Polymer Blends and Mixtures

NATO ASI Series

Advanced Science Institutes Series

A Series presenting the results of activities sponsored by the NATO Science Committee, which aims at the dissemination of advanced scientific and technological knowledge, with a view to strengthening links between scientific communities.

The Series is published by an international board of publishers in conjunction with the NATO Scientific Affairs Division

A	Life Sciences	Plenum Publishing Corporation
B	Physics	London and New York
C	Mathematical and Physical Sciences	D. Reidel Publishing Company Dordrecht and Boston
D	Behavioural and Social Sciences	Martinus Nijhoff Publishers Dordrecht/Boston/Lancaster
E	Applied Sciences	
F	Computer and Systems Sciences	Springer-Verlag Berlin/Heidelberg/New York
G	Ecological Sciences	

Series E: Applied Sciences – No. 89

Polymer Blends and Mixtures

edited by

D.J. Walsh
J.S. Higgins
A. Maconnachie

Imperial College
London, UK

1985 **Martinus Nijhoff Publishers**
Dordrecht / Boston / Lancaster
Published in cooperation with NATO Scientific Affairs Division

Proceedings of the NATO Advanced Study Institute on Polymer Blends and Mixtures, Imperial College, London, UK, July 2-14, 1984

Library of Congress Cataloging in Publication Data

NATO Advanced Study Institute on Polymer Blends and
   Mixtures (1984 : Imperial College)
   Polymer blends and mixtures.

   •(NATO ASI series. Series E, Applied sciences ; no. 89)
   "Proceedings of the NATO Advanced Study Institute on
Polymer Blends and Mixtures, Imperial College, London,
England, July 2-14, 1984"--T.p. verso.
   "Published in cooperation with NATO Scientific
Affairs Division."
   Includes index.
   1. Polymers and polymerization--Congresses.
I. Walsh, D. J. II. Higgins, J. S. III. Maconnachie, A.
IV. North Atlantic Treaty Organization. Scientific

   Affairs Division. V. Title. VI. Series.
   QD380.N377 1984   668.9   85-4922
   ISBN 90-247-3152-6

ISBN 90-247-3152-6 (this volume)
ISBN 90-247-2689-1 (series)

Distributors for the United States and Canada: Kluwer Academic Publishers,
190 Old Derby Street, Hingham, MA 02043, USA

Distributors for the UK and Ireland: Kluwer Academic Publishers, MTP Press Ltd,
Falcon House, Queen Square, Lancaster LA1 1RN, UK

Distributors for all other countries: Kluwer Academic Publishers Group, Distribution
Center, P.O. Box 322, 3300 AH Dordrecht, The Netherlands

Printed in The Netherlands

PREFACE

A couple of years ago a small group of people began discussing the possibility of running an advanced summer school in the area of polymer blends. There had been a number of recent advances in this field, and given the considerable interest in these new polymeric materials, we thought such a meeting would be well received both by industry and academia. We wanted it to contain a wide range of background science and technology and also up to date recent advances in the field. It became clear as the discussion progressed that the experts in the field were scattered over the length and breadth of Europe and North America and thus the cost of bringing them together for a summer school would necessitate a high registration fee which would deter many of the research workers we wished to attract. The NATO Advanced Study Institute programme enables a subject to be covered in depth and by giving generous funds to cover lecturers' costs ensures that a wide spectrum of research workers can attend. We decided to apply to NATO and this book contains the results of our request.

The ASI was funded under the 'Double-Jump' Programme which is not a new Olympic event but a way of supporting courses on subjects of direct industrial interest. The Institute was also backed by donations from several companies and approximately half those attending were from industrial organisations. Twelve of the sixteen NATO countries were represented by participants or lecturers.

The Institute was designed to illustrate ways of studying polymer blends. These include preparative and characterisation methods as well as techniques for measuring properties important in end uses. It was also designed to consider correlations which can be made between the chemical structures of the component polymers, the methods of preparation of blends, their morphology, and their physical properties.

It can be seen from the relevant lectures that the chemical and physical techniques for controlling the morphology of the blends are very well developed, and that the methods used for studying the morphology and measuring the mechanical properties of the blends are also satisfactory. The experimental methods available for studying the thermodynamics of polymer mixtures are however less satisfactory. On the whole they give only indirect information. At present it is not possible to directly and accurately measure the enthalpy and entropy changes of mixing over a range of temperature and composition.

With regard to the correlation between preparative methods, chemical structure, and blend morphology, two questions can be posed. Firstly, is it possible to predict which polymers will be

miscible or at least understand why miscibility occurs? It
is clear that this will be possible in the very near future.
Secondly, for two phase blends, can one relate the morphology
to the preparative procedure? In most cases these relationships are fairly well understood.

The correlation between morphology and physical properties for homogeneous blends can be reasonably understood in
terms of the properties of the pure components. For two
phase structures, however, we are a long way from being able
to make this correlation though progress is being made at
least for the simpler morphologies.

This picture of the state of the subject summarised here
became clear during the lectures and the discussions which
followed and we hope it is also equally clear from the pre-
sentations in this book.

We would like to thank all the lecturers for fulfilling
their part in putting the story together, the participants
for their contributions and for creating a lively environment
for the meeting, and NATO for making it all possible. Thanks
are also due to Montedison, General Electric, du Pont and Dow
Corning for financial assistance.

D.J. Walsh
J.S. Higgins
A. Maconnachie   London, December 1984

TABLE OF CONTENTS

THERMODYNAMIC THEORY AND EXPERIMENTAL TECHNIQUES FOR POLYMER BLENDS

D. R. Paul

Department of Chemical Engineering
  and Center for Polymer Research
The University of Texas
Austin, Texas 78712

INTRODUCTION

The phase behavior of polymer blends has become a topic of major scientific investigation during the past decade because of fresh insights into the issues involved and because of intensified technological interest in multicomponent polymer systems. The purpose here is to review the basic thermodynamic principles involved in polymer-polymer miscibility and to introduce some of the simpler methods for quantifying the interaction between the components which is a major factor. This will form a framework for developing relationships between blend phase behavior and the molecular structure of the component polymers. The approach will be restricted to classical theories of the Flory-Huggins type and to simple physical techniques which are generally available in most research laboratories. Subsequent chapters will deal with the newer and more advanced theories as well as sophisticated experimental techniques like neutron scattering. The utility and limitations of the solubility parameter as a means predicting phase behavior will be outlined briefly.

Several monographs and reviews (1-18) are available for further study.

GENERAL ASPECTS OF PHASE BEHAVIOR

Blends of two or more different polymers may exist in a completely homogeneous state where their segments are mixed at the most intimate level or they may segregate into distinct phases. Phase segregation may be the result of incomplete miscibility between two fluid or molten polymers or it may be caused by crystallization of one or more components from an otherwise homogeneous melt mixture. Obviously, both modes of phase segregation may exist simultaneously. A homogeneous amorphous phase upon cooling will eventually become a glass at a single temperature intermediate between the glass transitions of the pure components. The glass transition of the blend will depend on composition and reflect the mixed environment of the segments. This will be so even if phase segregation by crystallization has occurred provided there remains a homogeneous amorphous phase albeit of a different composition than the overall blend. On the other hand, blends comprised of separate amorphous phases will exhibit glass transitions characteristics of each phase. Thus, glass transition behavior can be a powerful tool for identifying the amorphous phase structure of blends as seen in subsequent chapters.

Homogeneous blends may experience liquid-liquid phase separation as the result of either raising or lowering the temperature (19) as suggested in Fig. 1. Generally, UCST behavior is characteristic of systems which mix endothermically while LCST behavior is characteristic of exothermic mixing and associated entropy effects. LCST behavior is rather common in polymer blends while UCST behavior is usually limited to cases where miscibility is the result of the low molecular weight of the components, e.g. mixtures of oligomers. In general, liquid-liquid phase equilibrium results in the presence of both components to some extent in each phase although this can be very limited in polymer blends. Because of partial miscibility,

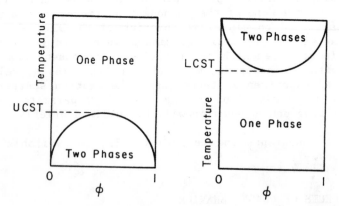

Figure 1. Liquid-liquid phase behavior for binary mixtures illustrating systems with an upper critical solution temperature (left) and a lower critical solution temperature (right). (By permission of the MMI Press. Copyright 1982.)

the separate glass transitions seen for phase separated blends
may not correspond to those of the pure components.

BASIC THERMODYNAMICS

In this section, the concept of miscibility and the possibility of liquid-liquid phase separation for binary polymer
blends will be discussed in terms of classical equilibrium
thermodynamics. To have miscibility, i.e. a homogeneous phase,
it is necessary for the free energy of mixing

$$\Delta G_{mix} = \Delta H_{mix} - T\Delta S_{mix} \tag{1}$$

to be negative; however, for this phase to be stable against
phase segregation the additional requirement that

$$\left(\frac{\partial^2 \Delta G_{mix}}{\partial \phi_i^2}\right)_{T,P} > 0 \tag{2}$$

must also be met. Here ϕ_i is the volume fraction of either
component - any other suitable measure of composition could be
used instead. The simplest model for describing the free energy
of mixing two polymers is based on an extension of results developed originally for polymer solutions by Flory and by Huggins
which assumes the only contribution to the entropy of mixing is
combinatorial in origin and is well-approximated by (20)

$$\Delta S_{mix} = -R(V_A + V_B)[\frac{\phi_A}{\tilde{V}_A} \ln\phi_A + \frac{\phi_B}{\tilde{V}_B} \ln\phi_B] \tag{3}$$

where $\tilde{V}_i$ is the molar volume of component i and V_A and V_B are
the actual volumes of these components comprising the mixture.
This model further assumes the heat of mixing is described by a
van Laar type of expression

$$\Delta H_{mix} = (V_A + V_B) B \phi_A \phi_B \tag{4}$$

where B is a binary interaction energy density characteristic of
mixing segments of components A and B. The latter is simply
related to the familiar Chi parameters by

$$\frac{B}{RT} = \frac{\chi_A}{\tilde{V}_A} = \frac{\chi_B}{\tilde{V}_B} = \frac{\chi_{AB}}{V_u} = \tilde{\chi}_{AB} \tag{5}$$

4

where V_u is the molar volume of an arbitrary unit of chain A, e.g. its repeat unit. While each of these measures of the interaction parameter is used in the literature, B is the least confusing since its basis, a unit of mixture volume, is always clear. In this model, the heat of mixing is independent of the component molecular weights while the two terms in the entropy of mixing are inversely related to the molecular weights of the corresponding components since $\tilde{V}_i = M_i/\rho_i$, ρ_i = mass density of pure i. For simplicity, each polymer is assumed here to be monodisperse. For polydisperse components, appropriate average molecular weights must be used in these and subsequent thermodynamic relations (21).

The free energy of mixing per unit volume of mixture for this model is thus obtained by combining equations 1, 3 and 4, i.e.

$$\frac{\Delta G_{mix}}{(V_A + V_B)} = RT \left[\frac{\rho_A \phi_A}{M_A} \ln \phi_A + \frac{\rho_B \phi_B}{M_B} \ln \phi_B \right] + B \phi_A \phi_B \qquad (6)$$

The discussion of equation 6 is simplified by considering the symmetrical case where molecular weights and densities are the same for both components, i.e. $M = M_A = M_B$ and $\rho = \rho_A = \rho_B$. In this case, the value of $BM/\rho RT$ will determine whether the necessary and sufficient conditions for having a stable, homogenous phase are met at a given temperature or not. Owing to the logarithmic terms, the entropy part of equation 6 is always negative thus favoring miscibility. However, depending on the sign of B, the enthalpy part may or may not favor miscibility. Applying the condition given by equation 2 to equation 6 for the symmetrical case reveals that a completely stable mixture exists for all proportions of the components only when

$$\frac{BM}{\rho RT} < 2 \qquad (7)$$

The ramifications of this result will be made clear by the following.

For the moment, it will be assumed that mixing is endothermic, i.e. $B > 0$, and that B does not depend either on composition or temperature. As molecular weight is varied at constant temperature, there will be a range where equation 7 is satisfied and stable, homogeneous mixtures exist for all ϕ_A. At a critical molecular weight M_{cr}, $BM_{cr}/\rho RT$ is exactly equal to two. Figure 2 illustrates what is happening to the entropy relative to the enthalpy as M is varied (22). An arbitrarily fixed heat of mixing is shown, and this does not vary with M. The lowest curve shows $-T \Delta S_{mix}$ when $M = M_{cr}$ — this curve shifts

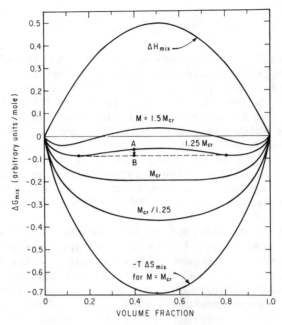

Figure 2. Free energy of mixing for polymers A and B with the same molecular weight M calculated from equation 6. (By permission of Academic Press. Copyright 1978.)

up or down from the one shown in inverse proportion to M. Algebraically adding the enthalpy and entropy parts yields the four free energy curves shown in the middle. When $M < M_{cr}$, the free energy of mixing satisfies equation 2 for all ϕ_A, and complete miscibility is the result. When $M > M_{cr}$, the free energy of mixing develops a negative curvature in the mid-composition region violating equation 2 and even becomes positive when M is sufficiently larger than M_{cr}. A homogeneous mixture at point A can achieve an even lower free energy by moving to point B corresponding to two separate phases having compositions given by the end points of the dotted line. This is not possible for free energy curves with $M < M_{cr}$. Thus, M_{cr} is the critical molecular weight below which all mixtures are homogeneous and stable against phase separation. When $M > M_{cr}$, phase separation occurs at equilibrium in the mid-composition region, and stable one-phase mixtures only exist at the extremities of the composition range, i.e. partial miscibility exists. As the ratio M/M_{cr} increases, the width of the miscibility gap broadens.

For relatively non-polar components, the interaction parameter B can be estimated from the solubility parameters of the two components, δ_i, using (22,23)

$$B = (\delta_A - \delta_B)^2 \tag{8}$$

When the solubility parameters differ by $1.0(cal/cm^3)^{\frac{1}{2}}$, M_{cr} is less than 1200 at 25°C assuming $\rho_A \sim \rho_B \sim 1$ g/cm^3. If the solubility parameters are matched more closely so that the difference is only 0.1 $(cal/cm^3)^{\frac{1}{2}}$, then M_{cr} is about 120,000. Thus, molecular weight is an influential factor affecting miscibility of components which mix endothermically, and very precise matching of component solubility parameters would be required to achieve miscibility of high molecular weight polymers. However, for polar polymers the solubility parameter approach breaks down, and mixing may become exothermic. In the latter case, molecular weight has only a secondary effect on phase behavior.

Next, it is of interest to examine the effect temperature has on phase stability. Figure 3 shows two possible ways the quantity $BM/\rho RT$ might depend on temperature. Using the condition given by equation 7, a value of two on the ordinate divides a region where all mixtures are one phase at equilibrium from a region where some compositions will separate into two phases. A critical solution temperature given by the following

$$T_{cr} = \frac{BM}{2\rho R} \tag{9}$$

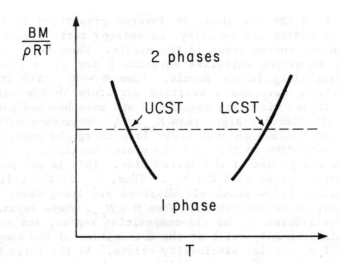

Figure 3. Thermodynamic conditions for UCST and LCST behavior.

divides these regions on a temperature scale. The intersection
the curve on the left makes with the dashed line defines an
upper critical solution temperature since two phases exist below
the T_{cr} but above this temperature all mixtures form one phase –
see Figure 1. A UCST may occur even if B is independent of
temperature since the temperature factor in the denominator can
cause the ordinate to drop below the critical value of two when
the mixture is heated sufficiently. Clearly, the curve on the
right must be the result of an interaction parameter which
increases with temperature and at a rate which offsets that of T
in the denominator. The intersection this curve makes with the
dashed line defines a lower critical solution temperature since
all mixtures are one phase below this value but phase separation
occurs on heating – see Figure 1. Thus, the simple theory
outlined here does not predict LCST behavior since it does not
incorporate a priori such a temperature dependence for B. The
newer "equation-of-state" theories of mixtures can predict LCST
behavior (13,18,24,25) without requiring any temperature
dependence of the corresponding interaction parameters.

In the case where the molar volumes, or molecular weights,
of the two components are not equal as assumed above, the
entropy and free energy of mixing curves will not be symmetrical
about the mid-point of concentration. Instead, the minimum of
the $-T\Delta S_m$ curve will shift to the side of the component having
the lower molar volume while the enthalpy of mixing curve will
remain symmetrical so long as B is not composition dependent.
The free energy of mixing curve will be skewed in the same
direction as the entropy. For this case, equation 7 must be
replaced by the following

$$\frac{B}{RT} < \frac{1}{2}\left[\frac{1}{\sqrt{\tilde{v}_A}} + \frac{1}{\sqrt{\tilde{v}_B}}\right]^2 \tag{10}$$

The coordinates of the critical point are (20)

$$T_{cr} = \frac{2\,B\,\tilde{v}_A\,\tilde{v}_B}{R(\sqrt{\tilde{v}_A} + \sqrt{\tilde{v}_B})^2} \tag{11}$$

and

$$(\phi_B)_{cr} = \frac{\sqrt{\tilde{v}_A}}{\sqrt{\tilde{v}_A} + \sqrt{\tilde{v}_B}} \tag{12}$$

The thermodynamic model outlined above cannot account for
all of the complexities of real mixtures; however, it does
provide a useful framework for identifying the major factors

affecting phase behavior and for analyzing experimental data. When used for the latter purpose, it is important to note that the model contains only one adjustable parameter, viz. B, introduced in a very simplistic way to describe the heat of mixing. Recognizing all of the shortcomings of the model intro-duced by the simplifying assumptions about the enthalpy and entropy terms, it has become common practice for analyzing data to relax the original interpretation of the interaction parameter by accepting equation 6 as its defining equation rather than equation 4. That is, B becomes a parameter which is selected to force data to fit equation 6. Consequently when used in this way, B must be regarded as a free energy parameter rather than one of purely enthalpic origin (19,21,26), and all issues not included or improperly accounted for in the model are lumped into B. As a result, B may be a function of blend composition, component molecular weights, and temperature rather than being independent of these quantities as assumed in the construction of the model. The object of more advanced theories is to properly consider all of these issues and thereby have an interaction parameter which is not variable and which has a simple physical meaning.

Given the generalized definition of B mentioned above, standard thermodynamic relations

$$\Delta S_{mix} = - \left(\frac{\partial \Delta G_{mix}}{\partial T} \right)_p \tag{13}$$

$$\Delta H_{mix} = - T^2 \left[\frac{\partial (\Delta G_{mix}/T)}{\partial T} \right]_p \tag{14}$$

may be applied to equation 6 to break it into enthalpic and entropic contributions (20). Doing so gives the following as a replacement for equation 4

$$\Delta H_{mix} = (V_A + V_B)(B - T \frac{\partial B}{\partial T}) \phi_A \phi_B \tag{15}$$

and an excess entropy contribution

$$\Delta S_{mix}^{(e)} = - (V_A + V_B)(\frac{\partial B}{\partial T}) \phi_A \phi_B \tag{16}$$

which must be added to the approximation of the combinatorial entropy of mixing given by equation 3. Note that recombining the various terms according to equation 1 gives back equation 6 as expected. The generalized free energy parameter, B, may be thought of as having enthalpic and entropic parts related as follows

$$B = B_h - TB_s \tag{17}$$

$$\text{where} \quad B_s = -\frac{\partial B}{\partial T} \tag{18}$$

$$\text{and} \quad B_h = B - T\frac{\partial B}{\partial T} \tag{19}$$

In general, both B_h and B_s may be composition dependent.

To this point, it has been implied that mixing is generally endothermic, and the main conclusion from this thermodynamic analysis is that two polymers will not form miscible mixtures unless their molecular weights are very low or the interaction parameter is very small. However, mixing need not always be endothermic especially when the components are polar and capable of certain specific interactions (10,13,14,17). Exothermic mixing provides a driving force for miscibility and completely removes the restrictions mentioned above. Realization of this simple fact has resulted in the discovery of a substantial number of miscible polymer pairs over the past decade or more. This has led to new optimism and possibilities for forming interesting and useful multicomponent polymer systems. There is now a substantial body of literature on miscible blends (10,13, 14,17) some of which deals with understanding and quantifying the interactions responsible for miscibility in specific systems. The remainder of this contribution serves as an introduction to the more important techniques commonly used for this purpose. Methods for a priori prediction of this information are severely limited. As shown by equation 8, the solubility parameter approach cannot predict exothermic mixing - its utility is limited to relatively non-polar structures.

TECHNIQUES FOR STUDYING MISCIBLE BLENDS

The previous discussion has identified the main factors affecting miscibility of polymer-polymer mixtures. Even when the molecular weights of both components are very large, miscibility is expected if the interaction parameter B is negative. The purpose here is to describe some techniques by which this parameter can be quantitatively measured and to give some examples of its determination for selected systems. Quantitative information of this sort is essential for developing an understanding of the relationships between phase behavior of blends and the molecular structures of the components. Other reviews of such techniques are available (19,27), so the current discussion is limited to a few classical techniques which have proven especially useful. Neutron scattering is not included since it is treated more fully elsewhere in this volume.

Table 1 lists a series of aliphatic polyesters whose blends with poly(vinyl chloride) and the polyhydroxy ether of bisphenol-A (designated as Phenoxy) have been extensively studied in our laboratory (28-31). The statements about the miscibility of the various polyesters with PVC and Phenoxy were deduced by noting whether the blends showed one or two glass transitions. Examples from this table will be used to illustrate results obtained by some of the following techniques.

Melting Point Depression

When crystals of one component are in equilibrium with a mixed amorphous phase, the melting point will be lower than when the equilibrium is with a pure amorphous phase of the same component comprising the crystals (20). This fact can be used to gain valuable information about interactions between the components of a blend when experimental measurements are properly executed and analyzed (28-34). A simple analysis can be formulated from basic thermodynamics using the model from the previous section to describe the behavior of the homogeneous, amorphous mixture with which very large, perfect crystals of one of the components are in equilibrium. The chemical potential of the crystallizable component, e.g. A, in the amorphous mixture relative to pure amorphous A can be computed from equation 6 using

$$\mu_A^{(a)} - \mu_A^{\circ} = \left(\frac{\partial \Delta G_{mix}}{\partial V_A} \right)_{T,P,V_B} \tag{20}$$

where these potentials are expressed per unit volume of A. The chemical potential of perfect crystals of A at any temperature relative to the same reference state can be approximated by (20)

$$\mu_A^{(c)} - \mu_A^{\circ} = - (\Delta H_u / V_u)(1 - \frac{T}{T_m^{\circ}}) \tag{21}$$

where $(\Delta H_u / V_u)$ is the heat of fusion per unit volume of A and T_m° is the melting point of perfect crystals of A when in equilibrium with an undiluted melt of A. For the blend, equilibrium between A in the crystal and in the melt mixture exists when

$$\mu_A^{(c)} = \mu_A^{(a)} \tag{22}$$

Combining equations 20-21 gives the following expression for the extent the melting point of A is decreased by blending with polymer B

Table 1: Phase behavior observed for blends of various
polyesters with poly(vinyl chloride) and Phenoxy

Abbreviation	Structure of Repeat Unit	$\dfrac{CH_2}{COO}$	Poly(vinyl chloride)	Phenoxy
PES	$-(CH_2)_2-O-\overset{O}{\overset{\|}{C}}-(CH_2)_2-\overset{O}{\overset{\|}{C}}-O-$	2	immiscible	immiscible
PEA	$-(CH_2)_2-O-\overset{O}{\overset{\|}{C}}-(CH_2)_4-\overset{O}{\overset{\|}{C}}-O-$	3	immiscible	miscible
PBA	$-(CH_2)_4-O-\overset{O}{\overset{\|}{C}}-(CH_2)_4-\overset{O}{\overset{\|}{C}}-O$	4	miscible	miscible
PCL	$-(CH_2)_5-\overset{O}{\overset{\|}{C}}-O-$	5	miscible	miscible
PCDS	$-CH_2-\underset{}{\bigcirc\!\!\text{S}}-CH_2-O-\overset{O}{\overset{\|}{C}}-(CH_2)_2-\overset{O}{\overset{\|}{C}}-O-$	(5)	miscible	miscible
PHS	$-(CH_2)_6-O-\overset{O}{\overset{\|}{C}}-(CH_2)_8-\overset{O}{\overset{\|}{C}}-O-$	7	miscible	immiscible
PDOA	$-(CH_2)_{12}-O-\overset{O}{\overset{\|}{C}}-(CH_2)_4-\overset{O}{\overset{\|}{C}}-O-$	8	miscible	n.t.
PDS	$-(CH_2)_{10}-O-\overset{O}{\overset{\|}{C}}-(CH_2)_8-\overset{O}{\overset{\|}{C}}-O-$	9	miscible	n.t.
PDEDE	$-(CH_2)_{10}-O-\overset{O}{\overset{\|}{C}}-(CH_2)_{10}-\overset{O}{\overset{\|}{C}}-O-$	10	miscible	n.t.
PDODE	$-(CH_2)_{12}-O-\overset{O}{\overset{\|}{C}}-(CH_2)_{10}-\overset{O}{\overset{\|}{C}}-O-$	11	miscible below 170°C	n.t.
PDODO	$-(CH_2)_{12}-O-\overset{O}{\overset{\|}{C}}-(CH_2)_{12}-\overset{O}{\overset{\|}{C}}-O-$	12	miscible below 155°C	n.t.
PHEDO	$-(CH_2)_{16}-O-\overset{O}{\overset{\|}{C}}-(CH_2)_{12}-\overset{O}{\overset{\|}{C}}-O-$	14	miscible below 130°C	n.t.

n.t. = not tested; however, all polyesters having a CH_2/COO ratio of 7 or more
are expected to be immiscible with Phenoxy.

$$(T_m^o - T_m) = - RT_m T_m^o \left(\frac{V_u}{\Delta H_u}\right) \left[\frac{\ln \phi_A}{\tilde{V}_A} + \left(\frac{1}{\tilde{V}_A} - \frac{1}{\tilde{V}_B}\right)\phi_B\right]$$

$$- T_m^o (V_u/\Delta H_u) \ B \ \phi_B^2 \tag{23}$$

assuming B is independent of the concentration of the blend. The first term on the right in equation 23 stems from combinatorial entropy contributions and is nearly zero if the molecular weights of both A and B are large. The second term on the right results from enthalpy of mixing effects.

The analysis given above assumes the crystals of B are infinitely large which may not be valid depending on the conditions of their formation. This condition can be assured experimentally by using the extrapolation procedure of Hoffman and Weeks (31,34,35). This consists of measuring T_m for a series of samples crystallized at different temperatures T_c and extrapolating T_m versus T_c to the line $T_m = T_c$. This is recommended for data to be used in melting point depression analyses (34); however, failure to make this correction only introduces errors in the analysis if crystal thickness actually varies with blend composition.

For blends of high molecular weight polymers, the melting point depression (after appropriate correction for crystal size if needed) can be analyzed for B by plotting versus ϕ_B^2. As suggested by equation 23, this should result in a straight line, as seen in Figure 4, whose slope is related to B. The interaction parameters deduced for blends of the three polyesters with Phenoxy from the data in Figure 4 are plotted in Figure 5 versus the ratio of aliphatic carbons per ester group in their structures. The line drawn through these points reflects the fact that the polyesters PES and PHS are not miscible with Phenoxy and, therefore, their interaction parameters must be positive although their values are not known quantitatively.

Similar data are shown in Figure 6 for various polyesters in Table 1 which are miscible with PVC (28,31). Note that the polyesters PES and PEA are not miscible with PVC so their interaction parameters must be positive. The interaction parameters shown in Figures 5 and 6 were deduced from the depression of the melting points of the crystalline polyesters since both Phenoxy and PVC do not crystallize.

For all of the above systems, plots of $(T_m^o - T_m)$ versus ϕ_B^2 were linear within experimental error implying that B is independent of blend concentration. Morra and Stein (34) have reported highly curved plots for the system poly(vinylidene

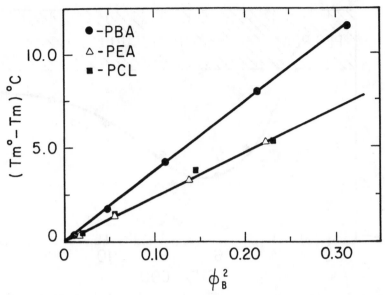

Figure 4. Melting point depression for three polyesters when blended with Phenoxy. (By permission of John Wiley and Sons, Inc. Copyright 1982.)

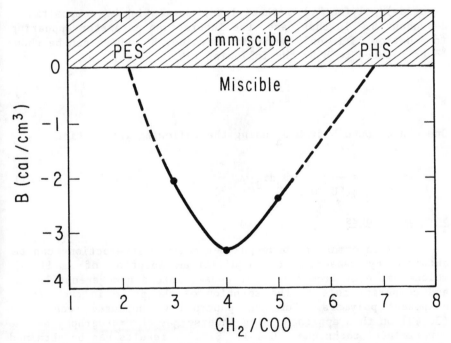

Figure 5. Blend interaction parameters computed from Figure 4 shown as a function of polyester structure.

14

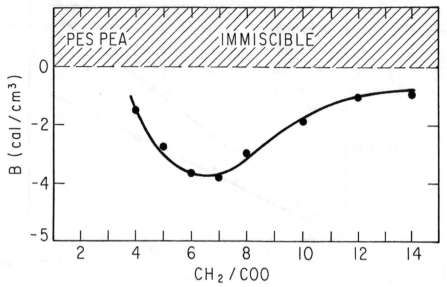

Figure 6. Interaction parameters for poly(vinyl chloride)/
polyester blends deduced from melting point depression.

fluoride)/poly(methyl methacrylate) suggesting a composition
dependent B. In this case, the interaction parameter appearing
in equation 23 is only an apparent value, B_a, which can be shown
to be related to B by the following (26)

$$B_a = B + \phi_B \frac{\partial B}{\partial \phi_B} \qquad (24)$$

One can compute B from B_a using the following integration

$$B = \frac{1}{\phi_B} \int_0^{\phi_B} B_a d\phi_B \qquad (25)$$

Sorption Probes

Some information about polymer-polymer interactions can be
obtained by measuring the equilibrium sorption of small
molecules, e.g. from the vapor phase, into a homogeneous blend
and comparing this to the sorption of the probe in the pure
component polymers. Numerous papers have appeared recently
(36-43) on this approach using inverse gas chromatography as the
experimental technique (40-43). Similar results can be obtained
using quartz springs, electronic microbalances, quartz crystal
oscillators, etc. Data analysis can be made from a ternary

formulation of the basic theory described above. In the limit where Henry's law describes the sorption, C, of the probe, species 1, into component i

$$C = k_{1i}\, p_1 \tag{26}$$

at probe partial pressure p_1, the result is (39)

$$\ell n k_{1b} = \phi_A\, \ell n k_{1A} + \phi_B\, \ell n k_{1B} + (B\tilde{V}_1/RT)\, \phi_A \phi_B \tag{27}$$

Here A and B denote the two polymeric components as before and b denotes a blend of composition ϕ_A treated as a single component. Figure 7 shows the Henry's law coefficient for CH_2Cl_2 sorption into miscible blends of Phenoxy and PEA using semi-logarithmic coordinates as suggested by equation 27. The dashed line shows the additive result given by the first two terms on the right while the solid line represents the best fit of equation 27 to the data resulting in the value of the interaction parameter shown. Other data of this type will be discussed in the next section.

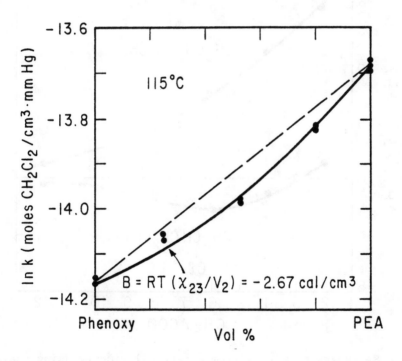

Figure 7. Henry's law solubility coefficient for methylene chloride in Phenoxy/PEA blends. Interaction parameter shown was computed by fitting data to equation 27.

Some investigators have found interaction parameters determined by this method to depend on the choice of probe used (38).
There is some reason to believe that a more sophisticated theory
of mixtures could remove at least a portion of this difficulty;
however, to date, all analyses have been based on equation 27 or
equivalent forms.

Direct Calorimetry Using Low-Molecular Weight Analogs

While calorimetry cannot be performed directly to obtain
the heat of mixing for two high molecular weight polymers because
of obvious difficulties, such measurements can be readily made
for low-molecular weight liquids whose molecular structures are
representative portions of the polymer chains (30,31,44,45).
Clearly, the movements, distances of separation, and

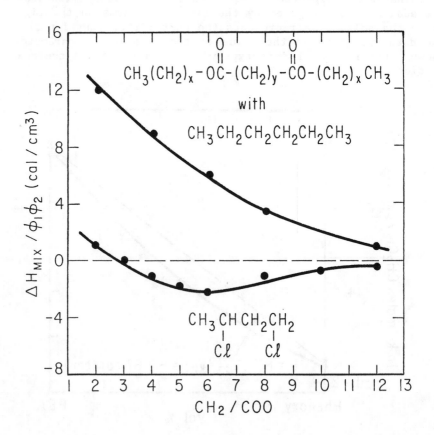

Figure 8. Heats of mixing from direct calorimetry using analogs
of polyesters and polyethylene (upper curve) and of polyesters
and poly(vinyl chloride) (lower curve). Ordinate is equivalent
to B.

accessibility for interaction of chemical groups are not the same in liquids as in polymers; however, if suitable analogs can be identified, this can be a powerful tool for understanding the thermodynamics of polymer blends in at least a semi-quantitative way.

Figure 8 illustrates this nicely using simple analogs for PVC and the polyesters described in Table 1. These results (31) are very similar in form to the interaction parameters deduced by melting point depression for the corresponding polymers shown in Figure 6. The values are not identical as might be expected; however, the trends with polyester structure seen in the two graphs are virtually the same. The immiscibility of PVC with polyesters having a low ratio of CH_2/COO (see Table 1) is predicted by the calorimetry results since mixing becomes endothermic in this range. At high CH_2/COO ratios the heat of mixing approaches zero as was also seen in Figure 6. Because of the weakening of the favorable interaction for miscibility of PVC with polyesters in this range, these blends undergo phase separation on heating as shown in Figure 9. Figure 8 also shows the heat of mixing of various esters with hexane. Mixing in this case is endothermic but approaches zero as the CH_2/COO ratio increases.

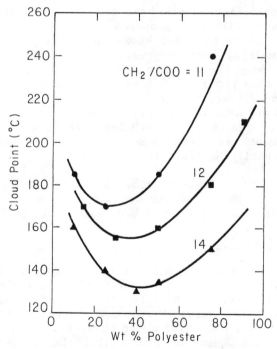

Figure 9.   Phase diagrams for blends of poly(vinyl chloride) with polyesters having the structures shown.

18

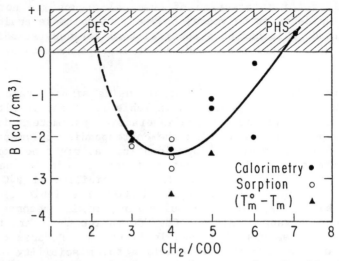

Figure 10. Comparison of interaction energies for blends of Phenoxy with various polyesters determined by three different methods.

Similar calorimetry data have been reported for various esters mixed with diphenoxy propanol selected as an analog for Phenoxy (30). These data are shown in Figure 10 along with results from melting point depression shown in Figure 5 plus data obtained using the sorption probe technique like that shown in Figure 7. This provides an interesting comparison of results using the three different techniques described here. Agreement among the various data is not perfect, and this may be attributed to a variety of reasons. Nevertheless, it is clear that the general agreement seen lends some confidence to using these methods for quantifying blend thermodynamics.

By use of Hess's law, calorimetric measurement of the heats of dilution for a blend and its components by a solvent can be used to estimate the polymer-polymer heat of mixing (46,47).

Spectroscopic Techniques

Obviously there is a great deal of interest in knowing more about the mechanisms responsible for exothermic mixing in various systems since this is essential to the full understanding of the relationship between phase behavior and molecular structure. On purpose, little has been said about this until now. The use of spectroscopic techniques and especially FTIR to identify the interacting chemical groups has recently become an extremely active area of investigation (48-55). Very illuminating

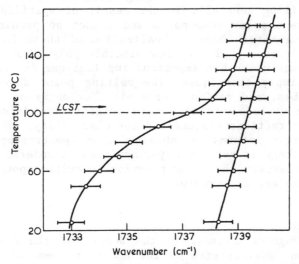

Figure 11. Frequency of carbonyl band in ehtylene/vinyl acetate copolymer (right) and after blending with a chlorinated polyethylene (left) by FTIR. Dotted line shows temperature where homogenous blend phase separates on heating. (By permission of Butterworth and Co. Copyright 1983.)

information is now emerging; however, it is beyond the present scope to review this literature. Likewise, a discussion of the various types of specific interactions which may be important in blends will not be given here since this will occur later in this volume.

Results in Figure 11 serve to illustrate the type of information now being developed by investigations of blends using FTIR techniques (55). The mixing of a chlorinated polyethylene with an ethylene/vinyl acetate copolymer causes a shift in the carbonyl peak of the latter to lower wave numbers as seen. However, upon heating this shift greatly diminishes near the point where the blend phase separates owing to the existence of an LCST. Future investigations of this type are expected to have a tremendous impact on understanding of blend phase behavior.

SUMMARY

A simple thermodynamic model for polymer-polymer blends based on the theories of Flory and Huggins leads to the important conclusion that to have miscibility in the limit of high molecular weights requires a negative interaction parameter.

Physically the latter means mixing must be exothermic.
Exothermic mixing generally is the result of specific interactions between polar components and cannot be predicted by
solubility parameter theory. Realization of these facts have
resulted in discovery of many new miscible polymer pairs during
the past decade which has important implications for commercial
products. Simple techniques like melting point depression,
sorption probes, and various forms of calorimetry as described
here provide useful ways to obtain quantitative information
about the interactions responsible for miscibility. When coupled
with mechanistic information obtained from spectroscopic techniques, the stage is set for major advances in understanding the
relationship between the phase behavior in multicomponent polymer
systems and molecular structure.

Acknowledgment is made to the donors of the Petroleum
Research Fund, administered by the American Chemical Society,
and to the Army Research Office for their support of this
research.

REFERENCES

1. Keskkula, H. (ed.): 1968, Polymer Modification of Rubbers and Plastics, Appl. Polym. Symp., Vol. 7.
2. Bruins, P. F. (ed.): 1970, Polyblends and Composites, Appl. Polym. Symp., Vol. 15.
3. Platzer, N.A.J. (ed.): 1971, Multicomponent Polymer Systems, Adv. Chem. Ser., Vol. 99.
4. Molau, G. E. (ed.): 1971, Colloidal and Morphological Behavior of Block and Graft Copolymers, Plenum Press, New York.
5. Sperling, L. H. (ed.): 1974, Recent Advances in Polymer Blends, Grafts, and Blocks, Plenum Press, New York.
6. Platzer, N.A.J. (ed.): 1975, Copolymers, Polyblends, and Composites, Adv. Chem. Ser., Vol. 142.
7. Klempner, D. and Frisch, K. C. (eds.): 1977, Polymer Alloys: Blends, Blocks, Grafts, and Interpenetrating Networks, Plenum Press, New York.
8. Manson, J. A. and Sperling, L. H.: 1976, Polymer Blends and Composites, Plenum Press, New York.
9. Bucknall, C. B.: 1977, Toughened Plastics, Applied Science Publishers, London.
10. Paul, D. R. and Newman, S. (ed.): 1978, Polymer Blends, Vols. I and II, Academic Press, New York.
11. Cooper, S. L. and Estes, G. M. (eds.): 1979, Multiphase Polymers, Adv. Chem. Ser., Vol. 176.
12. Klempner, D. and Frisch, K. C. (eds.): 1980, Polymer Alloys II, Plenum Press, New York.
13. Olabisi, O., Robeson, L. M. and Shaw, M. T.: 1979, Polymer-Polymer Miscibility, Academic Press, New York.
14. Paul, D. R. and Barlow, J. W., Polymer Blends (or Alloys), J. Macromol. Sci. - Rev. Macromol. Chem. C18 (1980) 109.
15. Solc, K. (ed.): 1982, Polymer Compatibility and Incompatibility: Principles and Practice, Vol. 2 in MMI Press Symposium Series.
16. Martuscelli, E., Palumbo, R. and Kryszewski, M.: 1979, Polymer Blends: Processing, Morphology, and Processing, Plenum Press, New York.
17. Barlow, J. W. and Paul, D. R.: Ann. Rev. Mater. Sci. 11 (1981) 299.
18. Sanchez, I. C.: Ann. Rev. Mater. Sci. 13 (1983) 387.
19. Paul, D. R.: pp. 1-23 in Ref. 15.
20. Flory, P. J.: 1953, Principles of Polymer Chemistry, Cornell University Press, Ithaca.
21. Koningsveld, R., Chermin, H.A.G., and Gordon, M.: Proc. Royal Soc - London A319 (1970) 331.
22. Paul, D. R.: pp. 1-23 in Ref. 15.
23. Krause, S.: Ch. 2 in Ref. 10.
24. Sanchez, I. C.: Ch. 3 in Ref. 10.
25. Sanchez, I. C.: pp. 59-76 in Ref. 15.

26. Koningsveld, R. and Staverman, A. J.:   Adv. Colloid and Interface Sci. 2 (1968) 151.
27. Robeson, L. M.: pp. 177-211 in Ref. 15.
28. Ziska, J. J., Barlow, J. W., and Paul, D. R.:   Polymer 22 (1981) 918.
29. Harris, J. E., Goh, S. H., Paul, D. R., and Barlow, J. W.: J. Appl. Polym. Sci. 27 (1982) 839.
30. Harris, J. E., Paul, D. R., and Barlow, J. W.:   Polym. Eng. Sci. 23 (1983) 676.
31. Woo, E. M.:   1984, PhD Dissertation, University of Texas at Austin.
32. Nishi, T. and Wang, T. T.:   Macromolecules 8 (1975) 909.
33. Paul, D. R., Barlow, J. W., Bernstein, R. E., and Wahrmund, D. C.:   Polym. Eng. Sci. 18 (1978) 1225.
34. Morra, B. S. and Stein, R. S.:   J. Polym. Sci.:   Polym. Phys. Ed. 20 (1982) 2243.
35. Hoffman, J. D. and Weeks, J. J.:   J. Res. Natl. Bur. Stand. U.S. 66 (1962) 13.
36. Kwei, T. K., Nishi, T., and Roberts, R. F.:   Macromolecules 7 (1974) 667.
37. Deshpande, D. D., Patterson, D., Schreiber, H. P., and Su, C. S.:   Macromolecules 7 (1974) 530.
38. Olabisi, O.:   Macromolecules 8 (1975) 316.
39. Masi, P., Paul, D. R., and Barlow, J. W.:   J. Polym. Sci.: Polym. Phys. Ed. 20 (1982) 15.
40. Walsh, D. J. and McKeown, J. G.:   Polymer 21 (1980) 1335.
41. DiPaola-Baranyi, G., Fletcher, S. J., and Degre, P.: Macromolecules 15 (1982) 885.
42. Walsh, D. J., Higgins, J. S., Rostami, S., and Weeraperuma, K.: Macromolecules 16 (1983) 391.
43. Doube, C. P. and Walsh, D. J.:   Eur. Polym. J. 17 (1982) 63.
44. Cruz, C. A., Barlow, J. W., and Paul, D. R.:   Macromolecules 12 (1979) 726.
45. Walsh, D. J., Higgins, J. S., and Zhikuan, C.:   Polymer 23 (1982) 336.
46. Weeks, N. E., Karasz, F. E., and MacKnight, W. J.:   J. Appl. Phys. 48 (1977) 4068.
47. Karasz, F. E. and MacKnight, W. J.: pp. 165-175 in Ref. 15.
48. Coleman, M. M., Zarian, J., Varnell, D. F., and Painter, P. C.:   J. Polym. Sci.:   Polym. Lett. Ed. 15 (1977) 745.
49. Coleman, M. M., and Zarian, J.:   J. Polym. Sci.:   Polym. Phys. Ed. 17 (1979) 837.
50. Ting, S. P., Pearce, E. M., and Kwei, T. K.:   J. Polym. Sci.:   Polym. Lett. Ed. 18 (1980) 201.
51. Roerdink, E., and Challa, G.:   Polymer, 21 (1980) 509.
52. Coleman, M. M., and Varnell, D. F.:   J. Polym. Sci.: Polym. Phys. Ed. 18 (1980) 1403.
53. Varnell, D. F., and Coleman, M. M.:   Polymer, 22 (1981) 1324.

54. Garton, A., Aubin, M., and Prud'homme, R. E.: J. Polym. Sci.: Polym. Lett. Ed. 21 (1983) 45.
55. Coleman, M. M., Moskala, E. J., Painter, P. C., Walsh, D. J., and Rostami, S.: Polymer 24 (1983) 1410.

GLASS TRANSITIONS AND COMPATIBILITY; PHASE BEHAVIOR IN COPOLYMER CONTAINING BLENDS

Frank E. Karasz

Polymer Science and Engineering Department
Materials Research Laboratory
University of Massachusetts
Amherst, Massachusetts   01003   USA

INTRODUCTION

The equilibrium phase state of a mixture of high molecular weight polymers may not be as obvious by visual inspection as is an analogous mixture of low molecular weight materials. Consequently the nature of this state and of the properties of macromolecular mixtures has become an important problem in polymer science. A binary polymer mixture at a given temperature and pressure can exhibit compatible (miscible) or, more often, essentially incompatible (immiscible) behavior. The former is characterized by the formation of a solution, homogeneous at the segmental level, which in principle is no different from that occurring in low molecular weight or polymer-solvent systems; the latter retains a discrete heterogeneity, even though it is nearly always possible to bring two polymers into a state of fine mutual dispersion, stable only because of kinetic barriers to phase separation. Nevertheless metastable incompatible mixtures are sufficiently important to warrant further subdivision into those systems which while thermodynamically unstable, possess mechanical integrity over a given temperature-time regime, and those which cannot retain mechanical integrity for any practical time period in the temperature regime of interest. Clearly such divisions are highly arbitrary and will be functions of the thermal, solvent and mechanical history of the mixture as well as the fundamental interactions between the two macromolecules.

As already indicated, to distinguish between the thermo-dynamically compatible systems and those in a state of fine dispersion maintaining some degree of mechanical integrity may not be straightforward. A relatively facile test for compatibility depends on the observation of a single glass transition, while in finely dispersed yet basically heterogeneous incompatible polymer mixtures, the individual blend constituent glass transitions will be retained. While this diagnostic test is in favorable cases easily applied and reliable, it is by no means infallible and may be compromised by one or more factors discussed below. Furthermore, it relies on the assumption that the manifestation of a single glass transition may be intrinsically equated with segmental level homogeneity, an assumption which while pragmatically unassailable, has not been established as having a theoretical validity.

Since the compatibility or incompatibility of a given polymer system is often ultimately related to a practical application it may also be noted that the advantage of compatibility with its concomitant single glass transition often relates to the possibility of thereby adjusting the rheological processing "window" lying between the glass transition and chemical degradation temperatures. Mechanical properties of compatible systems tend to deviate only to a relatively small extent from those anticipated on the basis of a linearly additive scheme. In contrast, the advantages of an incompatible blend tend to be concentrated in terms of an often spectacular improvement of mechanical properties at the subsequent use temperatures; the effects on processing characteristics, while present, tend to be secondary.

In this chapter we shall first discuss the glass transitions and their determination in the context of polymer compatibility. In the second part, recent developments in the study of compatibility in macromolecular mixtures containing random copolymers - based on experimental T_g studies for such systems - will be reviewed.

GLASS TRANSITION CRITERION FOR COMPATIBILITY

The traditional experimental criteria for compatibility have been extensively reviewed elsewhere, (1-3). In recent years the advent of novel and powerful scattering techniques have led to increased emphasis on methods which in principle lead to conclusions regarding the conformation of individual macromolecular chains and which therefore tend, as is proper, to minimize distinctions between polymer-polymer and polymer-solvent systems. Techniques such as neutron scattering have been especially valuable in this regard but are not readily applicable on a routine basis. For somewhat pragmatic reasons, therefore, the examination of glass transition behavior has largely prevailed as a diagnostic test of compatibility, especially so since the advent of commercially available instrumentation of increasing sensitivity which has made T_g measurements relatively routine. A diagnosis based on this technique is not infallible, as was indicated above, and often warrants careful verification, preferably by other techniques. Before addressing potential problems in this area, however, we shall first review some common methods of determining T_g's equally applicable to single as well as to multiple component systems. It is important to note in this context that the "single glass transition" criterion for compatibility in fact comprises a somewhat stronger set of requirements: the single transition temperature must be a monotonic function of blend composition, and must be concomitantly associated with the disappearance of the blend constituent T_g's.

Glass Transition Determinations

A very large variety of physical measurements have been proposed for the determination of glass transition temperatures. Many of these may however be regarded as tours-de-force of either unnecessary experimental complexity and/or inadequate sensitivity, and may be unimportant in a practical schema. In practice, therefore, the most significant techniques fall into the following categories:

(1) Calorimetric determination of heat capacities as a function of temperature.

(2) Dynamical mechanical (low strain) measurements of complex modulus as a function of temperature.

These two techniques together constitute by far the majority of contemporary tests of compatibility by T_g measurement. Their predominance has been assisted by the availability of, respectively, commercial differential scanning calorimeters (DSC's) and automated dynamical mechanical spectrometers.

Other non-negligible measurements that may have specialized applicability include:

(3) dielectric relaxation spectroscopy,
and

(4) dilatometry.

Standard references may be consulted for extensive lists of less widely used measurements.

A few comments may be offered with respect to these major techniques.

1. Differential Scanning Calorimetry. This technique has the advantage of small sample requirements, relatively rapid measurement capability and high sensitivity. The last, in fact, demands a cautious and conservative evaluation of results since artifactual phenomena are not infrequently encountered in practical application. The glass transition is distinguished by a discontinuity in the C_p vs. T curve, which may however be modified by the peaks associated with enthalpy relaxation phenomena. A recent development has been the use of a derivative presentation ie. dC_p/dT vs. T, which may assist in the evaluation of more problematical situations.

2. Dynamical Mechanical Spectroscopy. This technique has more stringent sample requirements than calorimetry, but is sometimes preferred because of a widely held belief that it is more sensitive than calorimetry, though this assertion does not appear to have been put to an objective test. The diagnostic measurement is nearly always that of the loss angle, tan δ, or the complex tensile modulus, E", and this can usually be obtained over a frequency range, typically two to four decades wide, which may be advantageous under some conditions.

3. Dielectric Relaxation Spectroscopy. This technique is similar in many respects to its mechanical analogue with the exception of the fact that a much wider frequency range - perhaps six or more decades - is normally available. Thus a time-temperature superposition treatment may be used in principle to obviate the necessity of heating the sample to the "normal" glass transition temperature regime, though this potential advantage is seldom if ever realized in practice. The sensitivity of this technique is a function of the chemical structure of the macromolecules involved since it is essentially a probe of electrical dipolar relaxation.

4. Dilatometry. The volumetric determination of the glass transition temperature in terms of a discontinuity of the expansion coefficient α (cf. C_p) was one of the earliest techniques used for the measurement of T_g's, in single as well as in multi-component systems, but now appears to have been almost eclipsed by the techniques cited above, largely because of pragmatic considerations. Dilatometry may also be used, in one of several available experimental variants, for the determination of the volume of mixing in binary polymer mixtures. It is a matter of record that compatible systems are associated with a densification or negative mixing volume (usually of the order of 1-3%), whereas the density of incompatible systems would be expected to display a strictly linearly additive behavior. However, it must be added that relatively few systems have been investigated in detail; moreover, measurements made at temperatures below the T_g may be compromised by well known time-dependent relaxation phenomena associated with the glass transition.

PROBLEMS ASSOCIATED WITH THE T_g TEST FOR COMPATIBILITY

As already indicated, the single T_g diagnosis for compatibility is subject to certain limitations, (4). Amongst these are the following:

1. Proximity of Component T_g's

Obviously if the component T_g's cannot be differentiated the single T_g test fails by itself to discriminate between the incompatible and compatible cases. A definitive statement of the necessary differentiation is not possible, as this will depend, amongst other factors, on the sharpness of the T_g's in the constituents, possible transitional smearing in the mixture, the proximity of the phase boundaries and on instrumentational characteristics. In practice one may anticipate problems if the T_g's lie within 20°C or less of each other. The problem may be somewhat alleviated by the use of the differential presentation dC_p/dT vs. T already alluded to, but insufficient evidence is as yet available regarding this point.

2. Sensitivity

The utility of the single T_g test will also be compromised if the potential detectability of the T_g of either component is reduced. This will occur, obviously, if one or other of the components is present in low concentration in the mixture. Since the T_g vs. blend composition curve for compatible systems is typically convex with respect to the abscissa, this problem may be enhanced if mixtures containing small concentrations of the higher T_g constituent are being studied. In the latter case the T_g of the compatible mixture may well be hardly distinguishable from that of the pure major component. Again, any numerical statement regarding this limitation is invariably arbitrary; nevertheless one may anticipate difficulties if the minor constituent is present in less than 10% concentration.

The problem is accentuated in mixtures where crystallization of either or both constituent polymers takes place. In this case not only is the intrinsic sensitivity of the technique reduced by the diminution of the amorphous phase concentration, but a distinct potential arises for confusion amongst glass and melting transitions inherent in this more complex situation. Thus the diagnosis of compatibility in crystallizable polymer blends is nearly always more difficult than in amorphous-amorphous systems.

3. Other Problems

Occasional discrepancies between results obtained by different experimental techniques for measuring T_g have been reported, (4). More common are apparent discrepancies in diagnoses where different criteria (e.g. scattering versus single T_g) have been applied. Reasons for such discrepancies can often be found by additional investigation but the cumulative effect is to encourage a cautious approach in reaching conclusions in the diagnosis of compatibility in macromolecular systems. Furthermore, it must be recalled that the T_g technique is often employed as a probe of the phase behavior at temperatures higher than T_g itself. The value of this diagnosis therefore may depend on the ability to <u>avoid</u> phase re-equilibration during the quench and the T_g measurement. This ability will be a function of several factors - the relative positions of the test and the glass transition temperatures, the kinetics of re-equilibration relative to the time necessary for measurement and so on.

We conclude that for other than the simplest amorphous-amorphous systems with well differentiated constituent T_g's which are either totally compatible or essentially incompatible within the temperature-composition regime of interest, problems may well occur.

PHASE BEHAVIOR

The rich variety of phase behavior of polymer mixtures as represented in classical temperature-composition diagrams is discussed elsewhere in this volume, (5), and has been extensively reviewed in standard treatments of the subject, (2,6). In principle, there is no necessity to differentiate between the behavior of mixtures of low molecular weight species and those of macromolecules. In practice, the presence of the glass transition may cause problems in obtaining information over a wide temperature regime. It will normally be impossible to achieve thermodynamic equilibrium below the glass transition of the mixture and this inability will therefore mask phase behavior in this regime. It may be that this effect is responsible for the near absence of definitive reports of UCST's (upper critical solution temperatures) in polymer-polymer blends, in contrast to the widespread observations of lower critical solution temperatures (LCST's) in these systems, though theory tends to suggest, in addition, that the former behavior is intrinsically less likely to occur in macromolecular mixtures, (7).

TEMPERATURE-COPOLYMER COMPOSITION DIAGRAMS

Recent interest in compatibility of random copolymer/copolymer and copolymer/homopolymer systems has led to the introduction of a non-classical representation of phase behavior, the so-called temperature-copolymer composition diagram, (8). In T-c plots domains of equilibrium miscibility and immiscibility are shown for a given <u>blend</u> composition (typically, though not necessarily, 50/50 weight percent) as a function of annealing temperature and of <u>copolymer</u> composition.

The nature of the T-c diagram depends on the respective interactions of the constituent monomeric species present in the blend. For example, in a homopolymer/copolymer blend, A_n/$(B_{1-x}$ $C_x)_n$ equilibrium phase behavior will depend on the three interaction parameters x_{AB}, x_{AC}, and x_{BC}, (8). Where the T-c plot covers a domain of partial miscibility (blends could also be either totally miscible or immiscible in a given regime) two types of T-c diagrams are encountered (Figs. 1 and 2).

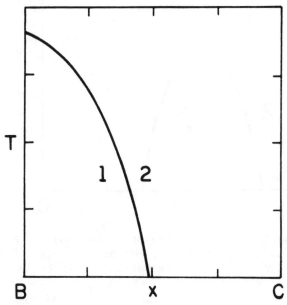

Figure 1: Schematic of phase behavior in constant composition mixtures of a homopolymer A_n with random copolymer $(B_x C_{1-x})_n$' as function of temperature and composition. In this figure $x_{AB} < 0$; x_{AC}, $x_{BC} > 0$; n,n' →∞. The intersection of the phase boundary with the ordinate approximately represents the LCST of the blend of A_n with B_n'. Above this temperature $x_{AB} > 0$.

In Fig. 1, homopolymer A_n interacts relatively favorably with homopolymer B_n(X_{AB} is negative) and unfavorably (positive X_{AC}) with homopolymer C_n ($x = 0$ and 1 respectively). If LCST behavior is displayed in these blends, as has been commonly observed, then the phase boundary represents, approximately, the loci of the LCST's of the blends of A_n with the respective copolymers, and necessarily has a negative slope as shown. A more complex situation may occur in the homopolymer/copolymer case if all three X's are positive, i.e. all possible combinations of homopolymers are immiscible. In this case for certain numerical combinations of the interaction parameters a "window of miscibility" may be observed as in Fig. 2.

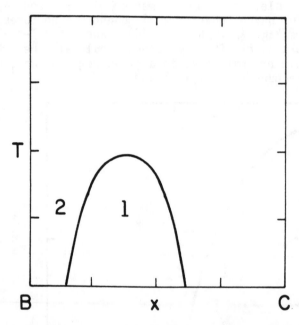

Figure 2: Schematic of phase behavior of homopolymer A_n with copolymer $(B_x C_{1-x})_n$, as in Fig. 1, but now all $X_{ij} > 0$ ($i,j \equiv$ A,B,C). The change in X's with temperature results in the maximum in the miscibility window.

The case of random copolymer blends of the type $(A_{1-x} B_x)_n/(A_{1-y} B_y)_{n'}$ is of interest. In this case the mean field theory prediction (8) is that the system is totally immiscible for all copolymer compositions other than x = y provided that n, n' are infinite and $X_{AB} > 0$. This condition is relaxed for systems of finite molecular weight, so that a "chimney of miscibility" is found, Fig. 3, (9). It must also be reemphasized that these diagrams represent equilibrium situations and hence are not necessarily valid for situations below T_g. Moreover the diagrams are predicated on the observance of high temperature phase separation; if the opposite behavior, i.e. with UCST's, also occurs, then the T-c diagrams will have a different appearance, e.g. Fig. 4. This more complex case has not yet been observed.

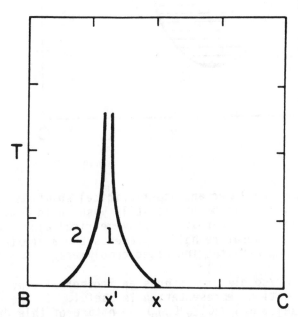

Figure 3: Schematic of a "chimney of miscibility" for random copolymers containing identical monomers A and B. Theory predicts miscibility should only occur for the condition x ≡ y for infinite molecular weight copolymer (see text). For finite molecular weights this condition is relaxed so that an approximately sym-metrical "chimney-type" phase boundary is predicted, centered about the copolymer composition x', where x' equals a given value of y,y'.

34

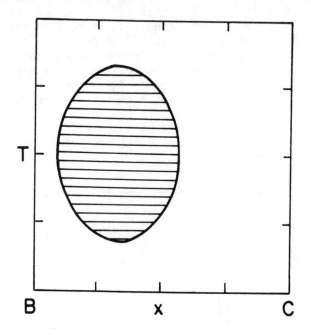

Figure 4:   When both lower and upper critical solution temperatures occur, the window of miscibility shown in Fig. 2 will be
joined to an inverted, approximately symmetrical window as shown,
where the lower temperature half represents the approximate loci
of the UCSTs of A_n with the indicated copolymers.

For copolymer/copolymer mixtures an isothermal copolymer-
copolymer composition representation is useful. In the most
general case, $(A_{1-x} B_x)_n / (C_{1-y} D_y)_n$, the nature of this diagram
will be a function of the six χ parameters corresponding to each
of the possible monomer-monomer interactions in such a system,
(8). A large range of behavior is predicted, e.g. Fig. 5. An
example of the special case of two copolymers with a common
monomer $(A \equiv C)$, is shown in Fig. 6.

ACKNOWLEDGEMENT

This work was supported by AFOSR Grant 84-0100.

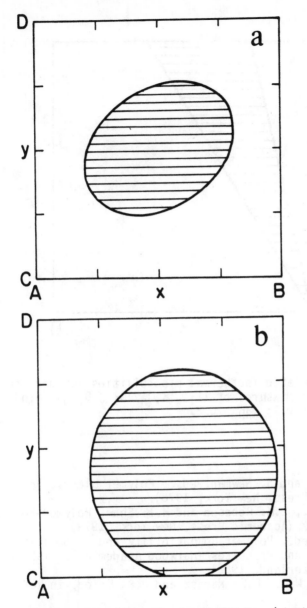

Figure 5: Calculated isothermal miscibilities (shaded areas) for
random copolymer mixtures, $(n,n' \to \infty)$, (8).
   (a) $x_{AB} = 0.3$; $x_{AC} = 0.1$; $x_{AD} = 0.2$; $x_{BC} = 0.2$; $x_{BD} = 0.1$; $x_{CD} = 0.4$.
   (b) $x_{AB} = 0.6$; $x_{AC} = 0.2$; $x_{AD} = 0.3$; $x_{BC} = 0.1$; $x_{BD} = 0.2$; $x_{CD} = 0.5$.

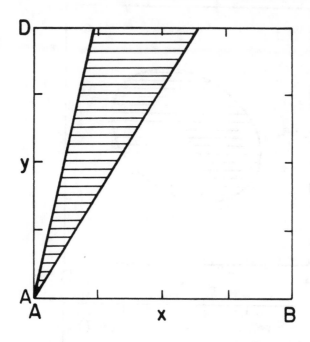

Figure 6: Calculated isothermal miscibilities (shaded area) for random copolymer mixtures of $(A_{1-x} B_x)_n/(A_{1-y} D_y)_{n'}$ $(n, n' \rightarrow \infty)$. $\chi_{AB} = 0.7$; $\chi_{AD} = 0.1$; $\chi_{BD} = 0.2$.

REFERENCES

1. Paul, D.R. and S. Newman, eds., Polymer Blends, Vol. 1, (Academic Press, New York, 1978).
2. Olabisi, O., L.M. Robeson and M.T. Shaw, Polymer-Polymer Miscibility (Academic Press, New York, 1979).
3. Šolc, K., ed., Polymer Compatibility and Incompatibility: Principles and Practices (Harwood Academic Publishers, Chur, Switzerland, 1982).
4. MacKnight, W.J., F.E. Karasz and J.R. Fried, in Reference 1, Chapter 5.
5. D.R. Paul, this volume.
6. Sanchez, I.C., in Reference 1, Chapter 3.
7. Sanchez, I.C., Ann. Rev. Mat. Sci. 13 (1983) 387-412.
8. ten Brinke, G., F.E. Karasz and W.J. MacKnight, Macromolecules 16 (1983) 1827.
9. Karasz, F.E. and H. Ueda, to be published.

MICROSCOPY AND OTHER METHODS OF STUDYING BLENDS

M. T. Shaw

University of Connecticut
Storrs, CT   06268

1   INTRODUCTION

Examination of a given polymer blend is most often directed at answering questions concerning the degree of miscibility and the morphology of the phases (if two or more are present). In some instances, additional information is sought concerning the composition of the phases and the degree of interaction of the components.

At a most basic level, confirmation of the presence or absence of two phases is desired. While in many cases the presence of two phases is obvious from strong scattering of light, instances arise when the situation is not clear cut. The most notable is when the refractive indices of two components match, but muddy cases also arise when the components are nearly completely miscible and the phases are very small and close in composition. For these reasons, a portfolio of methods has been assembled for characterizing blends.

In Table 1, some experimental methods for the characterization of blends have been listed. The division is according to the amount of information sought. As might be imagined, the difficulty of the analysis increases with the amount of information needed. Some of the methods which can be used for phase analysis can, of course, be used for detection of two phases, but are not often used for screening because of their complexity.

The influence of blending on the primary and secondary transitions in polymers has been covered in a previous chapter. These methods provide valuable and often easy-to-get information concerning the number of phases present in the blend and even the composi-

Table 1

Methods for Analyzing Polymer Blends

A. One or two phases?
    1. Opacity (light scattering)
    2. Glass transition
        a. DSC, DTA, TMA, TOA
        b. Dynamic Mechanical
        c. Dielectric
        d. Volumetric
        e. etc.
    3. Microscopy (visible, phase contrast, electron)
    4. Diffusion/Permeability
    5. Excimer fluorescence
    6. Mechanical, rheological response

B. Compositions of equilibrium phases
    1. NMR
    2. Glass transition
    3. Phase equilibria using any method in A. (Phase boundaries
        are found by varying T, P or concentration)

C. Thermodynamic Parameters
    1. Phase equilibria
    2. Inverse GC
    3. Vapor sorption
    4. Neutron scattering
    5. Heats of solution
    6. Freezing-point depression
    7. Ternary (polymer 1 + polymer 2 + solvent) phase behavior

tion of each phase. They are highly recommended as the first instrumental methods to use in most any blend study.

A drawback of glass transition methods is that they invariably
involve non-equilibrium conditions; typically temperature is swept
at 2 to 20°C/ minute. The results then reflect a combination of
the state of the blend at its preparation temperature and at the
measurement temperature. With high-Tg glassy polymers this usually
does not present much of a problem; equilibrium is established in
the melt during preparation of the specimen and the phase composition and structure are "frozen" in by rapid cooling below Tg. During subsequent rapid heating the blend has insufficient time to re-
equilibrate. With rubbery polymers, however the equilibration temperature may not be evident. Conceivably, the blend could have
been biphasic at the high mixing temperature, yet show one Tg after
sitting at room temperature for a period of time.

The important area of scattering has been discussed in a separate chapter. Scattering of electromagnetic radiation can give information on the size, shape and motion of molecules in a contrasting medium. Size, or coil expansion can be directly related to the interaction between the components of the blends.

In this lecture, a potpourri of methods will be discussed. Some of these may be applicable in special circumstances or because of the availability of equipment. Often they have proved indispensable for blend characterization.

2 MICROSCOPY

For the detailed characterization of the phase morphology in blends, microscopy is unmatched by any other technique. Modern shape analysis equipment eliminates the tedium and increases the accuracy of size, shape, and orientation analyses. These techniques as well as direct observation depend on the development of contrast between the phases. With polymers, this requirement often proves to be very bothersome, and efforts to increase contrast may lead to spurious results.

2.1 Origins of contrast

Optical (visible light) contrast can arise from a number of sources. Color, opacity, refractive index, orientation, absorption, or dichroic differences can all be exploited. With ultraviolet illumination, differences in fluoresence characteristics can be used. Even a difference in mechanical properties can lead to contrast; polishing the specimen and observation with reflected light can be exploited with some blends having sufficiently large and distinct phases.

With the transmission electron microscope, electron scattering differences are the primary source of contrast. With most polymers, adequate differences in scattering are achieved by electron etching of one of the phases to produce a thickness difference, or by reacting one of the phases with a heavy metal. The scanning electron microscope, on the other hand, depends primarily on surface texture for contrast; polishing and etching, or cryofracture are commonly used to generate a height difference between the phases. If a polished spectrum is used, both secondary and back-scattered electrons can be combined to yield contrast, the latter being generally superior. Differences in atomic composition are required, however, and "staining" is often needed for best results (1).

2.2 Visible Microscopy

Good optical microscopes are capable of resolutions of 0.2 μm
if the contrast is sufficient. Phase-contrast optics can enhance
the contrast in many blends by exploiting any refractive index difference between the components; however, resolution often suffers.
Dark-field techniques can provide excellent contrast and resolution
as does ultramicroscopy; with the latter the potential for shape
characterization is lost. Because the eye is capable of high
color resolution, staining the specimen may be helpful. However,
the observer should be aware that contact of the specimen with any
solvents could alter the phase equilibria and structures.

Optical microscopes can serve as an optical bench for a number
of useful experiments with blends. One example is thermal optical
analysis (TOA) where light is transmitted through a temperature-
controlled specimen; phase changes resulting in changes in the
transmitted light are followed with a photocell at the eyepiece.
Changes in orientation or crystallinity are recorded with polarizing optics. Using dark field optics and a photomultiplier, light
scattering at a fixed average angle can be measured, again as a
function of temperature or time.

2.3 Electron Microscopy

The electron microscrope is capable of much higher resolution
than the optical microscope, which opens both doors and traps. Of
the two, transmission electron microscopy (TEM) has seen as much
application to blends as the easier-to-use scanning electron microscope (SEM). Both are capable of high resolution, but both can
have problems with contrast because the elemental compositions of
many polymers are not very different. Attempts to improve contrast
by staining with heavy metals, etching, etc can often introduce
what appears to be structure even in single-component systems.
Nevertheless, TEM/SEM have proved valuable.

One example of an early application of TEM to blends was the
investigation of structures produced by the spinodal decomposition
of a PMMA/SAN blend (2). At room temperature this blend is miscible, but if the temperature is raised to above 200°C (depending
on composition and MW) a phase decomposition proceeds and the blend
gets cloudy. By quenching the blend down quickly from these temperatures after various amounts of time, McMaster (2) was able to
preserve the structure of the decomposing blend. These specimens
were examined by TEM after ultramicrotoming and mounting on a metal
grid. Fortunately, after some exposure to the electron beam, contrast developed; later investigation showed that the PMMA-rich
parts were preferentially etched away by the beam and became thinner
transmitting more electrons. Clearly this system was carefully (or

adventitiously) chosen to be as free from artifacts as possible.
If the blends' or components' Tg's were lower, structure preserva-
tion would have been much more difficult.

Examination of miscible blends by electron microscopy, particularly TEM, has been attempted often for the purpose of defining
the scale of the mixing. With miscible systems, replicas may have
to be made to obtain the desired resolution. Replicas (e.g. a thin
polymer or carbon coating often shadowed with a metal) can have
artifactual structures either from the replication process itself
or from the method for preparing the blend surface.

With SEM, replication or thin sections are unnecessary because
the surface of the blend is examined directly. Contrast, however,
depends, for the most part, on surface texture which corresponds
(hopefully) to the blend structure. Getting sufficient contrast
may be as straightforward as breaking the specimen, the differences
in mechanical properties of the two phases providing the necessary
differences in texture. As with light microscopy, metallographic
polishing techniques have proved useful, each phase taking a different degree of polish. The trouble begins when attempts are made
to improve contrast by solvent or chemical etching. Clearly, if
one phase is completely inert to the treatment while the other
phase reacts strongly (e.g. PP/Nylon 6 etched with formic acid (3)),
the technique probably will work. However, most organic solvents
will affect to some extent most organic polymers. This influences
not only the phase equilibria but may destroy any metastable structures which were present. Almost any form of etching which deposits
energy on the surface can have the same undesirable effect unless
the rate of energy deposition is low compared to the rate which
thermal energy can be removed from the surface. Electron- beam and
ion-beam etching are examples of such techniques.

2.4 Examples

2.4.1 An immiscible blend, HDPE/PET (8). Examination by SEM.
Fracture surfaces of the components, especially PET, were found to
be highly textured; therefore, the blends were etched to improve
contrast. Xylene was used for extracting PP; a mixture of phenol
and tetrachloroethane for the PET. 50/50 blends were found to be
co-continuous when a block copolymer was used as a dispersant; the
PP was continuous when an EPR was used as the dispersant.

2.4.2 Blend of LDPE/PP (5). Blend was made using a two-roll mill
and studied with a polarizing light microscope equipped with a hot
stage. Films 100 μm thick were obtained by pressing the blends
between a glass slide and a cover slip. Blends were found to contain large spherulites of PP; this was identified as being the low-
melting β form and was created as a result of the admixture with
LDPE.

2.4.3 Block copolymers of styrene. Butadiene and Styrene-Isoprene
(6). Phase contrast using SEM, staining not necessary. The material
property of importance is "mean inner potential" which ranges from
6.3-7.3V for hydrocarbon polymers. This quantity is analogous to
refractive index for phase contrast in the optical microscope; it
provides the phase shift. Images obtained using the phase-contrast
technique were not as sharp as those using OsO_4 staining.

2.4.4 Phase decomposition of PVME/PS [7]. Bright-field microscopy
was used with the blend sandwiched between a glass slide and cover
slip. To increase contrast, the aperture diaphragm was nearly
closed. The authors report this caused loss of resolution and
shadowing. A 50/50 blend of PS (MW = 100,000) and PVME (MW = 44600)
showed an interconnected structure after 10 min at 423 K, typical
of spinodal decomposition. After 60 minutes, the phases coarsened.

2.4.5 Phase separation in EVA/chlorinated PE [8]. Light microscopy, used in the shearing mode (a method of interference contrast),
was used to examine blends heated to induce separation. Phase separation after heating at 90°C for 3 hours was clearly visible.

2.4.6 Staining with Mercuric Trifluoroacetate for TEM contrast
(9). To demonstrate contrast development in a blend, a laminate of
alternating layers of PPO and PS was stained with a 10% solution of
H_2O in trifluoroacetic acid. The H_g preferentially attaches to the
PPO and good contrast develops in 60 minutes. Obviously, the same
stain could be employed for SEM and imaging using backscattered
electrons or x-rays (energy dispersive x-rays, EDAX).

2.4.7 Optical Phase Contrast and Interference Contrast (10). Two-
phase blends of PS/PVME were heated at temperatures below the LCST.
Dissolution of the structure was followed by phase-contrast microscopy. Resolution was about 2 μm.

3   SPECTROSCOPY

3.1 Overview

Various types of energy absorption or fluorescence spectro-
scopies have been proven immensely useful for understanding the
structures and dynamics of organic molecules. It is only natural
that these methods should be applied to polymer blends. While it
cannot be stated that spectroscopy has answered all our questions,
it has provided some information which would be difficult to obtain
any other way.

Nuclear magnetic resonance (NMR), infrared absorption (IR),
and UV/Visible fluorescence spectroscopy have seen wide application

in blend studies. X-ray photoelectron spectroscopy (XPS or ESCA) and energy dispensive x-ray analysis (EDAX) have been also used.

3.2 Nuclear Magnetic Resonance

The main application of NMR in polymer blends, until recently, has been as a probe of the physical state of the phases.. In this mode, NMR might be considered as a supplement to dynamic mechanical analysis. NMR does have some advantages, however, in that is sensitive to smaller phases; is very fast (can be used to follow phase separation processes); and can be used, under favorable circumstances, to diagnose how much of each component is in each phase. Pulse techniques are commonly employed; relaxation of magnetization is influenced by the physical environment of the nuclei. Mobile protons will show very slow spin-spin (T_2) relaxation, while protons in a rigid matrix will relax more rapidly. The strength of these relaxations is proportional, as a first approximation, to the number of protons involved. Similar arguments apply for T_1, the spin-lattice relaxation process, but T_1, generally exhibits a minimum as the mobility of the matrix is varied.

It should be cautioned that the mere presence of multiple relaxation processes does not demonstrate the presence of two phases; the protons in a miscible polymer system will experience a broad range of local environments. Experiments designed to follow the change of the relaxation behavior with temperature, concentration, pressure, etc. are usually needed to determine the location of phase boundaries.

Solid state ^{13}C NMR, because of its high resolution, should be capable of revealing a great deal about the individual environments of the carbon atoms in a blend. Chemical shifts quantify the average magnetic environment while relaxation experiments can characterize the dynamics of motion. In principle chemical shifts will be perturbed by the presence of a second component in a miscible system, but generally it is the line shapes which tend to be changed most significantly. These changes are quantified readily by examining the influence of proton polarization on ^{13}C polarization in what is called the $T_{1\rho}$ (H) process.

Following are some examples of the application of NMR to blends:

3.2.1 Wide line ^1H NMR of PVC/EVA blends (11). Key experimental result is k/k_s where k is the ratio of amplitude of the narrow and broad lines (mobile vs. immobile protons) and k_s is the same ratio for a standard sample. The purpose of a standard is to account for instrument drift. Mixing at RT yields higher k/k_s values than mixing at 110°. k/k_s dropped with mixing time as well, indicating increased incorporation of the soft EVA in the hard PVC phase.

3.2.2 Wide line ^{1}H and ^{19}F NMR of PE/PTFE blends [12]. Displayed were the derivative of the absorption and second moments of the absorption lines. Proton lines at 20°C were broadened by addition of PTFE indicating immobilization of PE molecules. The second moment of the line emphasizes changes in the "tails" of the line.

3.2.3 Magic angle ^{13}C NMR of PPO/PS (13). Aliphatic main-chain resonance of the PS in the blend was broadened, as was the protonated aromatic carbon resonance of PPO.

3.2.4 $T_{1\rho}$ (H) experiments on PS/PVME (14). $T_{1\rho}$ (H) refers to the spin-lattice relaxation in the rotating frame for protons. By observing the ^{13}C resonance, the relaxation of spins for protons most directly coupled with each carbon can be examined individually. For PS carbons, the $T_{1\rho}$ (H) values dropped from around 3.9 to 2.4 ms as the blend was compatibilized, indicating close association.

3.3 Excimer Fluorescence

Certain aromatic groups are capable of absorbing energy in the UV and releasing this energy in a number of ways including fluorescence. The fluorescence spectrum is characteristic of the particular aromatic group.

Two fluorescence methods have been used for the study of blends. One is termed "nonradiative energy transfer" in which the amount of energy transferred from a donor group to an acceptor group is assessed. Because the transfer is nonradiative, the donor and acceptor must be very close; thus if two polymers are labeled with donor and acceptor, a large amount of energy transfer would imply miscibility. A napthyl group might be used or a donor while an anthryl group could be used as the acceptor.

In a second technique, the intensity of emission from a fluorescent group (e.g. napthyl) is compared with the intensity from dimers of this group. The dimers are termed "excimers". The key experimental observation is I_D/I_M, the ratio of the intensities from the dimer and monomer. As the miscibility of the photo-sensitive polymer in the host increases, the formation of dimers will be inhibited by coil expansion and the presence of inert material. I_D/I_M will then decrease.

The supreme advantage of these methods is that the proximity of groups at the several Angstrom level can be unambiguously concluded. The disadvantage is the lack of generality; the polymer must contain a fluorescent group. While tagging can remove this objection to some extent, it can be argued that the tags may themselves associate in some nonrandom fashion, particularly if they are highly different from the rest of the polymer. All components used in these studies must be very free of fluorescent molecules, including antioxidants. For this reason, experiments at elevated temperature may be limited by thermal degradation.

Some examples of the application of these techniques to blends are as follows:

3.3.1 Nonradiative energy transfer in acrylate copolymer blends (15). Napthyl (N) was used as a donor, anthryl (A) as an acceptor. The I_N/I_A ratio is expected to decrease as the system becomes more miscible. In 50:50 blends of poly(methyl methacrylate) with poly (methyl methacrylate-co-butyl acrylate), I_N/I_A increased rapidly when the butyl acrylate content of the copolymer exceeded 0.2 mole fraction.

3.3.2 Study of spinodal decomposition of PS/PVME blends (16). The decomposition was followed by measuring the intensities of the excimer (dimer) and monomer. The ratio I_D/I_M at $423°K$ increased rapidly in the first 30 minutes.

3.4 Infrared Spectroscopy

The reasoning behind the use of infrared spectroscopy (IR) in blend studies is as follows: If the blend is immiscible, the absorption spectrum of the blend will be the sum of those for the components. If the blend is miscible because of an interaction between the two components which perturbs the bonding between atoms, then differences will be noted in the spectrum of the blend relative to the sum of those for the components.

In many blends, these changes may be exceedingly small; thus the high resolution of Fourier Transform IR (FTIR) is usually necessary. There is a reported danger in working with blends in that apparent changes in frequency can arise due to dispersion effects when the refractive indices of the components differ significantly (17). Additionally, any changes in atomic geometry forced upon the molecules by the blending will change the normal mode frequencies and the spectrum; an interaction need not be present. The latter is common with crystalline polymers (18). If all goes well, the IR investigation of a miscible blend will not only reveal the presence of an interaction, but will provide information on which groups are involved. This knowledge could prove valuable for designing components for new miscible systems.

One widely studied interaction is hydrogen bonding. Hydrogen bonding involves a distinct shift of electron density from the vicinity of the electron-rich atom (e.g. oxygen) to between this atom and the proton. The shift of electron density often lowers the vibrational transition energies associated with both old groups; considerable broadening is also usually noted.

Below are annotated examples of the application of IR to blends:

3.4.1 FTIR studies with PS/PVME [19]. With 50:50 blends cast from solvents to produce one-phase or two-phase films, the PVME absorption in the 1100 cm^{-1} region (involving the oxygen atom) showed strong changes. With polystyrene, a CH out-of-plane bending mode at 700 cm^{-1} proved most sensitive to the degree of miscibility.

3.4.2 PE/PTFE (12). Comparisons were made in the intensity of the CH$_2$ rocking mode (731 cm^{-1}) as the sintering time was increased. Blends at each end of the composition range showed little change with sintering time while a 50:50 blend showed a large increase, suggesting interdiffusion.

3.5 X-ray Photoelectron Spectroscopy

Nearly any type of spectroscopy which measures energy levels associated with translation, vibration, electronic transitions, or nuclear magnetic transitions can show solvation effects and therefore be applied to the problem of the physical mixing of polymers. Depending on the interaction suspected, some types are more appropriate than others; for example, UV/Visible spectroscopy might to chosen if electron transfer complexes could form between the components. In addition, any spectroscopy which can do compound or elemental analysis can be employed for the determination of phase compositions.

One example is x-ray photoelectron spectroscopy (XPS or ESCA). On irradiating a polymer surface with x-rays, electrons with characteristic energies are emitted; oxygen can be distinguished easily from carbon for example. By varying the grazing angle, the depth of sampling into the polymer can be varied; 10-50 Å is typical. With a penetration depth of only tens of angstroms, the composition of one _very_ thin polymer film cast upon another polymer can be determined as a function of time [20]. Proper analysis of this one dimensional diffusion problem will then provide the miscibility of the substrate polymer in the film polymer.

Some examples of the application of XPS to blends are given below:

3.5.1 Sintered PTFE/PE blends (21). Mixed powders of the two components were sintered and then cut or cleaved. The cut surfaces reflected the overall composition of the blend whereas the cleaved surfaces were high in PTFE. Changes of the composition of the cleaved surface with bulk concentration were taken as evidence that the PTFE/PE interface was _not_ the site of the cleavage.

3.5.2 Diffusion of PS into a PEO layer (20). Laminates were heated at 130°C for various times and the C_{1s} core level signals were monitored. An exponential decrease in the rate of appearance of the PS in the PEO surface layer was found, suggesting diffusion.

4   VOLUME OF MIXING

On mixing two polymers the total volume of the system will, in general, undergo some change. If the miscibility of the components is very low, the volume of mixing will be small. However, if the blend is miscible because of strong intercomponent interactions, a measurable negative volume of mixing might be expected.

Before discussing the thermodynamics of volume changes, it is important to settle on a definition for volume of mixing, ΔV^M, which will reflect only those changes due to interactions. In terms of total volumes:

$$\Delta V^M = V_B - \Sigma \ V_i \qquad (1)$$

where V_B is the total volume of the blend and V_i are the total volumes of each component as they would be alone but in the same physical state as they exist in the blend. Thus if component 1 is semicrystalline when pure, but an amorphous rubber when mixed, then the amorphous volume should be used. Conversely the rather common case of a rubbery component causing the crystallization of a glassy polymer should be guarded against.

There is no easy direct connection between volume of mixing and ΔG^M, the free energy of mixing. However, the empirical observation with low molecular weight compounds is that a negative ΔV^M is often associated with a negative (exothermal) ΔH^M. (The change of miscibility with pressure must also be known to make a quantitative connection.) An early example with polymers involved the miscible system PS/PVME, where densification was found (22). A more recent example, involving PVC/poly(α-methyl-α-n-propyl-β-proprolactone) also showed a small densification (23).

5   TERNARY METHODS

The examination of interactions in a polymer-polymer system can often be simplified by using a third component, usually a volatile solvent. The activity of a volatile solvent is nearly always easy to measure accurately; and, if present in sufficient quantity, the solvent promotes equilibration and analysis of phases. The drawback is that the solvent will, in general, modify the interactions between the polymer segments to a degree which is not readily determinable or eliminated. This modification does not necessarily disappear smoothly with solvent concentration; thus, extrapolation to 100% polymer may not lead to a valid answer.

Below we discuss two high (polymer) concentration ternary methods and two low-concentration methods.

5.1 Vapor Sorption:

The general idea behind vapor sorption is that the activity of a volatile solvent in a polymer blend will be increased over the rule-of- mixtures value if the two polymers interact strongly. Solvent activity is readily available by direct vapor pressure measurement over the blend. Typically a sample will be exposed to a certain vapor pressure of solvent in a vacuum chamber and the uptake of the solvent will be determined by weighing the sample. Application of the Flory-Huggins equation can provide an interaction parameter for the two polymers, but the individual solvent-polymer interaction parameters must also be measured. Extrapolation to zero solvent concentration should follow; additionally a demonstration of the independence of the answer on solvent nature would be a wise precaution. Note that the usual treatment assumes the validity of the Flory-Huggins theory; however, more thorough theories could, in principle, be applied.

An simplified version of the type of equation used for the analysis of vapor sorption in blends is given by Spencer and Yavorsky (24).

$$\chi_{23}' = \frac{\ln k_{1b} - w_2 \ln k_{12} - w_3 \ln k_{13}}{w_2 w_3} \qquad (2)$$

where χ_{23}' is $\chi_{23} V_1/V_2$ (V_i = molar volume), k_{1b} is the Henry's-law constant for vapor (component 1) over the blend, k_{12} and k_{13} are the Henry's-law constants for pure polymers 2 and 3, and w_i are weight fractions in the blend. If the densities of the two polymers are not close, then volume fractions must be used. The above equation states that if χ_{23}' is zero, then the proper straight-line "rule-of-mixtures" will result from a plot of $\ln k_{1b}$ vs weight fraction of component 2 in the blend. A deviation below this line (less solubility of the vapor in the blend) will signify a negative χ_{23}' (favorable interaction between the polymers).

Some examples using vapor absorption techniques are outlined below:

5.1.1 Sorption of CO_2 in a blend of polycarbonate and a copolyester of 1,4-cyclohexanedimethanol with mixed iso- and terephthalic acids (25). CO_2 absorption isotherms were measured and found to follow a combined Henry's and Langmuir's law; thus some manipula-

tion was required to get the Henry's-law constants, k. With a
50:50 blend, the value of k was nearly 10% below the rule-of-
mixtures value, indicating a favorable interaction between the components.

5.1.2 Blends of PVME/PS examined with propane (24). χ_{23}' was
found to be positive at low PVME fraction, going through zero at
about 30 wt % PVME. The expected increase of χ_{23}' with temperature
was observed.

5.1.3 Sorption of decahydronapthalene by blends of PS/PoClS (26).
χ_{23} decreased with increasing PoClS content, becoming negative between 50 and 75%.

5.2 Inverse Gas Chromatography (IGC)

Gas chromatography is often a convenient method for measuring
the interaction of a solvent with a polymer. "Inverse" comes from
the fact that the test material (the polymer) is the stationary
phase, while the probe (the solvent) is injected into the carrier
gas. With a blend, the probe will be swept through the column
faster than expected from the rule-of-mixtures if strong interactions between the polymers are present.

While convenient, IGC is not without its difficulties (27).
The most severe problem may be the state of the polymer blend on
the column packing. If the blend is single phase and coats the
support uniformly, few problems with the analysis should arise.
However, if the polymer coating is too thick, diffusion effects
will become important and only a small fraction of the total polymer will contribute to the slowing of the probe. Clearly, the
diffusion of the probe in and out of the coating will be influenced
by the physical state of the polymer as well. Any crystallinity
will remove polymer from interaction with the probe and exacerbate
the diffusion problem. On the other hand, incomplete coverage of
the packing material may lead to severe adsorption effects with
some probes. For two-phase polymer systems the retention volume of
the probe will depend on the morphology of the system, and will, in
general be unpredictable.

The basic equation for analyzing IGC data is

$$V_g^o = \lim_{Q \to o} (t_s - t_a) \frac{Qj}{w} \frac{273.}{T} \tag{3}$$

where V_g^o is the retention volume per gram of polymer, Q is the gas
flow rate, t_s and t_a are the retention times for solvent and air
respectively, and T is the column temperature. j is a correction
factor for the difference in inlet and outlet pressures in the
column. For analyzing for the binary interaction parameter χ_{12},

$$X_{12} = \ln \frac{273.2 \, R \, v_{sp,2}}{P_1^{\,0} \, v_g^{\,o} \, V_1} - \frac{P_1^{\,0}}{RT} (B_{11} - V_1) \; -1. \qquad (4)$$

where $v_{sp,2}$ is the specific volume of the polymer, V_1 is the molar volume of the solvent, B_{11} is its second virial coefficient, and $P_1^{\,o}$ its vapor pressure. All these numbers are to be taken at the column temperature. For a blend

$$X_{23} = \frac{V_2}{\phi_3} [\frac{1}{\phi_3} \ln \frac{v_{sp,2}}{v_{g,2}^{\,o}} + \frac{1}{\phi_2} \ln \frac{v_{sp,3}}{v_{g,3}^{\,o}} -$$

$$\frac{1}{\phi_2 \phi_3} \ln \frac{w_2 \, v_{sp,2} + w_3 \, v_{sp,3}}{v_{g,23}^{\,o}}] \qquad (5)$$

where ϕ_i is volume fraction.

Doubé and Walsh (27) have demonstrated the use of IGC for detecting the phase transition in polymer blends. In these studies, they found a sharp discontinuity in retention volume vs. temperature; the temperature of the discontinuity was in good agreement with that found by other methods for the cloud point.

Some recent examples of the application of IGC to blends are cited below:

5.2.1 PS/Polycarbonate blend (28). X_{23} was measured over the entire composition range and at 465K and 475K. (The T_g of PC is ~425K). The value of X_{23} dipped below zero at two points, one at low concentration, the other at high. (Because no data points are included in the figure, it is not possible to judge the possible significance of the observed complexities.)

5.2.2 IGC investigation of PS/Poly(n-butyl methacrylate) blends [29]. IGC experiments using a wide variety of probes (solvents) were conducted at 140°C. The results varied considerably, e.g. at 58% PS (high MW) the range of $X_{23}' = X_{23} V_1/V_2$ was from -0.08 to 0.24. A simple average showed that X_{23}' could be expected to be positive for this immiscible system. With a low MW PS, the X_{23}' values showed a complex variation with composition. (Conceivably these variations could be due to problems with column packing at some of the compositions.)

5.3 Solution Viscosity

Most polymer laboratories have equipment for measuring the solution viscosities of polymers. It is not surprising, therefore, that many have attempted to apply solution viscometry to the investigation of polymer-polymer interaction.

Intrinsic viscosity of a polymer is the normal concentration-independent quantity obtained from solution viscosity. It is a measure of coil size in the solvent. If the intrinsic viscosity of a blend is measured, one might expect the value to be in between that of the components if there is little interaction and above the average if there is a strong positive interaction causing aggregation.

An unfavorable interaction might show up at higher concentration as a lowering in viscosity because of a collapse of polymer coils. The data from such experiments have been interpreted in terms of the expression

$$\eta_{sp}/c = [\eta] + bc \tag{6}$$

where η_{sp} is the specific viscosity of the blend solution, $[\eta]$ is the zero-concentration limit (intrinsic viscosity), and b the slope of the usual η_s/c vs c plot. Deviations of b from a rule-of-mixtures value is taken as evidence of positive or negative interaction. More complicated expressions have been used (30) for example

$$\eta_{sp} = c_1[\eta_1] + c_2[\eta_2] + c_1^2 b_1 + c_2^2 b_2 + \tag{7}$$
$$2 c_1 c_2 b_{12} + \ldots$$

where b_{12} is considered to reflect the interactions between the components in the solution. One group [31] has simply interpreted deviations from a linear rule-of-mixtures for solution viscosity vs weight fraction in the blend (total solids in the solution = 2%) as a sign of interaction in PVC/poly(vinyl acetate) blends.

5.4 Ternary Phase Behavior

Perhaps the earliest method of checking the "compatibility" of two polymers was to examine the phase behavior of the two polymers in a common solvent. This method still finds use.

Scott, in 1949, extended the Flory Huggins theory to include ternary systems containing two polymers (32). Patterson and coworkers have done a great deal to update this theory.

The working relationship developed from the Scott equations is:

$$X_{23} = \frac{[\frac{1}{r_3} \ln (\frac{\phi_3''}{\theta_3'}) - \frac{1}{r_2} \ln (\frac{\phi_2''}{\phi_2'}) - (X_{13} - X_{12})(\phi_1' - \phi_1'')]}{(\phi_3'' - \phi_3' + \phi_2' - \phi_2'')}$$

where the prime and double prime refer to conjugate phases; ϕ_i is volume fraction; r_i is the molar volume of polymer i relative to the solvent; and subscripts 2 and 3 refer to the polymers, 1 to the solvent. The experiment thus requires an analysis of both phases (which, hopefully, will have different densities and separate into clean layers), a knowledge of the densities and molecular weights of each component, and the binary (polymer/solvent) interaction parameters. Because of the problem of separating the phases at polymer concentrations higher than ~10%, the X_{23} calculated with the relationship above can be expected to depend strongly on solvent.

A simplification of the above equation obtains if $r_3 = r_2 = r$, $X_{13} = X_{12}$, and the amounts of polymer are equal. Under these conditions the only experimental quantity required is the amount of solvent needed to form a single-phase solution (i.e. ϕ_1). Then

$$X_{23} = \frac{2}{r} \frac{1}{1-\phi_1} \qquad (9)$$

If the plait point (the ternary composition where the two phases merge toward the same composition) can be found then the Scott theory predicts that

$$X_{23} = \frac{1}{2} [\frac{1}{r_2^{1/2}} + \frac{1}{r_3^{1/2}}] \frac{1}{1-\phi_1} \qquad (10)$$

and only the molecular weights of each component must be known.

Analysis of the equilibrium phases follows conventional routes, with dual-detector gel permeation chromatography being particularly convenient (33,34).

One unconventional method of running the ternary experiment is to crosslink one of the polymers (component 3) and equilibrate the gel with solution of the second polymer (component 2) (35). None of the crosslinked polymer enters the solution, but the second polymer and solvent enter the gel until equilibrium is achieved. On assuming X_{12} is independent of concentration, the following expression obtains:

$$X_{23} = \frac{V_2}{V_1} \{X_{13} - X_{12} + 2X_{12} \frac{\phi_1 - \phi_1''}{\phi_3''} - \frac{1}{\phi_3''}[\ln(\frac{\phi_1'}{\phi_1''}) - \frac{V_1}{V_2} \ln \frac{\phi_2'}{\phi_2''}]\} \qquad (11)$$

Here the symbols are the same as the former expression; the double prime refers to the gel phase. ϕ_1'' and ϕ_2'' can be found by weighing the gel phase before and after drying off the solvent. If component 2 is high in molecular weight, the amount of weight gain will be small.

Some examples illustrating the application of ternary phase equilibria to blends are noted below:

5.4.1 Poly(vinyl acetate)/PS using the gel technique (35). The poly(vinyl acetate) gel was formed by air curing at 140°C for 3 hrs. Polystyrene had a degree of polymerization equal to 24; about 15 mg of polystyrene entered 1 g of gel when the PS/benzene solution contained 10% PS. χ_{23}' was calculated to be 0.5.

5.4.2 PS/Polybutadiene (34). Plait point compositions were found using GPC to analyze the phases. χ_{23} at the plait point depended on PS molecular weight, but was the same for the two solvents used (THF and toluene).

5.4.3 PS/Poly(vinyl acetate) (30). Ternary phase diagrams were obtained by turbimetric titration of blends of various ratios using THF as the solvent; plait points and the lines cannot be established with this method. (Since the two molecular weights were the same, the extreme symmetry of the phase diagram suggests $\chi_{12} = \chi_{13}$. Application of equation 9 above with r = 2.80, $\phi_1 = 0.92$ gives $\chi_{23} = 0.09$.) The authors found that by introducing small amounts of interacting ionomers, the miscibility increased.

6   MELT RHEOLOGY

The use of solution or melt rheology to look for interactions in blends is largely empirical in nature. As described in the discussion of ternary methods, the expectation is that polymers with strong favorable or unfavorable interactions will show viscosity vs. concentration responses that are largely non-linear   There really is very little theory to back up this expectation, however.

A typical expression is that described by Carley (36)

$$\ln \eta_b = (\ln \eta_1) w_1^3 + (\ln \eta_2) w_2^3 + B_1 w_1^2 w_2 + B_2 w_2^2 w_1 \quad (12)$$

which is capable of fitting quite complex blend viscosity behavior. In this equation, η_b is the blend viscosity, η_1 and η_2 are the viscosities of the components, w_i are weight fractions, and B_1 and B_2 are interaction terms. Deviations from a rule-of-mixtures (in this case from a $\log\eta - w_2^3$ rule) will be reflected in B_1 and B_2.

One of the more extensive analyses of thermodynamics vs. rheology has been described by Lipatov et al. (28). They found relationships between the phase diagrams for various blends and the viscosity-concentration behavior. In several instances a close correlation between the sign of χ_{23} from IGC and the sign of the deviation of viscosity from a rule-of-mixtures ($\log\eta$ vs w) was found. No explanations were offered.

Some have proposed that the shear-rate response of blends or their solutions can reveal something about the state of the mixture. As an example, Chitrangad and Middleman (37) found some irregularities in the viscosity-shear rate response of 12% PPO/PS solutions with varying ratios of the two components.

References

1. Hobbs, S. Y. and V. H. Watkins. The Use of Chemical contrast in the SEM Analysis of Polymer Blends. Journal of Polymer Science, Polymer Physics Edition 20 (1982) 651-658.
2. McMaster, L. P. Aspects of Liquid-Liquid Phase Transition Phenomena in Multicomponent Polymeric Systems. Advances in Chemistry Series 142 (1975) 43.
3. Liang, B.-R., J. L. White, J. E. Spruiell and B. C. Goswami. Polypropylene/Nylon 6 Blends: Phase Distribution Morphology, Rheological Measurments, and Structure Development in Melt Spinning. Journal of Applied Polymer Science 28 (1983) 2011-2032.
4. Traugott, T. D., J. W. Barlow and D. R. Paul. Mechanical Compatibilization of High Density Polyethylene-Poly(ethylene Terephthalate) Blends. Journal of Applied Polymer Science 28 (1983) 2947-2959.
5. Teh, J. W. Structure and Properties of Polyethylene-Polypropylene Blends. Journal of Applied Polymer Science 28 (1983) 605-618.
6. Handlin Jr., D. L. and E. L. Thomas. Phase Contrast Imaging of Styrene-Isoprene and Styrene-Butadiene Block Copolymers. Macromolecules 16 (1983) 1514-1525.
7. Gelles, R. and C. W. Frank. Phase Separation in Polystyrene/-Poly(vinyl methyl ether) Blends as Studied by Excimer Fluorescence. Macromolecules 16 (1983) 1448-1456.
8. Walsh, D. J., J. S. Higgins and S. Rostami. Compatibility of Ethylene-Vinyl Acetate Copolymers with Chlorinated Polyethylenes/Compatibility and Its Variation with Temperature. Macromolecules 16 (1983) 388-391.
9. Hobbs, S. Y., V. H. Watkins and R. R. Russell. TEM Studies of Polymer Blends Stained with Mecuric Trifluoroacetate. Journal of Polymer Science, Polymer Physics Edition 18 (1980) 393-395.

10. Davis, D. D. and T. K. Kwei. Phase Transformation in Mixtures of Polystyrene and Poly(vinyl Methyl Ether). Journal of Polymer Science, Polymer Physics Edition 18 (1980) 2337-2345.

11. Gevert, T. U., I. Jakubowicz and S. E. Swanson. Phase Relations in Blends of Poly(vinyl chloride) with Copolymer of Ethylene-Vinyl Acetate and the Effects of Thermal Stabilizers. European Polymer Journal 15 (1979) 841-847.

12. Nagarajan, S. and Z. H. Stachurski. A Study of the PE-PTFE System. I. IR and NMR Measurement. Journal of Polymer Science, Polymer Physics Edition 20 (1982) 989-1000.

13. Stejskal, E. O., J. Schaefer, M. D. Sefcik and R. A. McKay. Magic-Angle Carbon-13 Nuclear Magnetic Resonance Study of the Compatibility of Solid Polymeric Blends. Macromolecules 14 (1981) 275-279.

14. Kaplan, S. An NMR Study of Compatibility of Polystyrene/-Poly(Vinyl methyl ether) Blends. Polymer Preprints 25(1) (1984) 356-357.

15. Amrani, F., J. M. Hung and H. Morawetz. Studies of Polymer Compatibility of Nonradiative Energy Transfer. Macromolecules 13 (1980) 649-653.

16. Gelles, R. and C. W. Frank. Effect of Molecular Weight on Polymer Blend Phase Separation Kinetics, Macromolecules 15 (1982) 1486-1491.

17. Coleman, M. M. and D. F. Varnell. Fourier-Transform Infrared Studies of Polymer Blends. III. Poly(β-Propiolactone)-Poly-(vinyl chloride) System. Journal of Polymer Science, Polymer Physics Edition 18 (1980) 1403-1412.

18. Varnell, D. F., J. P. Runt and M. M. Coleman. Fourier Transform Infrared Studies of Polymer Blends. 6. Further Observations on the Poly(bisphenol A carbonate)-Poly(ϵ-caprolactone) System. Macromolecules 14 (1981) 1350-1356.

19. Lu, F. J., E. Benedetti and S. L. Hsu. Spectroscopic Study of Polystyrene and Poly(vinyl methyl ether) Blends. Macromolecules 16 (1983) 1525-1529.

20. Thomas, H. R. and J. J. O'Malley. Surface Studies on Multicomponent Polymer Systems by X-ray Photoelectron Spectroscopy: Polystyrene/Poly(ethylene oxide) Homopolymer Blends. Macromolecules 14 (1981) 1316-1320.

21. Nagarajan, S., Z. H. Stachurski, M. E. Hughes and F. P. Larkins. A Study of the PE-PTFE System. II. ESCA Measurements. Journal of Polymer Science, Polymer Physics Edition 20 (1982) 1001-1012.

22. Kwei, T. K., T. Nishi and R. F. Roberts. A Study of Compatible Polymer Mixtures. Macromolecules 7 (1974) 667-674.

23. Aubin, M. and R. E. Prud'homme. Miscibility in Blends of Poly(vinyl chloride) and Polylactones. Macromolecules 13 (1980) 365-369.

24. Spencer, H. G. and J. A. Yavorsky. Solubility and Diffusion of Propane in Blends of Polystyrene and Poly(vinyl methyl ether) at T > Tg, Journal of Applied Polymer Science 28 (1983) 2837-2446.

25. Masi, P., D. R. Paul and J. W. Barlow. Gas Sorption and Transport in a Copolyester and Its Blend with Polycarbonate. Journal of Polymer Science, Polymer Physics Edition 20 (1982) 15-26.

26. Zacharius, S. L., G. tenBrinke, W. J. MacKnight and F. E. Karasz. Evidence for Critical Double Points in Blends of Polystyrene and Poly(o-chlorostyrene), Macromolecules 16 (1983) 381-387.

27. Doubé, C. P. and D. J. Walsh. Studies of Poly(vinyl chloride)/Solution Chlorinated Polyethylene Blends by Inverse Gas Chromatography. European Polymer Journal 17 (1981) 63-69.

28. Lipatov, Yu. S., A. E. Nestorov, V. F. Shumsky, T. D. Ignatova and A. N. Gorbatenko. Comparison of Thermodynamic and Rheological Properties of Binary Polymer Mixtures, European Polymer Journal 18 (1982) 981-986.

29. DiPaola-Baranyi, G. and P. Degré. Thermodynamic Chracterization of Polystyrene-Poly(n-butyl methacrylate) Blends. Macromolecules 14 (1981) 1456-1460.

30. Lecourtier, J., F. Lafuma and C. Quivoron. Compatibilization of Copolymers in Solution Through Interchain Hydrogen Bonding. European Polymer Journal 18 (1982) 241-246.

31. Singh, Y. P. and R. P. Singh. Compatibility Studies on Solutions of Polymer Blends by Viscometric and Ultrasonic Techniques. European Polymer Journal 19 (1983) 535-541.

32. Scott, R. L. The Thermodynamics of High Polymer Solutions. V. Phase Equilibria in the Ternary System: Polymer 1-Polymer 2-Solvent. Journal of Chemical Physics 17 (1949) 279-284.

33. Narasimhan, V., D. R. Lloyd and C. M. Burns. Phase Equilibria of Polystyrene/Polybutadiene/Tetrahydrofuran Using Gel Permeation Chromatography. Journal of Applied Polymer Science 23 (1979) 749-754.

34. Narasimhan, V., R. Y. M. Huang and C. M. Burns. Polymer-Polymer Interaction Parameters of Polystyrene and Polybutadiene from Studies in Solutions of Toluene or Tetrahydrofuran. Journal of Polymer Science, Polymer Physics Edition 21 (1983) 1993-2001.

35. Sakurada, I., A. Nakajima and H. Aoki. Phase Equilibria in Ternary Systems Containing a Solvent, a Linear Polymer, and a Crosslinked Polymer. Journal of Polymer Science 35 (1959) 507-517.

36. Carley, J. F. A Simple, Accurate Model for Viscosities of Polymer Blends, Conference Proceedings, Society of Plastics Engineers, 42nd Annual Technical Conference 30 (1984) 439-442.

37. Chitrangad, B. and S. Middleman. Shear Rate Dependent Viscosities of PPO-PS Polymer Blend-Solvent Systems. Macromolecules, 14 (1981) 352-354.

PREPARATION OF BLENDS

M. T. Shaw

University of Connecticut

Storrs, CT   06268   USA

1   OVERVIEW

The preparation of a blend is important both from the point of view of its properties and its economics. In this section the discussion will center on the influence of preparation method on the properties of the blend. Further insight into blending technology will be provided in the lectures on Rubber Blends, Processing and Rheology, and Interfacial Agents.

The objective of mixing during the preparation of blends is to bring the component materials into close proximity, facilitating the relaxation of any non-equilibrium concentration gradients. Generally, mixing is aided by solvents, heat, or both. In addition, shearing of the mixture is required. The limit of mixing is determined mainly by thermodynamics: concentration equilibria and interfacial energies are of primary consideration. While clearly the strain energy in parts of the mixing device can be very high, it generally is difficult to increase permanently the fineness of a polymer-polymer dispersion by application of intensive mechanical input alone. As will be seen, however, mechanical energy can lead to formation of interfacial agents which influence blend dispersion.

2   MELT MIXING

2.1   Advantages

Mixing in the melt state is often the method of choice for the preparation of polymer blends. It offers the advantage of introducing no foreign components (e.g., solvents) into the blend. For this reason and because of the simplicity and speed of melt mixing,

it has economic advantages which make it the primary commercial blend-
ing method. By using proper equipment, it is possible to obtain excel-
lent dispersion and equilibration of the components. Temperature,
time and environment for the mixing can usually be carefully con-
trolled.

2.2 Limitations

The primary disadvantage of melt mixing is that both the compo-
nents must be in the molten state, which can mean that temperatures
may be high enough to cause degradation. Researchers doing basic
studies involving melt mixing should check for degradation even if
precautions are taken to prevent it. Removal of oxygen, while help-
ful, is not a guarantee against degradation; shear stresses during
melt mixing are high enough to cause chain scission, for example.

While most mixing with commercially available equipment pro-
ceeds satisfactorily, one occasionally runs across pairs of polymers
which are difficult to mix. In most cases, a large difference in
the melt viscosity of the components explains the difficulty.

Mixing equipment for melts invariably involves metal surfaces
which move in opposition; for this reason, bearings and seals are
always required. Unless care is taken to keep the machinery clean,
preferably by complete disassembly and cleaning between each mix,
contamination of the blend is very likely. The smaller the mixer,
the greater the potential for a problem.

A final disadvantage of melt mixing is the cost of the equipment
involved. Also, even laboratory-size mixing equipment generally
works well only with large amounts of material; e.g., 50 grams or
more. If mixing of quantities of less than 1 g is required, melt
mixing is usually not feasible.

2.3 Procedures

For laboratory-scale mixing a number of devices are available.
An electrically heated two-roll mill can be useful, and sizes are
available from 2-1/2 cm x 7 cm (roll size) on up. By extending
the guides (the plows which keep the melt from rolling off the ends
of the rolls) toward the center of the rolls, very small amounts of
polymer can be mixed. Roll mills are completely open to the air
and dust, a disadvantage, but they are the easiest mixing device to
clean. The mixing effectiveness of a two-roll mill can vary from
good to very poor depending upon the rheology of the components and
the skill of the operator. As opposed to many designs, the mixing
action of many roll units can be controlled by the operator; roll-
gap, roll temperature, amount of material on the rolls, order of
mixing and the method of adding the components can be varied on
most mills. On some units, the speed of the rolls and the speed
differential can also be varied.

Two techniques which can be applied to difficult-to-mix systems using a two-roll mill are fugitive plasticization and sequencing. (Other mixing devices can also be used.) The former is useful when one of the components has a much higher glass transition or melting temperature than the other. This material can be softened by the addition of a small amount of volatile solvent. This softened component is then added to the fluxed second component; as mixing proceeds the solvent escapes. The other technique, sequencing, is similar in purpose but uses heat instead of solvent. Here the high Tg material is fluxed first using a high roll temperature. After fluxing has occurred, the second component is added <u>slowly</u> to the first and the temperature of the rolls is dropped accordingly. In addition, the roll gap must be adjusted continuously to accommodate the increasing volume of the blend.

A very respected mixing device in the industry is known as the Banbury mixer. As with the two-roll mill, this device has two counter-rotating rolls, but they are lobed to move the blend axially on the rolls. (On a two-roll mill, the operator must provide axial mixing by removing and turning the melt.) In addition, the rotors are enclosed which produces more areas of high shear and also allows the melt to be forced against the rolls by a ram. Mixing in a Banbury can be impressively rapid and efficient, but it is critical to have exactly the right volume of melt. Small Banburys are available, but most still require kilograms rather than grams. The closest equivalent for small mixing jobs is the Brabender mixer (and similar devices); its design is very close to that of the Banbury and includes a ram. Importantly, the Brabender can be blanketed with inert gas or even placed in a vacuum chamber. Minimum volume is about $30 \ cm^3$.

Available even in small sizes are a wide variety of extruders which can be adapted to laboratory mixing. An extruder is almost a must if enough blend is to be made for injection molding, etc. (Banbury mixers come in large enough sizes, but also required are a two-roll mill for sheeting the melt and a grinder for forming pellets.) Extruders are fairly easy to clean and some are small enough to handle about a kilogram of blend. Of all the mixers, however, an extruder is most independent of the operator; consequently the quality of mixing is highly dependent on screw design. For this reason, most laboratories using extruders have a selection of screws or screw parts to handle different types of blends. While fugitive plasticization and sequencing techniques are possible with an extruder, they require more extensive equipment or multiple runs.

An interesting and different mixing device is available which, in principle, can mix very small amounts. Invented by Prof. Maxwell of Princeton University, it is called the Mini-Max mixer. The melt is mixed by torsional flow between two heated plates, but mixing across flow lines can also be accomplished. This is done by periodically changing the gap of the plates. Confinement of the melt can be credited to normal stresses; thus contamination can be controlled *easily. To simulate the Mini-Max geometry, any rotational rheometer accommodating parallel plates can be used. With custom fixtures of high precision and small size, melt mixing of milligram quantities is not out of the question using a rheometer.*

2.4 Examples

2.4.1 PVC/Poly(ethylene-co-vinyl acetate) (1). Mixing was accomplished with a two-roll mill. Heat treating after mixing appeared to decrease the miscibility.

2.4.2 Polycarbonate/Polyester blending (2). The polyester, a copolymer of 1,4-cyclohexanedimethanol and iso-/terephthalic acids, is amorphous if quenched from the melt. ($Tm = 265°C$, $Tg = 87°C$). Solvents would induce crystallization. Mixing was accomplished with an extruder equipped with a film die. Quenching the film in ice water prevented crystallization.

2.4.3 PS/Poly(styrene-b-butadiene) (3). Melt mixing of the low MW (2400-3500) PS with the block copolymer was accomplished in a glass tube containing a glass rod stirrer. The tube was sealed off under vacuum. Actuation of the stirrer was through a magnet acting on a piece of iron attached to the glass rod.

3   CASTING FROM COMMON SOLVENTS

3.1   Advantages

Casting of a blend from a common solvent is the simplest mixing method available and is widely practiced. Very small quantities of experimental polymers can be handled easily. The resulting product, a film, is immediately useful for thermal analysis, dynamic mechanical analysis, mechanical testing, microscopic examination, infrared analysis, etc. If pure solvents and clean glassware are used, contamination can be precluded. In most cases, temperatures never exceed ambient, so degradation is not a problem.

3.2   Limitations

Not all polymers are readily soluble in common, safe solvents. Aromatic polyesters and amides are examples. Many crystalline polymers require warming of the solution and control of the temperature during casting. Some polymers hold on tenaciously to some solvents and residual solvents can influence the results of an analysis. Generally it is difficult to make thick shapes by casting; a molding operation using the films is required.

The most severe problem with casting is the influence of the solvent and the casting history on the resulting product. In spite of the fact that most of the solvent can be removed from a cast film, the nature of the film can depend strongly on the solvent used and the conditions during casting. The classical example is the system PVME/PS. With toluene as a solvent, the films cast from these polymers are single phase, which is presumably the equilibrium condition for this system at room temperature. With chloroform as a solvent, the cast films are cloudy; apparently the chloroform is able to modify seriously the interaction between PS and PVME, even at low concentrations. Similar problems have been noted for the system EVA/chlorinated PE by Walsh et al. (4). Obviously, if castings are performed with water-sensitive solvents such as THF, dioxane, acetone, etc., the absorption of water from the air could adversely affect the blend.

Crystalline polymers present particular difficulties because the nature of the solvent has a strong influence on the development of crystals. An extensive study of some of these problems, using the system SAN/PCL, was undertaken by Runt and Rim (5). They found that the extent of PCL crystallization depended strongly on the solvent and also on details of the casting technique such as initial concentration, temperature, and film thickness. A similar study on polycarbonate/PCL has also been reported (6).

Glassy systems with high-Tg components will be very susceptable to cast-in memory of any irregular ternary phase behavior. As solvent evaporates from these systems the Tg climbs, finally approaching that of the casting temperature. If the ternary system is biphasic under these conditions, then the cast film will also be biphasic even though the blend should be miscible. These ideas have been explored more fully by Gashgari and Frank (7).

3.3 Procedures

To avoid concentration and temperature gradients during the removal of solvent, casting is best done in thin films and with slow solvent removal. Both also help produce bubble-free films, but removal of air from the casting solution is also important. Solvents must be free of water.

There are a number of methods for casting thin uniform films from polymer solutions. One easy method is to spread the solution over a glass plate by rolling a rod wound with wire through a puddle of the syrup. Most researchers simply cast dilute solutions into flat dishes. If films of very uniform thickness are needed, the solution can be cast onto mercury. Covering the cast solution with an inverted dish slows the evaporation of the solvent and promotes uniform films. In the case of water-sensitive solvents, the dish also helps keep water vapor away. If the film sticks to glass, a silane treatment may help (4). Soaking in water or methanol may aid in removing films from glass.

If solvent is used for the sole purpose of mixing the polymers, it can be removed more quickly and with less chance of gross phase separation under non-quiescent conditions. Solvent removal from syrups can be accomplished on a two-roll mill, or even in a vented extruder. Foaming the syrup in a vacuum oven also often works well; the dried foam can be subsequently compression molded. Solvent can also be removed by adding the solution dropwise to a well-stirred non-solvent. Small drops can be formed using a vibrating nozzle.

To remove traces of solvent from the casting, high temperatures are invariably needed; protection of the polymer then becomes a consideration. Flowing inert gas or vacuums are typically used for this purpose and to reduce the partial pressure of the solvent over the film. It should be noted that the vacuums in typical laboratory ovens are not high enough to prevent oxidation of sensitive polymers.

Residual solvent can be most readily detected by dissolving the film in another solvent and injecting the solution into a gas chromatograph. Glass transition, elemental analysis, and weight loss have also been used.

3.4 Examples

3.4.1 PS/PVME from toluene (8). In this study, toluene solutions were coated onto slides by dip casting, the slide being raised at 1 mm/s. The atmosphere around the operation was dried with phosphorous pentoxide. The casting thickness was a smooth function of solution concentration; a 10% solution gave about a 1-μm film.

3.4.2 PS/PMMA from toluene (9). 4% solutions were cast into Petri dishes and, reportedly, the solvent was evaporated in a vacuum oven. (To prevent foaming most of the solvent must be removed slowly.) The films were then dried at 80°C for 36 hours, again in vacuum. The resulting films were broken and compression molded at 140°C.

3.4.3 PS/Poly(2,6-dimethyl-p-phenylene oxide (10). Blends were prepared by precipitation of chloroform solutions in methanol. The precipitates were dried and molded at 270°C for 10 minutes.

4 FREEZE DRYING

4.1 Advantages

Freeze drying has some advantages over casting which may be critical for blending work. With freeze drying, a solution of the two polymers is quenched down to a very low temperature and the solvent is frozen. Ideally, the polymers will have little chance to aggregate, but will collect randomly in regions throughout the frozen solvent. Thus the state of the dilute solution is somewhat preserved. Solvent is removed cleanly by sublimation; no changes can occur because of the solid nature of the mixture. To a large extent, therefore, the resulting blend will be independent of the solvent, if the solution is single phase before freezing and the freezing occurs rapidly.

4.2 Limitations

Freeze drying seems to work best with solvents having high symmetry; i.e., benzene, dioxane, naphthalene, etc. Water also works well. The solvents must be pure for best results, although mixtures of solvents with similar melting points have been used successfully.

The quantities of polymer which can be processed by freeze drying are not large. Dilute solutions must be used and the solution volume must be kept low for good heat transfer. The powdery form of the blend after solvent removal is usually not very useful and further shaping must be performed.

Of more fundamental concern is the state of blend upon freeze drying. Clearly the components will not be at equilibrium on removal from the dryer; in fact, Shultz and Mankin (11) have shown that a metastable miscible state can be achieved by freeze drying of PS/PMMA mixtures. The blend thus should be annealed to allow equilibration at the desired temperature.

While not complex, freeze drying does require a good vacuum system for low-boiling solvents and some attention to detail. It is not a fast blending method.

4.3 Procedures

The key to freeze drying is heat transfer. Fast removal of heat is needed during the freezing process, while during sublimation, heat transfer to the sample must not exceed that which can be removed by sublimation. To achieve rapid cooling, solutions can be poured on chilled metal plates or stirred into cold non-solvents. For low melting point solvents, liquid nitrogen is preferred; the blend solution can be dropped slowly into the well-stirred bath and collected by filtration. For high-melting-point solvents (e.g., naphthalene, p-dichlorobenzene, biphenyl), ice water is suitable. Clearly, high-melting-point solvents will be easier to handle and are preferred.

The freeze dryer must prevent the frozen solution from melting by insulating it from the environment. However, it must also hold the solution at a high enough temperature so that the activity of the frozen solvent is greater than the solvent which collects in the dryer's cold trap; otherwise drying will not occur. For low-melting solvents, heat transfer to the solution is controlled by providing a vacuum around the frozen solution, insulating any points of necessary mechanical contact and limiting any radiation to the sample. Drying of solutions made from high-melting-point solvents can be more casual; a vacuum oven works just fine. Even air circulating ovens can be used for some systems.

4.4 Examples

4.4.1 PS/PMMA (12). *5% blend solutions in naphthalene were prepared at 100°C and added slowly to ice water in a blender. The frozen droplets of solution were filtered from the solution and dried on the filter with air for 24 hours at room temperature. Two methods were tried for removing the residual naphthalene: 1. one week in high vacuum and 2. nine more days of air drying. Both were equally effective.*

4.4.2 PMMA/Poly(ethyl methacrylate) (13). Freeze drying was done from 5-10% solutions in benzene by swirling a small amount of solution in a flask immersed in dry ice-acetone. The flask was then connected to a rotary evaporator immersed in an ice bath. After 8 hours, a nitrogen stream was passed over the powder to remove traces of solvent. Casting of the same blends from benzene produced more heterogeneous mixtures.

5 EMULSIONS

5.1 Advantages

Handling of polymers as emulsions has many of the same advantages as the use of solvents. Films can be cast; mixing requires no expensive equipment; no high temperatures are needed.

5.2 Limitations

Emulsions of polymers are not always available or easy to make. While emulsion polymerization is highly advanced, it is not applicable to all monomers.

The state of blends prepared from emulsion mixing and coagulation has not been extensively studied. Using emulsions of poly(ethyl acrylate) and poly(methyl methacrylate), Bauer (14) produced a blend with evidence of a metastable miscible phase. On the other hand, Hughes and Brown (15) found that mixed emulsions of poly(methyl acrylate) and poly(vinyl acetate) gave products which were biphasic, whereas solution-cast films were miscible.

6 MIXING VIA REACTION

6.1 Overview

Co-crosslinking and IPN formation are specialty methods for forming blends which will be discussed in detail in subsequent lectures. With both of these methods, and with a host of proprietary intensive mixing techniques, the idea is to force a degree of miscibility by reactions between the polymers. Other methods involve the polymerization of a monomer in the presence of a polymer and the introduction of groups onto the polymer chain which will react or interact strongly (16). Exchange reactions between polyesters or polycarbonates have also been examined (17).

6.2 Co-crosslinking

Co-crosslinking may not necessarily result in a thermoset network, but merely in copious amounts of grafting. It is practiced both in the rubber industry (where thermosets are the desired end product) and in connection with the manufacture of polyolefin-based thermoplastic elastomers (usually blends of PP with various elastomers (18)). The key in this process appears to be simultaneous mixing and crosslinking, initiated usually by a curative such as a peroxide (19).

6.3 Mixing Starting with a Polymer/Monomer Solution

By starting with a polymer dissolved in a monomer, the unlike segments of each component are already intimately mixed. If the solution is gelled or viscous enough, or if grafting occurs during polymerization of the monomer, some very intimate mixtures may result. If the polymer is crosslinked, IPN's may be formed.

6.4 Examples

6.4.1 LDPE/PS (19). A 50:50 mixture of the polymers was mixed in a torque rheometer (similar to a Brabender mixer) at 170°C in N_2. 1% cumene hydroperoxide was added and the mixing was continued for up to 60 minutes. Extraction of unbound PS with solvent showed that the amount of combined PS had increased monotonically with time.

6.4.2 Polycarbonate/dicyanate bisphenol A SIPN (20). A SIPN is a semi-interpenetrating network formed by one linear and one cross-linked polymer. A 50:50 mixture of polycarbonate (Tg = 152°C) and the dicyanate of bisphenol A was heated between 250°C and 300°C to polymerize the dicyanate. (Dicyantes form trifunctional crosslinks consisting of a triazine ring). The Tg of the clear mixture was 195°C.

6.4.3 Polystyrene/Poly(ethyl methacrylate) (21). The polystyrene was dissolved in ethyl methacrylate which was polymerized with AIBN. Single-phase mixtures could be achieved with ∼10000 MW PS at 5% in PEMA. Tests for grafting were negative.

6.4.4 Poly(ethylene-co-vinyl acetate)/silicone (22). A 50:50 mixture of these two immiscible polymers was blended in a Brabender mixer at 185°C. The silicone, which was modified with methylvinyl-siloxane units, began to crosslink and react with mechanically degraded polyethylene. A 75% gel fraction could be achieved.

REFERENCES

1. *Gevert, T. U., I. Jakubowicz and S. E. Swanson. Phase Relation-ships in Blends of Poly(vinyl chloride) with Copolymer of Ethylene-Vinyl Acetate and the Effects of Thermal Stabilizers. European Polymer Journal 15 (1979) 841-847.*

2. *Masi, P., D. R. Paul and J. W. Barlow. Gas Sorption and Trans-port in a Copolyester and Its Blend with Polycarbonate. Journal of Polymer Science, Polymer Physics Edition 20 (1980) 15-26.*

3. *Roe, R-J. and W-C Zin. Phase Equilibria and Transition in Mix-tures of a Homopolymer and a Block Copolymer. 2. Phase Diagram. Macromolecules 17 (1984) 189-194.*

4. *Walsh, D. J., J. S. Higgins and S. Rostami. Compatibility of Ethylene-Vinyl Acetate Copolymers with Chlorinated Polyethyl-enes. 1. Compatibility and Its Variation with Temperature. Macromolecules 16 (1983) 388-391.*

5. *Runt, J. and P. B. Rim. Effect of Preparation Conditions on the Development of Crystallinity in Compatible Polymer Blends: Poly-(styrene-co-acrylonitrile)/Poly(ε-caprolactone). Macromolecules 15 (1982) 1018-1023.*

6. *Varnell, D. F., J. P. Runt and M. M. Coleman. Fourier Transform Infrared Studies of Polymer Blends. 6. Further Observations on the Poly(bisphenol A carbonate)-Poly(ε-caprolactone) System. Macromolecules 14 (1981) 1350-1356.*

7. *Gashgari, M. A. and C. W. Frank. Excimer Fluorescence as a Molecular Probe of Blend Miscibility. 4. Effect of Tempera-ture in Solvent Casting. Macromolecules 14 (1981) 1558-1569.*

8. *Reich, S. and Y. Cohen. Phase Separation of Polymer Blends in Thin Films. Journal of Polymer Science, Polymer Physics Edition 19 (1981) 1255-1267.*

9. *Singh, Y. P. and R. P. Singh. Compatibility Studies on Solid Polyblends of Poly(methyl methacrylate) with Poly(vinyl ace-tate) and Polystyrene by Ultrasonic Technique. European Polymer Journal 19 (1983) 529-533.*

10. *Stejskal, E. O., J. Schaefer, M. D. Sefeik and R. A. McKay. Magic-Angle Carbon-13 Nuclear Magnetic Resonance Study of the Compatibility of Solid Polymer Blends. Macromolecules 14 (1981) 275-279.*

11. *Shultz, A. R. and G. I. Mankin. Freeze-Dried Poly(methyl meth-acrylate)-Polystyrene Blends: Inference of Molecular Mixing. Journal of Polymer Science, Polymer Symposia No. 54 (1976) 341-360.*

12. *Shultz, A. R. and A. L. Young. DSC on Freeze-Dried Poly(methyl methacrylate)-Polystyrene Blends. Macromolecules 13 (1980) 663-668.*

13. *Jachowicz, J. and H. Morawetz. Characterization of Polymer Chain Interpenetration in Solution by Fluorescence after Freeze-Drying. Macromolecules 15 (1982) 1486-1491.*

14. Bauer, P., J. Henning and G. Schreyer. *Dynamomechanical Studies on Mixtures of Poly(ethyl acrylate) and Poly(methyl methacrylate).* Angewandte Makromolekulare Chemie 11 (1970) 145-157.

15. Hughes, L. J. and G. L. Brown. *Heterogeneous Polymer Systems. I. Torsional Modulus Studies.* Journal of Applied Polymer Science 5 (1961) 580-588.

16. Zhou, Z-L. and A. Eisenberg. *Ionomeric Blends. II. Compatibility and Dynamic Mechanical Properties of Sulfonated cis-1, 4-Polyisoprenes and Styrene/4-Vinylpyridine Copolymer Blends.* Journal of Polymer Science, Polymer Physics Edition 21 (1983) 595-603.

17. Paul, D. R. and J. W. Barlow. *Polymer Blends (or Alloys).* Journal of Macromolecular Science, Reviews of Macromolecular Chemistry C18 (1980) 109-168.

18. Kresge, E. N. *Rubbery Thermoplastic Blends,* in D. R. Paul and S. Newman, eds. *Polymer Blends, Vol. 2.* (New York, Academic Press, 1978), pp. 293-318.

19. Hajian, M., C. Sadrmohaghegh and G. Scott. *Polymer Blends - IV Solid Phase Dispersants Synthesized by a Mechanochemical Procedure.* European Polymer Journal 20 (1984) 135-138.

20. Wertz, D. H. and D. C. Prevorsek. *SIPNs: A New Class of High-Performance Plastics.* Plastics Engineering, April (1984) 31-33.

21. Nicolai, K. A., D. J. Shaw and E. V. Thompson. *Morphology and Phase Relationships of Polystyrene/Poly(alkyl methacrylate) Systems. Low-Molecular-Weight Polystyrene in Poly(ethyl methacrylate).* Polymer Preprints 21(1) (1980) 167-168.

22. Falender, J. R., S. E. Lindsey and J. C. Saam. *Silicone-Polyethylene Blends.* Polymer Engineering and Science 16 (1976) 54-58.

LIGHT, NEUTRON AND X-RAY SCATTERING TECHNIQUES FOR STUDYING POLYMER BLENDS

Dr. J.S. Higgins

Department of Chemical Engineering and Chemical Technology,
Imperial College,
London SW7 2BY, England.

INTRODUCTION

It is not the purpose of this chapter to enter into the details of scattering theory but rather to outline the principles and explain the applications to study of blends. All scattering experiments are based on the existence of variations in the homogeneity of the scattering medium. These may arise either because a few molecules of one type are embedded in a matrix of others, as in dilute solutions, or because of concentration fluctuations in a high concentration mixture or from density fluctuations in a single component system. If these inhomogeneities also cause variation in the physical property relevant for scattering a particular type of radiation, a scattering pattern may result. These relevant properties are the polarisability (or refractive index) for light, the electron density for X-rays and the scattering length density (a nuclear property) for neutrons.

The diagram in Figure 1a demonstrates that if radiation is scattered from two points in a medium a distance D apart then the path difference of the two scattered rays is $2P \sin(\theta/2)$ where $P = D \sin(\phi-\theta/2)$ is the projection of D onto the plane bisecting the incident and scattered directions and θ is the scattering angle. For convenience we will refer to this projection P as the 'dimensions' of the scattering object. There may then be interference between the rays depending on the relationship between $2P\sin(\theta/2)$ and the wavelength λ. If there is no differentiation between positions within the medium then the net sum of all the rays will cancel. If however there is a relationship between the scattering properties of a series of A and B pairs and these

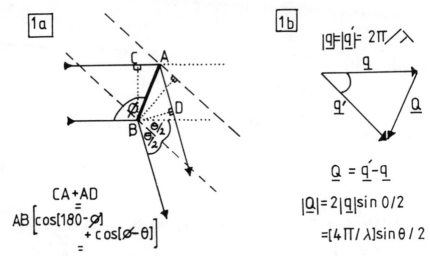

Figure 1 (a) Extra path length travelled by two rays scattered at points A and B. (b) Relationship between $\underline{q}$, $\underline{q}'$, $\underline{Q}$ and θ.

differ from surrounding points an interference pattern may result in the scattered radiation and this will be characteristic of the shape of the scattering object. The condition for this is that the path length $2\,P\,\sin(\theta/2)$ should be of the same order of magnitude as the wavelength of the scattered radiation

i.e. $\dfrac{2P\sin(\theta/2)}{\lambda} \simeq 1$ (1)

In scattering experiments it is usual to describe events in terms of the wave vector of the radiation. This is a vector, $\underline{q}$, pointing along the direction of travel and of length $2\pi/\lambda$. Figure 1b shows the relationship between the incident wave vector $\underline{q}$, the scattered wave vector $\underline{q}'$ and the change $\underline{Q} = \underline{q} - \underline{q}'$. The quantity we are interested in is

$Q = |\underline{Q}| = (4\pi/\lambda)\sin(\theta/2)$ (2)

When describing the scattering of light the refractive index, $\tilde{n}$, of the scattering medium compared to vacuum must be taken into account. The refractive index governs the velocity of light – and thus the wavelength in the medium, $\lambda' = \lambda/\tilde{n}$.

The wavelength change Q for scattered light is thus $Q = (4\pi\tilde{n}/\lambda)\sin(\theta/2)$ in terms of the incident wavelength, λ, in vacuo. For neutrons and X-rays there is no equivalent effect – the 'refractive index' compared to a vacuum is unity.

In a Bragg diffraction experiment from an ordered array of objects the projection P becomes the distance d between planes (indicated by the dashed lines in Figure 1a). The condition for constructive interference is then the familiar

$$2d \sin \theta/2 = n\lambda \qquad (3)$$
$$d = 2n\pi/Q \qquad (4)$$

and a series of peaks will be observed in the scattering pattern. Even a rather random array of objects (such as a gas) with an average separation d will cause at least one peak in the scattering at Q_m such that $Q_m d = 2\pi$.

In summary, then, the scattering pattern which arises if there are inhomogeneities in a medium may contain two pieces of information. There will be a pattern corresponding to the shape and dimensions P of the inhomogeneities and there may be an interference peak corresponding to their average spacing d. In both cases the angle of observation θ is such that P/λ or d/λ is of order $\sin \theta/2$, and it is necessary that the inhomogeneities are reflected in the scattering property for the particular radiation employed.

CONTRAST FACTORS AND SCATTERED INTENSITY

Dilute Solution

The condition $\sin \theta/2 \simeq P/\lambda$ indicates that in experiments in which scattering from macromolecules themselves in solution or from fluctuating concentrations within macromolecular mixtures are to be observed there will be a fundamental difference between light and X-rays and neutrons. Typical dimensions of P will be 10^2–10^3 Å. For light λ is of order 5×10^4 Å and θ is likely to range anywhere up to π, while for the other two radiations ($\lambda \simeq$ ~1–10 Å) θ will be very small.

In order to compare the different scattering techniques we consider the intensity I scattered per unit solid angle per unit sample volume from a beam of intensity I_0 incident on a solution of particles of mass M and dimension P with a number density N/V.

Light scattering. Following Flory (1) it can be shown that for $P \ll \lambda$

$$\frac{I}{I_0} = \frac{8\pi^4\alpha^2}{\lambda^4} (1 + \cos^2\theta) \frac{N}{V} \qquad (5)$$

where α is the excess polarisability of particles over solvent. In terms of the refractive indices $\hat{n}$ and $\tilde{n}_0$

$$\alpha = \frac{V}{4\pi N} (\tilde{n}^2 - \tilde{n}_0^2) \qquad (6)$$

and for dilute solution $\tilde{n}^2 - \tilde{n}_o^2 \simeq (\frac{d\tilde{n}^2}{dc})_o c$

$$\simeq 2\tilde{n}_o c(\frac{d\tilde{n}}{dc})_o \tag{7}$$

$$\alpha = \frac{V}{4\pi N} 2\tilde{n}_o c \left(\frac{d\tilde{n}}{dc}\right)_o = \frac{m}{2\pi N_{AV}} \tilde{n}_o \frac{d\tilde{n}}{dc} \tag{8}$$

since $c = \frac{NM}{N_{AV}}$ where N_{AV} is Avogadro's number

$$\frac{I}{I_o} = \frac{2\pi^2}{\lambda^4 N_{AV}} \tilde{n}_o^2 \left(\frac{d\tilde{n}}{dc}\right)^2 (1 + \cos^2\theta) \frac{NM \cdot M}{VN_{AV}} \tag{9}$$

$$= \frac{2\pi^2}{\lambda^4 N_{AV}} \tilde{n}_o^2 \left(\frac{d\tilde{n}}{dc}\right)^2 (1 + \cos^2\theta) \, Mc$$

I/I_o is the so-called Rayleigh Ratio R^o_θ, (the superscript indicating $P \ll \lambda$) and thus

$$R^o_\theta = K^* (1 + \cos^2\theta) \, Mc \tag{10}$$

$$\text{where } K^* = \frac{2\pi^2}{\lambda^4 N_{AV}} \tilde{n}_o^2 \left(\frac{d\tilde{n}}{dc}\right)^2 \tag{11}$$

It should be noted that

(1) Because of the properties of light the scattered radiation from an initially unpolarised beam will be partially polarised in a manner depending on θ. For $\theta = 90^0$ the light is plane polarised and the polarisation vanishes as $\theta \to 0$).

(2) If we allow P to increase towards the dimensions of λ, the shape of the scattering particles will modulate the intensity as a function of θ (i.e. we have a scattering pattern as described in the introduction). However, as $\theta \to 0$ this interference must disappear so that equation 10 is retrieved if the intensity of scattering is extrapolated to the forward angle. We can thus take the Rayleigh ratio R^o_θ either as indicating $P \ll \lambda$ or $\theta \to 0$.

X-rays. In this case $P > \lambda$ and we can only eliminate the effect of the interference function by extrapolating to the forward scattering angle. The term $(1 + \cos^2\theta)$ tends to a constant for all SAXS experiments and the equivalent of equation 5 is then (2)

$$\frac{I}{I_o} = I_e^2 \, \Delta\rho_e \, \frac{N}{V} \tag{12}$$

where $I_e = (\frac{e^2}{mc^2})$ is the Thomson factor (the so-called classical radius of the electron). I_e^2 is the effective cross-section of an electron for scattering an X-ray.

$$\Delta\rho_e = (\rho_e^P - \rho_e^S) \frac{M}{\rho} \tag{13}$$

where ρ_e^P is the mean electron density of the particles and ρ_e^S that of the solvent $\left[\rho = \frac{\Sigma \text{ atomic numbers}}{\Sigma \text{atomic weights}} \times \rho. \quad \rho \text{ is the density} \right]$.

The equivalent of equation (9) is then

$$\frac{I}{I_o} = I_e^2 \left(\frac{\rho_c^P - \rho_e^S}{\rho^2} \right) N_{AV} \, cM \tag{14}$$

$$\text{and } K^*_{X-ray} = I_e^2 \left(\frac{\rho_c^P - \rho_e^S}{\rho^2} \right)^2 N_{AV} \tag{15}$$

Neutrons. Again $P > \lambda$ so we extrapolate to forward scattering. Now

$$\frac{I}{I_o} = (\underset{p}{\Sigma}b - \frac{V_p}{V_s}\underset{s}{\Sigma}b)^2 \frac{N}{V} \tag{16}$$

where b is the neutron scattering length of a nucleus, $\underset{p}{\Sigma} b$ and $\underset{s}{\Sigma} b$ are the sums of these lengths over the nuclei in the particle and the solvent molecule respectively. V_p/V_s is the ratio of specific volumes for particle and solvent.

The equivalent of equations 9 and 14 now becomes:

$$\frac{I}{I_o} = \left\{ \frac{\underset{p}{\Sigma}b - \frac{V_p}{V_s}\underset{s}{\Sigma}b}{M} \right\} N_{AV} \, Mc \tag{17}$$

$$\text{and } K^*_{neutrons} = \left\{ \frac{\underset{p}{\Sigma}b - \frac{V_p}{V_s}\underset{s}{\Sigma}b}{M} \right\}^2 N_{AV} \tag{18}$$

$$= \left\{ \frac{\rho_b^P - \rho_b^S}{\rho} \right\}^2$$

if ρ_b^P and ρ_b^S are scattering length densities ($\rho_b = \Sigma b/M\rho$) and ρ the density.

Table I lists the values of b for a number of nuclei commonly found in polymeric systems. Unlike the electron density there is no systematic variation through the periodic table. This means that light atoms are no longer at a disadvantage, but even more importantly the large difference between hydrogen and deuterium allows isotopic labelling of molecules or parts of molecules.

This effect is shown up in Table II in which the values of K* for the different radiations are compared for solutions of polystyrene in benzene, deuterated benzene and deuterated polystyrene. Although the value of K* for light looks very small it should be remembered that even conventional light sources are many orders of magnitude 'brighter' than all neutron sources and most X-ray sources.

In many experiments only the relative values of K* are relevant, but when absolute intensities are required (e.g. to obtain the molecular weight from equations 10, 14 or 17) it is usual to calibrate against a standard sample rather than to attempt to determine I_o directly (2,3,4,5,).

Blends and Mixtures

Suppose now, instead of a dilute solution of particles we have a concentrated solution or mixture in which there are

TABLE I

Neutron Scattering Lengths

Nucleus	^1H	^2H	^{12}C	^{14}N	^{16}O	
b	−374	+.667	+.665	+.93	+.580	$\times 10^{-12}$ cm

TABLE II

Values of K* for light, neutrons and X-rays

$$K^* (cm^2 g^{-2} \text{ mole})$$

	light(4358Å)	X-rays	neutrons
Polystyrene in benzene h_6	2.4×10^{-7}	3.6×10^{-4}	7.8×10^{-6}
polystyrene in benzene d_6	"	"	2.4×10^{-3}
polystyrene h_8 in polystyrene d_8	0	0	3.8×10^{-3}

concentration fluctuations. Again, following Flory (1) the particles described above are replaced by elements of volume V. We
now require the equivalent of $\alpha^2 N/V$ in equation 5
the number density $N/V = \frac{1}{\delta V}$ and $\delta \alpha = \frac{\delta V}{4\pi} (\tilde{n}^2 - \tilde{n}_o^2)$

$$\text{thus } \delta \alpha = \frac{\delta V}{4\pi} \frac{d\tilde{n}^2}{dc} \overline{\Delta c} = \frac{\delta V}{4\pi} 2\tilde{n}_o \frac{d\tilde{n}}{dc} \overline{\Delta c} \qquad (19)$$

where $\overline{\Delta c}$ is the average concentration fluctuation. Following the
route from equation 5 onwards

$$\text{we obtain } R_\theta^o = K*(1 + \cos^2\theta) \delta V \overline{\Delta c}^2 N_{AV} \qquad (20)$$

Now the thermodynamic theory of fluctuation developed by
Einstein (6) shows that

$$\overline{\Delta c}^2 = \frac{kT}{(\partial^2 F/\partial^2 c)} = \frac{kTV}{\delta V/(-\partial\mu_1/\partial c)} \qquad (21)$$

where c is the mean concentration in a volume element δV, F is the
free energy, μ_1 is the chemical potential of component 1 (for a
solution, the solvent) and V_1 is its molar volume.

$$\text{Thus } R_\theta^o = K*(1 + \cos^2\theta) \frac{RT V_1 c}{(-\partial\mu_1/\partial c)} \qquad (22)$$

In a similar way the quantity $RTV_1 c/(-\partial\mu_1/\partial c)$ replaces Mc in
equations 14 and 17.

The forward intensity in a scattering experiment, which for
dilute solutions is simply governed by the size and number of
particles present (Mc), in an interacting concentrated system is
governed by the thermodynamic properties via the concentration
dependence of the chemical potential $(-\partial\mu_1/\partial c)$. Clearly, this
already indicates some applications (to be described below) for
scattering from blend systems. For example, since $\partial\mu_1/\partial c \to 0$ at
the spinodal the forward intensity (equation 22) will tend to
infinity and hence provide a method for obtaining the spinodal
conditions. As well as the forward intensity, the angular variation in the scattering pattern may contain useful information, but
further discussion is left to particular applications described
below.

LIGHT SCATTERING APPLICATIONS

As has already been indicated equation 22 immediately suggests a method of obtaining the spinodal of a partially compatible
blend by extrapolation to the concentration/temperature conditions
in which the forward scattered intensity tends to infinity (8).

(A detailed discussion of critical opalescence (7) must take into account the singularities that occur at the spinodal. However, since the divergences only occur within 10^{-1} ^{0}C of the spinodal temperature the extrapolation techniques work well (8)).

As Figure 2 indicates, starting from the single phase region, the extrapolation has to be made over a very wide temperature gap to avoid taking the sample into the shaded, metastable region. The elegant PICS (pulse induced critical scattering) method (9,10) overcomes this problem for solutions or low molecular weight polymers with high molecular mobility.

If a sample is simply cooled along the path indicated in Figure 2 impurities will almost certainly provoke nucleated phase separation before the spinodal is reached. The concentration in the sample will then begin to develop inhomogeneities rather than fluctuations (as shown in Figure 2b). The scattered intensity may be described in the form of equation 20 and will be governed both by the size, V, of the inhomogeneities and the concentration difference c. As the particles increase in size and the concentration variation becomes more extreme the sample may become cloudy. (This depends on the value of K* for the mixture). The cloudiness is relatively easy to detect either as a reduction in transmitted light, or an increase in scattered light, though the exact location of its onset (somewhere between the binodal, B, and the spinodal, S, in Figure 2a) is usually dependent on a number of factors such as sample purity, heating or cooling rates, molecular mobility, etc. Nevertheless, cloud point curves are widely used as indications of the miscibility limits on phase diagrams of partially compatible polymers. Figure 3 shows a typical turbidimeter, designed to observe the scattered intensity as a function of temperature from four samples (11). Low molecular weight

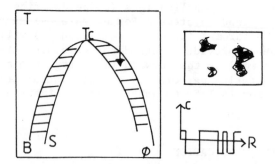

Figure 2   Phase separation at a UCST. The arrowed path would result in nucleation and growth and an irregular structure as shown.

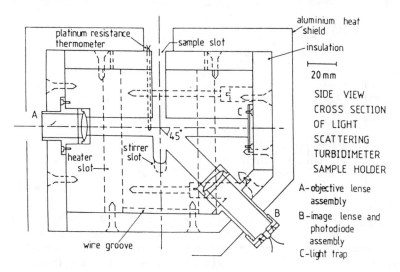

Figure 3   A light scattering turbidimeter (taken from reference 11).

samples are stirred between each measurement and heating or cooling cycles are compared at different rates until reproducibility is achieved. High molecular weight blends (showing LCST behaviour) are usually used in the form of thin films and will generally not remix with ease on cooling after heating into the two phase region. In this case, measurements are always made starting from the single phase region until reproducible cloud points are obtained. Figure 4a shows the variation with temperature of the intensity of light scattered from a blend of ethylene-vinyl acetate copolymer with solution chlorinated polyethylene (12). Inset are the results of a series of experiments in which the sample was heated to the indicated temperature and maintained there. In this way it is possible to locate the cloud point curve shown in Figure 4b within a few degrees. For a number of systems the method proves unsuitable. This is usually because K^* is too small, but occasionally if the inhomogeneities are very short range, δV becomes too small for detection.

The angular dependence of the scattering which develops when the sample is within the two phase region is itself of considerable interest. The type of structure which appears is characteristic of the region of the phase diagram (13,14). Within the spinodal a process called spinodal decomposition takes place (15). The system is completely unstable and one particular wavelength, 1, among the concentration fluctuations becomes favoured. The amplitude of the fluctuation grows and the regularity of the structure gives

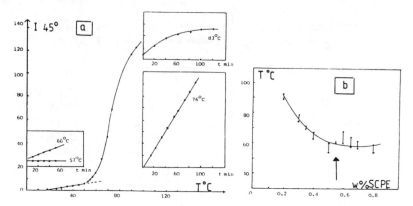

Figure 4 (a) Development of light scattered intensity as functions of temperature or time at 45^0 from a sample of ethylene vinyl acetate copolymer mixed with chlorinated polyethylene and (b) cloud point curve with the sample in (a) arrowed.

rise to a "diffraction" ring of maximum intensity in the scattered radiation such that $Q1 = 2\pi$. The initial position and growth rate of this diffraction ring and the subsequent increase in the domain size evidenced by the movement of the ring to smaller scattering angles (smaller Q) can all be related to the thermodynamic properties. van Aartsen (16) used the observed growth rate as a function of temperature to locate the spinodal. Figure 5 shows the development of the diffraction ring for a blend of EVA/SCPE as in Figure 4. In this case the sample was quickly heated to within the spinodal. If a sample phase separating by this process were observed at fixed angle the intensity might first increase and then decrease as the scattered intensity moves to smaller Q values.

In the metastable region between the binodal and the spinodal nucleation processes initiate phase separation and since these are not as regular in structure a diffraction ring does not generally result. Debye et al (17) showed that scattering from a randomly arranged two-phase system with sharp boundaries is described by

$$\frac{I(Q)}{I_o} = K* \overline{\Delta c}^2 \frac{a_c^3}{(1+Q^2 a_c^2)^2}$$ (23)

where a_c is the average correlation distance. Thus, a plot of $(I_o/I(Q))$ v Q^2 should yield a straight line, and the ratio of slope to intercept gives a value for a_c. For fully two phase systems a_c is of the order of light wavelengths, but in partially miscible systems a_c maybe quite short and require X-ray or neutron scattering measurements.

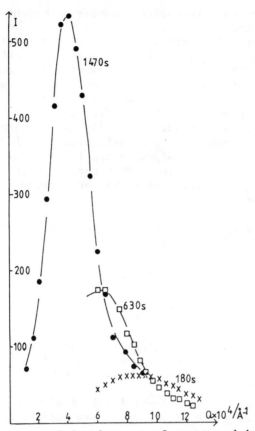

Figure 5   Later stages of development of scattered intensity of a sample similar to that in Figure 4, but with a different degree of chlorination which has been suddenly heated to 54.5^{0}C (within the spinodal for this blend).

NEUTRON SMALL ANGLE SCATTERING FROM BLENDS

While in principle the SANS technique can also be used to observe the miscibility limits in the equivalent of a cloud point measurement the relative inaccessibility of neutron sources precludes their use as an alternative to light or X-rays. It is only when neutrons have a unique advantage (as in the deuterium labelling technique) that it becomes sensible to deploy them.

The most common application to blends is in the study of molecular conformation of labelled chains in the single phase, but there may be phase separating systems where K* is unfavourable for light or X-ray observation or when the domain dimensions are closer to neutron than light wavelengths and neutrons then provide unique information.

The reader is referred to a number of recent articles for details of neutron scattering experiments and theory (4,18,19) applied to polymers – here we only cover the important principles. Figure 6 shows a diagram of a small angle scattering apparatus on the reactor at the Institut Laue–Langevin in Grenoble (20). It is fairly typical of such apparatus, the main feature being a large (64 cm x 64 cm) area detector which increases the count rate and compensates to some extent for the low fluxes of neutron sources compared to photon sources. Generally, as already mentioned, the interest in such experiments lies in obtaining information about the conformation of the polymer molecules and the interactions between them. We must therefore further develop equation 17 to include the angular variation in the scattering arising from the molecular shape.

For a single polymer molecule with a Gaussian configuration (i.e. very dilute solution)

$$\frac{I(0)}{I_o} = K*Mc \left\{ \frac{2}{Q^2 R_g^2} \left(e^{-QR_g} - (1-QR_g) \right) \right\} \qquad (24)$$

where R_g is the radius of gyration

For $QR_g \ll 1$, $I/I_o \to K*Mc(1-Q^2 R_g^2/3)$. At any finite concentration of a polymer mixture we must include the effect of the concentration fluctuations as in equation 22. Two routes may then be followed.

Small concentrations of polymer (2) in a mixture with polymer (1). In this case the expressions may be manipulated to give

$$K*c \left\{ \frac{I(0)}{I_o} \right\}_{QR_g \ll 1} = \frac{1}{M_2} \left\{ 1 + \frac{Q^2 R_{g2}^2}{3} \right\} + 2A_2 c \qquad (25)$$

where A_2, the so-called second virial coefficient is essentially a function of $(-\partial \mu_1 / \partial c)$ in equation 22. If we use the Flory-Huggins (21,22) description of the free energy of the polymer mixture then

$$A_2 = \frac{V_2^2}{V_1^2 M_2^2} \left(\tfrac{1}{2} - Z_1 \chi_{12} \right) \qquad (26)$$

where V_i and Z_i are the molar specific volume and degree of polymerisation of species (i). χ_{12} is the Flory interaction parameter (1).

Exploitation of equation 25 allows both R_{g2}, M_2 and χ_{12} to be obtained. The method is the so-called Zimm plot originally

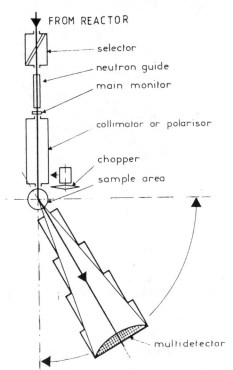

FROM REACTOR

selector

neutron guide

main monitor

collimator or polarisor

chopper

sample area

multidetector

Figure 6   A small angle neutron scattering spectrometer at the
Institut Laue-Langevin at Grenoble (20).

developed for light scattering (23) from polymer solutions.   Generally, in order to give a reasonable value for K* in the neutron
experiments from blends one of the components is deuterated.
Figure 7 shows a Zimm plot for a series of low concentrations of
deuterated poly(2,6-dimethyl,-1,4-phenylene oxide) dispersed in
polystyrene (24).   Extrapolation to zero c at fixed Q produces a
line described by the first term in equation 25 whose slope
gives a value for R_{g2} = 86Å.   Extrapolation to zero Q at constant
c gives a line described by a second term in equation 25 whose
slope gives A_2 = 6.39 x $10^{-4}g^{-2}$ cm^{-3} mol and an interaction parameter χ_{12} = -0.076.   The value of the intercept obtained at Q = 0,
c = 0, gives M_2 = 3.29 x 10^5 (the value of M for this polymer
obtained from light scattering measurements is 3.75 x 10^5).

Such measurements confirm the molecular dispersion of the
components via the values of R_g and M obtained and give values of
χ_{12} which may be compared with other measurements (25,26,27).

The restriction to low concentrations does impose an inconve-
nient limitation, particularly in systems with specific interac-

82

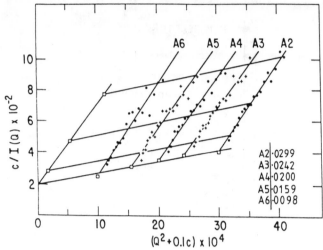

Figure 7   Zimm plot for deuterated poly(2,6-dimethyl-1,4-phenylene
oxide) in polystyrene at the concentrations shown.
Reproduced with permission from (24) Maconnachie et al
(Macromolecules 1984). Copright 1984, American Chemical
Society.

tions where χ_{12} may show unusual concentration dependence.

Finite Concentration Mixtures

In this case equation 25 can no longer be used and full
combination of equation 24 with 22 is required. A recent
derivation of this (28) gives for a mixture of polymer 1 in polymer 2 at concentration c, with a fraction f of polymer 1 chains
deuterated

$$\frac{I(Q)}{I_o} = K^* \left\{ \frac{(fd+(1-f)h-\beta b)^2}{F_1^{-1}(Q)+\beta \, F_2^{-1}(Q)-2\chi_{12}} \right. \tag{27}$$

$$\left. + f(1-f)(d-h)^2 F_1(Q) \right\}$$

where $F_i(Q) = c_i M_i \left(1 - \frac{Q^2 R_{gi}^2}{3}\right)$, the c_i are the volume fractions
and $\beta = \frac{V_1}{V_2}$, the ratio of specific segment volumes for the two
polymers.

Again simple Flory-Huggins theory has been used, and here χ_{12}
is explicitly defined per segment mole of polymer 1. d, h and b

are the sums of neutron scattering lengths for segments of polymer 1 (hydrogenous, h and deuterated, d) and polymer 2 (b).

The first term in equation (27) arises from the concentration fluctuations, the second gives the shape and size of the labelled chains. The two terms may be separated by varying f independently of c. The molecular dimensions are obtained from the second term together with a check on the correctness of the separation in the value of M_1 obtained. Figure 8 shows the results of such an experiment on methoxylated poly(propylene glycol) (PPGM M_n = 2,100) with methoxylated poly(ethylene glycol) (PEGM M_n = 600) (29). In this case the PPGM molecules were labelled by deuteration. The values of R_g and M obtained from the experiment has been divided by those obtained for deuterated PPGM in a mixture with hydrogenous PPGM. The molecular dispersion in the blend is confirmed and a slight tendency may be traced for the dimensions to decrease at low PPGM concentrations.

Extrapolation to Q=0 of the scattered intensity in equation 27 leads to

$$\frac{I(Q)}{I_o}\Bigg|_{Q\to 0} = \frac{(fd+(1-f)h-\beta b)^2}{(c_1M_1)^{-1}+\beta \; (c_2M_2)^{-1}-2\; \chi_{12}} \tag{28}$$
$$+ (d-h)^2 f(f-1)c_1M_1$$

so that χ_{12} can be obtained directly from the forward scattered intensity. In these experiments on PPGM/PEGM the value of χ_{12}

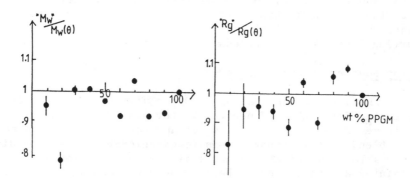

Figure 8   Comparison of the radius of gyration and the molecular weight of methoxylated poly(propylene glycol) (d_6) (M_n = 2100) when blended with methoxylated poly(ethylene glycol) (M_n = 600) with the same quantities obtained when mixed with methoxylated poly(propylene glycol) (h_6). Reproduced with permission from (29) Higgins and Carter (Macromolecules 1984). Copyright 1984, American Chemical Society.

obtained from neutron experiments was +0.07, to be compared with
the value obtained from heats of mixing data of +0.13. These
positive χ_{12} values are unfavourable for mixing. The latter is in
fact so unfavourable that the Flory-Huggins entropy term is not
large enough to allow a net free energy favourable for mixing.
Since the equation of state contributions are very small (30) the
mixing is explained in terms of an extra, favourable entropic
contribution to the interaction. This is not observed in the
purely enthalpic heats of mixing experiment, but since the intensity in the scattering experiment arises from $(-\partial\mu_1/\partial c)$ it is
included in the scattering value (a scattering measurement is a
free energy measurement).

All the neutron experiments rely on the labelling by deuterium causing no perturbation of the scattering system. This is of
course not strictly true. Deuterium, as a heavier atom, has a
different vibrational frequency and amplitude from hydrogen. This
in turn perturbs the partition function (31) and might be imagined
as giving a different effective specific volume to deuterated
segments of molecules. The effect in some blend systems is to
shift the spinodals by tens of degrees – deuterating polystyrene
or polybutadiene in blends of low molecular weights shifts the
UCST upward (32). Deuterating the polystyrene in high molecular
weight blends with poly(vinyl methyl ether) also raises the LCST
(33). On the other hand, for the PPGM/PEGM system described above
and for high molecular weight poly(methyl methacrylate) with solution chlorinated polyethylene there was no detectable effect of
deuteration. Based on these limited observations it seems that
systems where miscibility is caused by strong specific interactions are less likely to show effects of deuteration. It is
important to check this point before embarking on neutron experiments because manipulation of the coefficients in equation 27 by
varying f becomes meaningless if χ_{12} varies with f as well.

X-ray Scattering Experiments

X-ray techniques can be used for all the measurements so far
described for light and neutrons. But again, for reasons of convenience, X-ray observation is not applicable. If one component
in a blend contains heavy atoms (e.g. chlorine) thus providing a
reasonable value of K*) then the dilute solution methods for
obtaining R_g and χ_{12} described for neutrons can also be applied in
X-ray small angle scattering (25). This application has not been
widely used mainly because of experimental difficulties – the Q
range for SAXS is not as small as for SANS (λ X-rays = 1.5 Å, λ
neutrons = 5-15 Å) so that the limit $QR_g < 1$ is difficult to
achieve for reasonable molecular weight polymers.

Figure 9 shows a Kratky SAXS camera in which the single

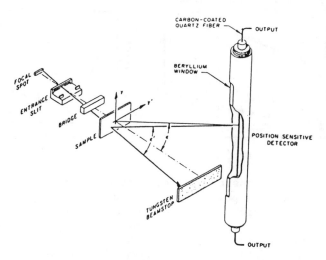

Figure 9   A small angle X-ray camera of the Kratky type but with a
linear detector. Reproduced from Russell & Stein (25)
by permission of John Wyley & Sons. Inc.

detector has been replaced by a linear position sensitive detector. The (relatively recent) employment of position sensitive
detectors has greatly speeded up the experiments, and, together
with intense rotating anode generators or synchroton sources made
possible live experiments on phase separating systems.

One typical example of the use of X-rays for studying blends
is seen in Figure 10a which shows the scattering from a series of
blends of polycaprolactone with polyvinyl chloride) analysed
according to equation 23 (25). Figure 10b shows the correlation
lengths a_c thus obtained. These very small values of a_c were
taken to indicate a high degree of miscibility in this case.

SUMMARY

All three radiations are important for the study of blends,
though it has only been possible here to sketch lightly the many
possibilities. The choice of radiation depends on three factors:
(1) the correspondence of the dimensions of the inhomogeneities to
be observed with the incident wavelength, (2) that a reasonable
value of the contrast factor K* for the chosen radiation arises
from these inhomogeneities and (3) convenience. This last is by
no means insignificant - the 'well-found laboratory' will
certainly provide optical equipment, may possess a small angle
X-ray camera but will certainly not possess a neutron source.

86

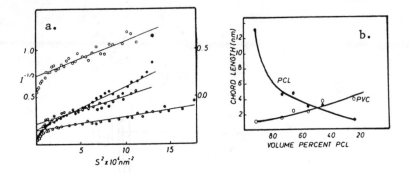

Figure 10   (a) Intensity scattered from blends of polycaprolactone with poly(vinyl chloride) plotted according
to equation 23 for several compositions and the (b)
resulting correlation lengths as a function of
composition (s = Q/2π). Reproduced from Russell &
Stein (25) by permission of John Wyley and Sons Inc.

REFERENCES

1.   Flory, P.J. Principles of Polymer Chemistry, (Ithaca, New
     York, Cornell University Press, 1953).

2.   Glatter, O. and O. Kratky. Small Angle X-ray Scattering,
     (London, Academic Press, 1982).

3.   Huglin, M. Light Scattering from Polymer Solutions
     (London, Academic Press, 1972).

4.   Higgins, J.S. and A. Maconnachie. In Methods in
     Experimental Physics: Neutron Scattering ed. K. Skold and
     D.L. Price, (New York, Academic Press, to be published
     1985).

5.   Jacrot, B. Rep. Prog. Phys. 39   (1976) 911.

6.   Einstein, A. Am. Phys. 33, (1910) 1275.

7.   Debye, P. J.Chem.Phys. 31   (1959) 680.

8.   Scholte, T.G. J.Pol.Sci. Part C 39, (1972) 281.

9.   Derham, K.E., J. Goldsborough and M. Gordon. Pure Appl.
     Chem. 38, (1974) 97.

10. Koningsveld, R., L.A. Kleintjens and H.M. Schoffeleers.
     Pure Appl. Chem. 39, (1974) 1.

11. Chong, C.L. Ph.D. Thesis, Department of Chemical Engineering, Imperial College (1981).

12. Walsh, D.J., J.S. Higgins, and S. Rostami. Macromolecules 16, (1983) 388.

13. McMaster, L.P. Adv. Chem. Ser. 1432, (1975) 43.

14. Snyder, H.L., P. Meakin and S. Reich. Macromolecules 16, (1983) 753.

15. Cahn, J.W. Trans. Metabl. Soc. AIME 242 (1968) 166 .

16. van Aartsen, J.J. Eur. Pol. J. 6, (1970) 919 .

17. Debye, P., H.R. Anderson and H. Brumberger. J. Appl. Phys. 28, (1957) 679.

18. Richards, R.W. In Polymer Characterisaion I, ed. J.V. Dawkins (Appl. Sci. Publishers, 1980).

19. Sperling, L.H. Polymer Engineering and Science 24, 1 (1984).

20. Neutron Beam Facilities Available to Users, Scientific Secretariat, Institut Laue Langevin, 156X Centre de Tri, 38042 Grenoble, France.

21. Flory, P.J. J. Chem. Phys. 9, (1941) 660 and 10, 51 (1942).

22. Huggins, M.L. J. Phys. Chem. 46, (1942) 151
                   J. Amer. Chem. Soc. 64, (1942) 1712.

23. Zimm, B.H. J. Chem. Phys. 16, (1948) 1093.

24. Maconnachie, A., R.P. Kambour, D.M. White, S. Rostami and D.J. Walsh. Accepted by Macromolecules.

25. Russell, T.P. and R. Stein. J. Pol. Sci. Pol. Phys. 20, (1982) 1593.

26. Kruse, W.A., R.G. Kirste, J. Haas, B.J. Schmitt and D.J. Stein. Makromol. Chem. 177 (1976) 1145.

27. Jelenic, J., R.G. Kirste, B.J. Schmitt and S. Schmitt-Strecker. Makromol. Chem. 180 2057 (1979).

28. Warner, M., J.S. Higgins and A.J. Carter. Macromolecules 16, (1983) 1931.

88

29. Higgins, J.S. and A.J. Carter. Accepted by Macromolecules.
30. Allen, G., Z. Chai, C.L. Chong, J.S. Higgins and J. Tripathi. Polymer 25, (1984) 239.

31. Buckingham, A.D. and H.G.E. Hentschel. J. Pol. Sci. Pol. Phys. 18, (1980) 853.

32. Atkin, A.L., L.A. Kleintjens, R. Koningsveld and L.J. Fetters. Polymer Bulletin 9, (1982) 347.

33. Yang, H., G. Hadziioannou and R.S. Stein. J. Pol. Sci. Pol. Phys. 21, (1983) 159.

LIQUID-LIQUID PHASE EQUILIBRIA IN POLYMER BLENDS

R. Koningsveld and L.A. Kleintjens

DSM Research & Patents, P.O. Box 18, 6160 MD Geleen, Netherlands

ABSTRACT

Liquid-liquid equilibrium phase relations in polymer mixtures are analysed in some detail with the aid of classic thermodynamical considerations. Polymolecularity in the constituents causes cloud-point curves not to represent coexisting phases and their extrema not to be identifyable with critical points. Such features should be taken into account when molecular theories are compared with experimental data. Multiple phase separations can be brought about by polymolecularity as well as by a special dependence of the inter-action function on concentration. Molecular origins of the latter may be: 1) different numbers of nearest neighbour contacts between the various repeat units, 2) mutual influence of the flexibility of the polymer chains, 3) non uniform segment density, 4) free volume. Statistical copolymers have the extra complication of a distribution of chemical composition. The latter enhances the tendency towards multiple phase equilibrium, chemical composition and chain length governing details sensitively. In addition, differences in size between the repeat units appear to be of primary importance. Experimental methods suitable for thermodynamic measurements on the highly viscous mixtures in hand are briefly reviewed.

1. INTRODUCTION

Equilibrium in a liquid polymer mixture is characterised by a minimal free enthalpy (Gibbs free energy) of mixing ΔG, which is either equal to zero or negative. If $\Delta G=0$ we have total immiscibility the free enthalpy of the mixture being precisely equal to the sum of the free enthalpies of the constituent phases, which are the

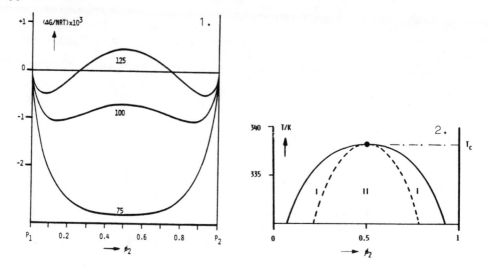

Fig. 1. Free enthalpy of mixing at 340K and 360K, calculated with
eq.(1) for a binary polymer mixture P_1/P_2. Chain lengths
($m_1=m_2$) indicated. Interaction function:
$g = 0.02 + 6(T/350-1)^2$ (ref.1)

Fig. 2. Binary phase diagram calculated with the ΔG-function used in
fig.1, $m_1=m_2=75$. Drawn curve: binodal (locus of coexisting
phase compositions); dashed curve: spinodal; ●: critical point

individual polymers. Complete miscibility goes with $\Delta G < 0$, and the
ΔG(concentration) curve or surface is everywhere positively curved.
If the system is partially miscible we have a plait on the ΔG-surface
(or curve) and there is a locus, called spinodal, where the curvature
in one direction is zero. Though concave/convex ΔG-curves may pass
through positive values between the two points of inflexion, the
relevant concentration ranges usually refer to unstable one-phase
states which can be ignored in equilibrium considerations.

Fig. 1 summarizes the situation for a strictly-binary system. For
the two higher molar masses double tangents can be constructed to the
concave/convex $\Delta G(\phi_2)$-curves and the tangent points determine the
compositions of the coexisting phases for the temperature considered.
Plotting these compositions and those of the inflexion points in a
$T(\phi_2)$-graph we obtain the familiar binary phase diagram (fig. 2) in
which the spinodal curve separates the metastable (I) and unstable
(II) regions within the miscibility gap. Upon reaching the critical
or consolute temperature T_c the two coexisting phases become identical
and beyond T_c we have complete miscibility.

Fig. 2 represents an upper critical miscibility gap which can be
shown usually to be accompanied by a positive enthalpy of mixing (ΔH)

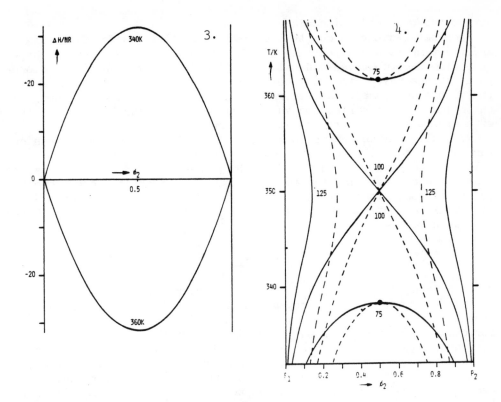

Fig. 3. Enthalpy of mixing at 340K and 360K calculated with
$\Delta H/NR = - T^2 \partial(\Delta G/NRT)/\partial T$ and eq.(1)

Fig. 4. Separated upper and lower critical miscibility and coalescence
into an hour-glass miscibility gap, calculated with the data
in fig.1. Chain lengths indicated. Dashed curves: spinodals;
drawn curves: binodals; •: critical points

in the temperature range of the gap. The reverse situation is known
to occur if the enthalpy of mixing is negative. Then the cloudpoint
curve demarcating the gap is upside down and we have lower critical
miscibility behaviour.

If the enthalpy of mixing shows a suitable temperature dependence
both phenomena may occur in the same system (fig.3). The location of
the two gaps being very sensitively determined by the molar masses
of the two polymers, it is not at all seldom found possible to let
the two gaps merge. One set of chain lengths exists at which the two
critical points just coalesce. The point of coalescence is not a
critical point in general, the two branches of spinodal and binodal
intersect with a finite slope (fig.4). At a further increase of the
chain lengths we obtain a single hourglass-shaped miscibility gap

with a two-branched spinodal.

Such phase diagrams may thus be rationalized in terms of the sign and the temperature dependence of ΔH. However, more complicated shapes have also been observed and call for less obvious explanations. Cloud-point curves have not seldom been reported that show deviations from the simple dôme shape. Shoulders or two extremes may develop when chain lengths are varied and such behaviour may occur in conjunction with upper and lower critical miscibility.

We shall analyse the various aspects with the aid of an expression for ΔG that formally covers almost all equations resulting from molecular theories. It is the Flory-Huggins-Staverman (2-4) expression that was shown by Tompa (5) to arise from elementary thermodynamic considerations without reference to a specific model. For a quasi-binary mixture of two homopolymers, each with a molar-mass distribution, we have (2-7)

$$\Delta G/N_\phi RT = \sum \phi_{1i} m_{1i}^{-1} \ln \phi_{1i} + \sum \phi_{2j} m_{2j}^{-1} \ln \phi_{2j} + g \phi_1 \phi_2 \tag{1}$$

where $N_\phi = \Sigma n_{1i} m_{1i} + \Sigma n_{2j} m_{2j}$, RT has its usual meaning, $n_{kl}, m_{kl}, \phi_{kl}$ = the number of moles, number of repeat units (chain length) and volume fraction of polymer kl, resp., $\phi_1 = \Sigma \phi_{1i}$, $\phi_2 = \Sigma \phi_{2j}$. If the system is strictly binary the two sums in eq.(1) have only one term each. Molecular models can often be expressed by eq.(1) where the interaction function g is used to accommodate dependencies on concentration, temperature, molar mass, etc., as the model may require. Virtually every theory contains the first two terms on the right-hand side for the combinatorial entropy of mixing.

The temperature dependence of g follows from elementary classical thermodynamics. Assuming c_p, the specific heat at constant pressure, to be linear in T, a first approximation often found, and Δc_p to show Van Laar-type concentration dependence, we can write

$$\Delta c_p = (c_o + c_1 T) \phi_1 \phi_2 \tag{2}$$

As a consequence, we obtain by integration of $\partial \Delta H = \Delta c_p \partial T$ and $T \partial \Delta S = \Delta c_p \partial T$:

$$g = g_o + g_1/T + g_2 T + g_3 \ln T \tag{3}$$

and

$$\Delta H/N_\phi R = (g_1 - g_3 T - g_2 T^2) \phi_1 \phi_2 \tag{4}$$

where g_o and g_1 arise from integration constants ($c_o \propto -g_3$; $c_1 \propto -2g_2$; 8,9). We note in passing that the last term in eq.(3), often neglected or combined with the first because of its weak T-dependence, may show up markedly in ΔH.

It is frequently found that $\Delta H(\phi_2)$ is not simply symmetric in ϕ_2 around $\phi_2 = \frac{1}{2}$, so that the term in brackets in eq.(4) must be concentration-dependent (10,11). We could write

$$g_3 = h_3 \phi_2^n; \quad g_2 = h_2 \phi_2^m \tag{5}$$

which accordingly causes g to depend on concentration. The parameters n and m can be chosen so as to suit the situation in hand. For instance, with n=1 and m=2 we have a quadratic series for $g(\phi_2)$ which we shall use later on.

According to Gibbs (12) the spinodal is defined by

$$J_s = 0 \tag{6}$$

where

$$J_s = \left|\left| \partial^2(\Delta G/N_\phi RT)/\partial \phi_k \partial \phi_1 \right|\right|_{p,T} \tag{7}$$

and ϕ_k and ϕ_1 represent the various independent concentration variables. Eqs.(6) and (7), applied to ΔG-function (1), lead to

$$(1/\phi_1 m_{w1}) + (1/\phi_2 m_{w2}) + \partial^2(g\phi_1\phi_2)/\partial\phi_2^2 = 0 \tag{8}$$

The critical state is determined by (12)

$$J_c = 0 \tag{9}$$

where J_c is the determinant formed from J_s upon replacement of the elements in any line by $\partial J_s/\partial\phi_{11}$, $\partial J_s/\partial\phi_{12}$, ..., $\partial J_s/\partial\phi_{21}$,.... For eq.(1) we find

$$m_{z1}/m_{w1}^2 \phi_1^2 - m_{z2}/m_{w2}^2 \phi_2^2 + \partial^3(g\phi_1\phi_2)/\partial\phi_2^3 = 0 \tag{9'}$$

In eqs.(8) and (9) m_{wk} and m_{zk} stand for the mass- and z-average chain lengths of polymer k.

We note that eq.(1) can be transformed into the familiar Flory-Huggins equation for a polymer solution. For a strictly-binary system, for instance

$$\Delta G'/N_\phi RT = m_1 \Delta G/N_\phi RT = \phi_1 \ln\phi_1 + \phi_2(m_2')^{-1}\ln\phi_2 + g'\phi_1\phi_2 \tag{10}$$

where $m_2'=m_2/m_1$ and $g'=gm_1$. It is seen that now m_1 is the unit of chain length, and g is scaled by a factor m_1. In the following we may and will draw examples from data on polymer solutions that thus can elucidate phase relationships in polymer mixtures as well.

The relevant literature is vast and we cannot nearly be complete. The preceding chapters and their lists of references should be consulted as well as refs 13-19 stated here.

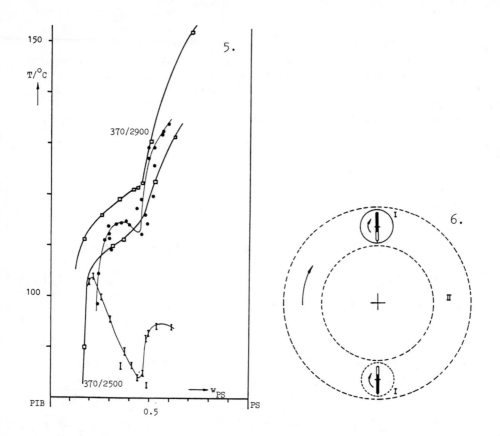

Fig. 5. Spinodals (**I**,•) and cloudpoints curves for polyisobutene(PIB)/
anionic polystyrene (PS) mixtures (molar masses indicated;
g/mole). Curves drawn by hand; w_{PS}= mass fraction PS

Fig. 6. Principle of the centrifugal homogeniser. The PICS capillaries
are mounted in segments I onto a rotating track II that can be
heated and thermostated up to 200°C. There are four segments,
each holding five capillaries. The segments can be made to
rotate very quickly in the sense indicated, thus superposing
a planetary motion on the rotation of track II

2. EXPERIMENTAL METHODS

Techniques applied to date to gather information about the
thermodynamic state of polymer blends may be split into two main
groups. Preceding chapters have dealt with microscopic, glass-transit-
ion- and other methods that roughly fall into one of the groups, which

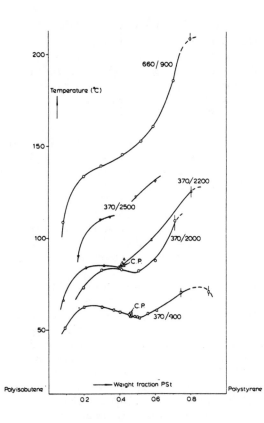

Fig. 7. Cloud points and critical points (C.P.) in short-chain liquid polymer mixtures. System: polyiso-butene/polystyrene, molar masses (g/mole) indicated

may be characterized by the following aspects: 1) the temperature at which the measurements are made or the samples are prepared may diffe: from those of interest, 2) the information gained usually refers to cloudpoints that are determined by details of the molar mass distri-butions and therefore very difficult to interpret quantitatively, 3) whether thermodynamic equilibrium is achieved remains questionable

There are also methods in which the thermodynamic state of the systems is changed essentially by the addition of a mutual solvent or replacement of part of a polymer by labelled chains. In both cases an unambiguous interpretation is complex and the amount of data neede: is large (20,21).

A second group of methods is based on equilibrium behaviour of the polymer blend itself. In the last decennium methods have been develop- ed to determine equilibrium data like cloud points, spinodals and critical points on molten polymer blends.

Spinodal points can be measured by pulse-induced critical scatter- ing (PICS) of polymer mixtures, recently also with sizeable chain lengths in the samples (22). Cloud points can also be obtained and such scattering methods thus supply useful information provided the refractive indices of the various constituents differ enough. Fig.5

gives some representative examples obtained by PICS.

PICS requires the preparation of homogeneous mixtures of two polymers in a capillary. To achieve this with highly viscous systems Gordon et al. (23,24) developed the centrifugal homogenizer. The principle of the apparatus is shown in fig.6. Of course, there is a limit to the viscosities that can be managed and, hence, to the chain lengths that can be handled. If higher molar masses must be investigated the mutual solvent method can be used. In contrast to its disadvantage in the usual situation (limited concentrations of the mixed polymer) we only need relatively small amounts of solvent to make the centrifuge method work. The amount of measurements needed is larger, obviously, but the apparatus can handle 20 systems at a time.

The centrifugal homogenizer can also be used to carry out phase-volume-ratio measurements. After homogenization the rotor tempera-ture is adjusted to a required temperature within the two-phase region. If, after thermal equilibration, the planetary motion is stopped, the centrifugal force will segregate the two phases. To follow the proces of phase separation, a stroboscope was fitted to the centrifuge which flashes at the rotor speed. With an optical system and/or a TV camera one can record the phase volumes of the systems in the four cell holders separately. After remixing the systems can go through the whole cycle again to record the phase volumes at another temperature. Such measurements yield the critical concentration and cloud points (25).

The compositions of the coexisting phases can be studied with the DSM-Spinco centrifuge (26), which supplies such information via diffusion measurements. The equilibrium methods discussed here (fig.7 gives some more examples) can be applied to systems showing upper- as well as lower critical miscibility behaviour. Restrictions only lie in refractive index differences, viscosities, thermal stabilities of the polymers and, sometimes, the ratio of domain size and wavelength of the light used.

Some typical cloudpoint curves and critical points obtained with these techniques are shown in figs 6 and 7. Critical points are usually found to be shifted away from the maximum of the cloudpoint curve, which mostly has an irregular shape. The theoretical back-ground for these phenomena is discussed in the next two sections.

3. POLYMOLECULARITY

It is virtually always wrong to treat a polymer as if it consists of a single component. In the literature this problem is often either ignored altogether or without checking assumed to involve negligible errors. The possible errors may be very large, however, in particular with wide distributions of molar mass. Under such circumstances a comparison between experiment and molecular theory becomes highly questionable and can at best be only be qualitative.

Apart from chain-length dispersity other aspects may complicate

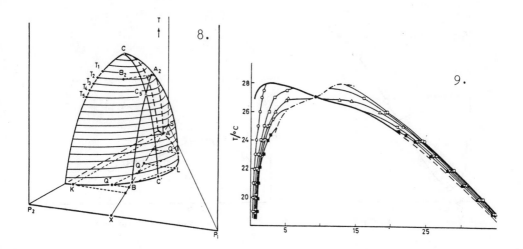

Fig. 8. Miscibility gap in solutions of a binary polymer (P_1/P_2) in a single solvent S. AA_2C_5B: quasi-binary section (cloudpoint curve); A_2: precipitation threshold, CC_5C': critical curve

Fig. 9. Quasi-binary phase diagram for solutions of a sample of poly-styrene in cyclohexane. Data from Rehage et al. (34). Cloud-point curve: ▬▬ ; coexistence curves at (from left to right on the left-hand branches): 2, 6, 10, 15 and 20 % overall polymer concentration by wt: ▬▬▬ ; shadow curve: —·—·—; critical point: ●

the picture (chain branching, copolymer composition). Here we discuss linear homo- and copolymers and first investigate the implications of molar mass distributions in homopolymers.

The primary consequences of an abundance of components for the usual two-dimensional phase diagrams have been known for many years (5, 12, 30, 31). A consideration of a ternary system already brings the main points to light. The extra component adds a degree of free-dom and we have a cloudpoint surface instead of a curve. It is seen in fig.8 that a cloudpoint curve does not represent coexisting phases and that its extremum (precipitation threshold, ref. 32) is not a critical point, except in very rare cases (see later). Cloud points coexist with incipient phases whose compositions lie outside the plane of drawing but can be projected onto it (e.g. K, coexisting with cloud point A). Figs 8 and 9 elucidate these aspects.

The cloudpoint curve encloses the (quasi-binary) miscibility gap and a system within it will separate into two phases whose composit-ions are both outside the plane of the diagram. Again, these concentrations can be projected onto it and we so obtain quasi-binary coexistence curves the location of which depends on the overall

98

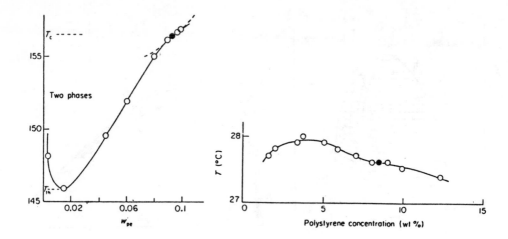

Fig. 10. Experimental cloudpoint curves and critical points (●).
Left: polyethylene (M_n=8x10^3;M_w=1.8x10^5;M_z=1.2x10^6). Dashed
curve indicates location spinodal; w_{pe} = mass fraction
polymer, solvent: n-hexane
Right: polystyrene (M_w/M_n=1.07;M_z/M_w=1.4); solvent: cyclohexane

composition of the mixture. They are located within the temperature-
overall composition region demarcated by the cloudpoint curve and its
coexistence curve (shadow curve, ref.33).

These general features have been amply verified by experiment
(33,34). Both critical temperature and concentration may deviate
considerably from the threshold values, polymolecularity always
being present. Examples are given in fig.10 which demonstrates the
possible fallacy of identification of critical and threshold conditions. Examples for polymer mixtures are shown in fig.11, which is
based on eq.(1) where the interaction function g is assumed not to
depend on concentration. We see that the critical point may shift
along either branch of the cloudpoint curve, depending on the m_z/m_w
ratios involved. Only in the rare cases where these ratios are equal
in the two constituent polymers the critical point occurs at the
extreme of the cloudpoint curve. This conclusion is only valid for
a concentration-independent g-function. In those rare cases where
critical points have been determined in polymer or, rather, oligomer
blends, they have been found far away from the threshold, fig.7
illustrates this.

It might seem inconsistent that a system outside the miscibility
gap could still phase separate (fig.9). Such systems, however, refer
to molar mass distributions that are shifted towards higher molar
mass, compared with the initial distribution. Hence, the polymers
they represent are less compatible with the solvent.

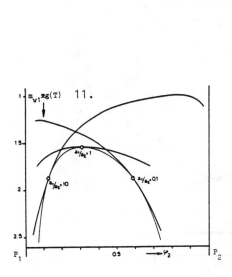

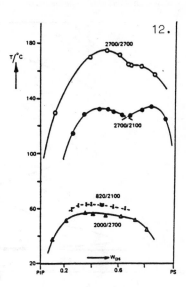

Fig. 11. Cloudpoint curves (bold curves), spinodal and critical
points (o), calculated with eq.(1) and $g \neq g(\phi_2)$, showing
the influence of polymolecularity. The ratio m_z/m_w for
polymer k is denoted by a_k, the mass-average chain lengths
differ by a factor 5

Fig. 12. Miscibility gaps in narrow-distribution polyisoprene/
polystyrene mixtures with indicated average molar masses.
The dashed curve is a spinodal by PICS

4. MULTIPLE PHASE SEPARATIONS

So far we considered separations into two liquid phases. Polymer
systems in particular do not seldom tend to split up into more than
two phases, three being the number occurring most frequently. Though
only rarely having been observed directly (35-37), multiple phase
separations have more often been detected indirectly by way of
irregularities in the shape of cloudpoint curves. More or less
pronounced deviations from the familiar dôme shape occur quite
often and are indicative for multiple phase behaviour. Dents, or
shoulders, have been reported by other authers than the present
(27-29). Such shoulders may develop into two extremes upon relatively
small changes in molar mass (see fig.12). These complexities can be
attributed to two possible causes, polymolecularity and concentration
dependence of the interaction function. Most of the considerations
following below go back to Korteweg (38).

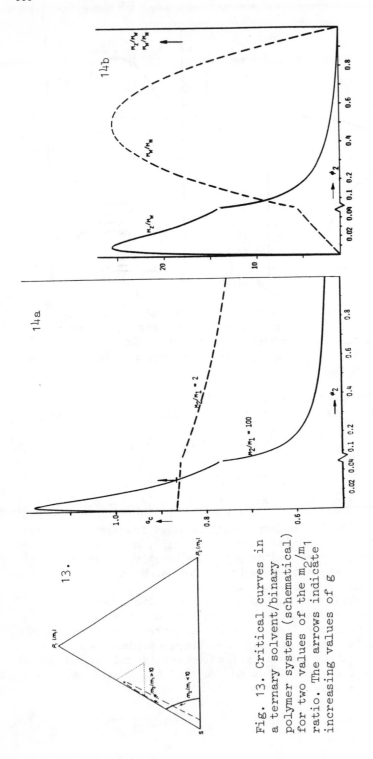

Fig. 13. Critical curves in a ternary solvent/binary polymer system (schematical) for two values of the m_2/m_1 ratio. The arrows indicate increasing values of g

Fig. 14. Critical value of g (g_c, fig.14a) and m_z/m_w and m_w/m_n ratios at g_c (fig.14b) as a function of the composition ψ_2 of the solvent-free polymer mixture. The m_2/m_1 ratios are indicated, fig.14b refers to $m_2/m_1=100$

4.1. Polymolecularity

Tompa (5,39) studied three-phase separations in solvent-polymer(1)-polymer(2) ternary systems where the two polymers differed by chain length only. He set g independent of concentration and found the course of the ternary critical curve to be quite different for small and large ratios of the chain lengths of the two polymeric components. If $m_2/m_1 < 10$ we have the obvious course from the largest critical g-value (g_{c1}) down to the smaller g_{c2} for m_2 (see fig.13). If $m_2/m_1 > 10$, however, the critical $g_c(\phi_1,\phi_2)$ curve passes through a maximum at $g_c > g_{c1}$. In that g-range there are two critical points for one g-value (i.e. at one temperature). One of these critical points may be stable, the other metastable or they may be metastable and unstable. In the latter case the cloudpoint curve has no stable and, hence, measurable critical point. It is seen in figs 14a and b that such behaviour can be expected if the binary polymer (1,2) mixture contains a very small amount of the large component 2. In terms of the usual polydispersity indices this implies m_z/m_w being larger than m_w/m_n, and the distribution being very asymmetric. Solc has thoroughly analysed the problem and derived general conditions for multicomponent polymers (40). His theoretical prediction on cloudpoint curves without a critical point was later verified by experiment (41).

Fig.15 shows the two critical points within the miscibility gap and their development into a stable/metastable set upon an increase of the concentration (volume fraction ψ_2) of the component with the longer chains in the solvent-free polymer mixture. When the metastable critical point has just appeared on the stable branch of the binodal (fig.16d) we have a critical phase in equilibrium with a third liquid phase. The other critical point may also develop such a situation (fig.16a) denoted by Korteweg as a heterogeneous double plait point (38). If other variables, like pressure, can be varied enough such heterogeneous plait points may be made to coincide in a triple critical point. In between the two double plait point situations there are three two-phase regions bordering on the three edges of a three-phase triangle. Cloudpoint curves (quasi-binary sections) will then consist of two intersecting branches showing up experimentally as dents or shoulders (40-43).

As remarked above, polymer mixtures and solutions can be viewed upon as behaving quite analogously. The relationship between dents in cloudpoint curves and three-phase separations has been established experimentally with polymer solutions only (42) but can reasonably be expected to be valid for polymer mixtures as well. The present three-phase mechanism appears to occur so close to the solvent-P_1 axis that it must be considered almost impossible to construct a quasi-binary section that intersects two of the two-phase regions neighbouring the three-phase triangle, and so have a bimodal shape. Such two-peaked cloudpoint curves will therefore not be found readily unless other factors dominate the location of meta-stable critical points.

It should be noted that the mechanism of three-phase separation

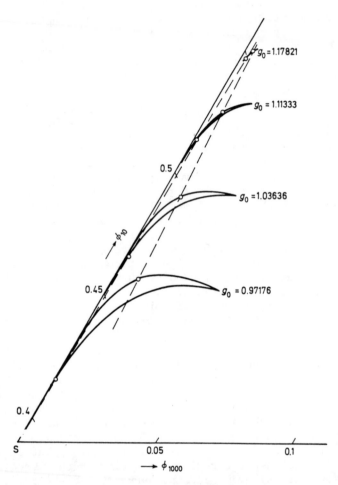

Fig. 15 Secundary binodals within the miscibility gap calculated
for for the ternary solvent (S)/polymer 1 (m_1=10)/polymer
2 (m_2=1000) system. Values of g (independent of concentration) are indicated. The dashed curve is the critical line.
This figure refers to part of fig.13 (indicated by dotted
lines)

by polymolecularity goes with a simple dôme-shaped spinodal. Eq.(8),
worked out for $g \neq g(\phi_2)$, m_{1i}=1 (solvent), yields

$$1/\phi_1 + 1/\phi_2 m_{w2} - 2g(T) = 0 \qquad (11)$$

which equation defines the spinodal curve in a $g(\phi_2)$ plane as a
parabola (ϕ_1=1-ϕ_2). Its location is determined by the mass-average

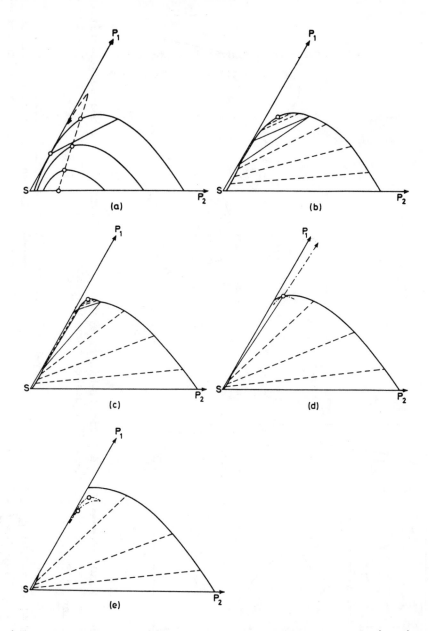

Fig. 16. Development and disappearance of a three-phase region in a
ternary polymer solution (schematic). Stable binodals:━━ ;
secundary binodals:━·—··—; critical points: o; tie lines:
━ ━ ━

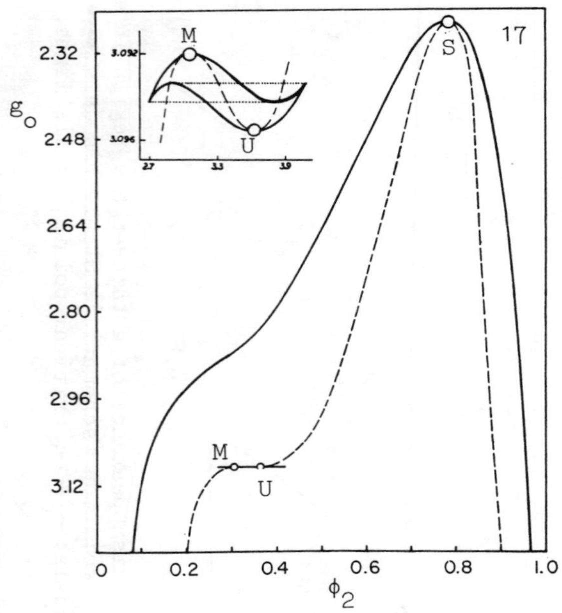

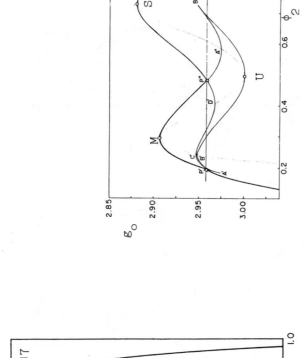

Fig. 17. Cloudpoint curve (———) for a strictly-binary system with a slightly perturbed double critical
point ($m_2/m_1=1$, $g_1=-2.0$, $g_2=2.531$). Dashed curve: spinodal, critical points: o (M, metastable;
U, unstable; S, stable)

Fig. 18. Composite cloud-point curve for a strictly-binary system with three single critical points
($m_2/m_1=1$; $g_1=-2.0$; $g_2=2.02$). Stable parts of binodals: ———; dotted curve: spinodal; critical
points: o (M,S, stable; U, unstable); P', P", P"', three-phase equilibrium. Coexisting phases
indicated by primed letters

chain length of the polymer, irrespective the spread in molar mass.

4.2. Concentration dependent interaction function

Multiplicity of critical points may also occur in strictly-binary systems. Within the scope of eq.(8) it is evident that the dependence of g on concentration must then be responsible. Recently, this possibility has been investigated in detail (43,44), we summarize some conclusions.

Critical points in binary systems are located at g-values where the spinodal has a horizontal tangent in $g(\phi_2)$ space. Multiplicity of critical points then calls for spinodal curves, more complex than a parabola with its single extreme. A simple way of producing such shapes provides a representation of $g(\phi_2)$ by a power series in ϕ_2. For the present purpose it suffices to ignore terms higher than the quadratic:

$$g = g_o + g_1\phi_2 + g_2\phi_2^2 \qquad (12)$$

Figs 17 and 18, taken from ref. 43, illustrate how separations into three liquid phases can thus develop in a binary system. In fig.17 two separate critical points (M, U) have developed from a double critical point, but they are still close together. The enlargement shows that the spinodal has a maximum at M, the meta-stable critical point, and a minimum at U, the unstable one. Note that the cloudpoint curve indicates this behaviour within the gap by a shoulder. If the coefficient g_2 in eq.(12) is changed, the secundary lense-shaped binodal may come to extend beyond the stable binodal which assumes a bimodal shape. Now we have two stable (M,S) and one unstable (U) critical point.

These theoretical considerations lead to the conclusion that cloudpoint curves with two extremes are very probably due to the interaction function being concentration dependent, rather than to polymolecularity.

5. IMPORTANT MOLECULAR PARAMETERS

Expressions for ΔG available in the literature have so far not been very successful for the purposes outlined above. For example, the original Flory-Huggins equation for the spinodal (eq.11) is quadratic in ϕ_2, so that there are only two roots in ϕ_2 at a given T and m_{w1}/m_{w2} set. Hence, two-peaked spinodals are outside the scope of this equation, no matter what the polymolecularities are. We have to deal with large effects caused by slight variations in molar mass (distribution) and other molecular parameters. The ΔG-function represents a subtle balance of various terms and predictions can only be highly unreliable and at best roughly qualitative, unless the nature of the individual contributions is understood.

Allowance for the fact that repeat units and molecules differ in size and shape provides a first step forwards. We follow Staverman's suggestion (45,46) and assume the numbers of interacting contacts per unit or molecule to be proportional to their specific surface area. The regular solution approximation then leads to a Van Laar type concentration dependence in terms of surface fractions and, expressed in volume fractions the g-function then is

$$g\phi_1\phi_2 = \{A + B(T)/(1 - \gamma\phi_2)\}\phi_1\phi_2 \tag{13}$$

where B(T)= the interaction free enthalpy, $\gamma= 1 - \sigma_2/\sigma_1$, σ_i = interacting contact surface area of unit i. There is ample evidence for the adequacy of eq.(14) in improving descriptions of thermodynamic properties (44,45,47-51). However, the concentration dependence it provides is not strong enough to produce two-peaked spinodals (52). Apart from this we are still forced to accept a purely empirical free enthalpy correction term $A\phi_1\phi_2$.

Huggins' later thermodynamic theories (50,51) contain corrections of the combinatorial entropy of mixing for the influence of the surroundings on the average randomness of orientation of a repeat unit relative to the orientation of the preceding segment in the chain. This could be interpreted freely as the flexibility of polymer chains embedded in a matrix of their own kind being changed by the addition of a second polymer with repeat units of a different chemical structure. Interpreting the empirical term $A\phi_1\phi_2$ in this way we can write

$$A\phi_1\phi_2 = -\phi_1\ln\{1+k_1(1-\gamma)\phi_2/(1-\gamma\phi_2)\} - \phi_2\ln\{1+k_2\phi_1/(1-\gamma\phi_2)\} \tag{14}$$

For a system like polyisobutene/polystyrene (PIB/PS) this type of reasoning would demand a positive k for less flexible (PS) chains and a negative value for flexible molecules (PIB). This is exactly a combination that may produce a bimodal spinodal (see fig.19). It is worth while noting that the interacting surface area ratio $(1-\gamma)$ must be assumed to differ from unit to obtain this result.

If the orientational entropy correction is to be an important molecular parameter it should show up significantly in the concentration dependence of the radius of gyration of each of the two polymers. With the aid of Silberberg's approach (53) one can rearrange Huggins' expression (14) and produce the concentration dependence of the radius of gyration in a qualitatively correct manner (54-56).

There is still another reason why the coil extensions could be expected to play an important role. The properties of very dilute polymer solutions are known to be determined by the nonuniformity of the polymer segment density. In polymer mixtures we have that situation on both ends of the concentration scale. To deal with this problem Stockmayer et al. (57,58) suggested an interaction function for polymer solutions that bridges the two concentration regimes

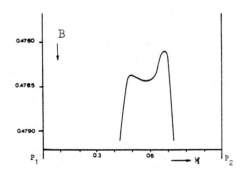

Fig. 19. Spinodal calculated with Huggins' orientational entropy correction (14) for $k_1=-0.69$; $k_2=+0.91$; $\sigma_2/\sigma_1=0.8$; $m_1=7.2$; $m_2=45$

continuously. This treatment can be extended to polymer blends and we write (18,19)

$$g = g_1^d P_1 + g_2^d P_2 + g^c(1 - P_1 - P_2) \tag{15}$$

where g_i^d represents the interaction function appropriate for the concentration range dilute in polymer i, and g^c to the uniform segment density range. For instance, g^c could be represented by eq.(12) or (13). P_i is a weighting factor standing for the probability that a given volume element does not fall within any coil domain i:

$$P_i = \exp(-\rho_i v_i) = \exp(-\lambda_i \phi_i) \tag{16}$$

where $\rho_i =$ the number of coils i per unit volume and v_i their domain. The latter is related to the radius of gyration which relation can be expressed in a proportionality to ϕ_i (57). We have shown elsewhere (18,19) that this model may yield bimodal spinodals for reasonable values of the parameters and also that appearance and disappearance of bimodality sensitively relates to polymolecularity, a phenomenon found experimentally on the system squalene/polystyrene (19).

Free volume treatments are known to be important, particularly when lower critical miscibility is being dealt with. Polymer mixtures have also been discussed from such a point of view, e.g. we have McMasters' (59) and Olabisi's (60) applications of Flory's equation of state theory (48,49). Olabisi could show that this theory is capable of describing two-peaked spinodals in the case of lower critical miscibility. Work by Walsh et al., who extended Flory's theory, should also be mentioned here (61,62). Other approaches have been suggested, for instance Sanchez' lattice-fluid model (63,64). Recently, Simha et al have begun exploring the potential applicability Simha's corresponding states theory to mixtures, including polymers (65-67).

The present authors have developed a simple free-volume treatment which is based on Trappeniers' two-component lattice-gas model (68-72). We extended Trappeniers' model by the introduction of Staverman's concept of interacting surface areas and found the free energy equation thus obtained to be very versatile. Nonpolar as well as polar molecules can be dealt with, in liquid, gaseous and near-critical fluid states, pure as well as mixed. Polymers also fall within this scope, references 68-70 give details and illustrations. These examples, and those presented in the other, similar treatments prove beyond doubt that free volume is a molecular parameter of primary importance and cannot be disregarded.

Because of the obvious similarity it might be worth-while to compare Sanchez' lattice-fluid model briefly with our mean-field lattice-gas treatment (MFLG)(73). The lattice-fluid approach is a true theory in that it does not contain empirical parameters. The MFLG treatment does have such parameters and is probably therefore more versatile at the cost of leaving questions as to their physical meaning open. The critical ratio $p_c V_c/RT_c$ comes out too high with Sanchez, in the MFLG it can only be correct since critical data are à la Van der Waals used to calculate some of the parameters. For mixtures Sanchez needs a combination rule for the cell volume, the MFLG is versatile enough to keep this parameter constant for all pure substances and their mixtures without losing accuracy. If the latter feature is viewed upon as not pointing to physical depth, it certainly is practically useful.

6. STATISTICAL COPOLYMERS

Worth-while a chapter in its own right, statistical copolymers should at least receive some attention in the present context. Some of the peculiarities presented in the preceding sections are exhibited emphatically by systems containing statistical copolymers. Not only do differences in numbers of nearest-neighbour contacts per repeat unit make themselves noticeable but also an interplay between the two polymolecularities appears to supply ample possibilities for multiple phase behaviour (74).

Binary mixtures of copolymers built up of the same two repeat units but differing in chemical composition may show limited miscibility. Such behaviour was predicted by Scott (75) who analysed copolymers with the Flory-Huggins model. The spinodal for such systems reads

$$1/m_1\phi_1 + 1/m_2\phi_2 - 2g_{ab}d_{12}^2 = 0 \tag{17}$$

where ϕ_1 and ϕ_2 are the volume fractions of copolymers 1 and 2, m_1 and m_2 their chain lengths, g_{ab} the interaction parameter for repeat units a and b contacts and d_{12} the difference in chemical composit-ion between the two copolymers (ϕ_{a1} and ϕ_{a2}; $d_{12} = \phi_{a1} - \phi_{a2}$).

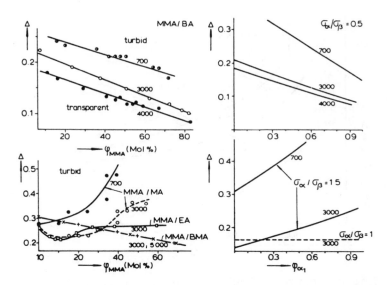

Fig. 20. <u>Left</u>: Tolerable difference for miscibility in 1:1 mixtures of
acrylate copolymers $P_{\alpha\beta}$ as a function of the chemical
composition $\phi_{\alpha 1}$ of copolymer 1 ($\Delta = -d_{12max}$). Mass-average
degree of polymerisation of the mixture is indicated. MMA=
methyl methacrylate, MA= methyl acrylate, EA= ethyl acrylate,
BA= butyl acrylate, BMA= butyl methacrylate (76,77)
<u>Right</u>: Tolerable difference for miscibility (spinodal values)
calculated for indicated values of the mass-average chain leng†
length of $P_{\alpha\beta 1}/P_{\alpha\beta 2}$ 1:1 mixtures, and interacting surface
areas σ_α and σ_β for the copolymer repeat units

Molau later demonstrated that homogeneous mixtures of styrene/acrylo-
nitrile (SAN) copolymers would allow a difference in AN content of
about 4 mole % (36). Beyond that value only phase-separated systems
are found.
    An extension of Scott's treatment proved necessary in order to
deal with observations by Kollinsky and Markert (76,77) on mixtures
of acrylic copolymers. Whereas Scott's eq.(17) predicts Δ, the
maximum tolerable difference d_{12} for miscibility not to depend on the
chemical composition ϕ_a of the individual copolymers, Kollinsky and
and Markert's detailed study revealed otherwise (fig.20). A simple
extension of eq.(17) to include a disparity in size between the repeat
units and, consequently, the numbers of nearest-neighbour contacts
they can make already suffices to cover the experimental findings
in a qualitative manner (78). Fig.20 shows this, the observed trends
are reproduced, obviously interacting surface areas are needed for
a first improvement of the simple model.
    A similar conclusion can be drawn from Molau's data on SAN mixtures

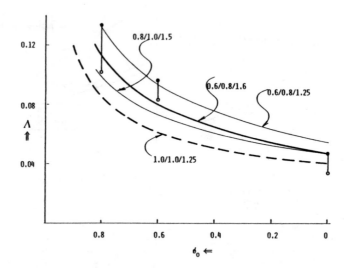

Fig. 21. Spinodal curves calculated for indicated sets of σ_a/σ_o; σ_b/σ_o; g_{ab} values (74). Tolerable composition difference Λ as a function of solvent concentration ϕ_o. Data for poly(acrylonitrile(a)-styrene(b)) in methyl ethyl ketone (o) by Molau (36). Filled circles: phase separated; open circles: homogeneous systems

dissolved in methyl ethyl ketone. The appropriate equation for the spinodal is complex and contains three interaction functions, relating to solvent/repeat unit a, solvent/repeat unit b and repeat unit a/b interactions. Yet, with reasonable values for g_{oa} and g_{ob} (both >0) it is only possible to pass calculated spinodals through the Λ regions where they must be expected to be located if the various interacting surface areas are supposed to be different (fig. 21).

Fig.22 shows calculated spinodals and critical points for an arbitrarily chosen (a/b) copolymer in a 50% solution. It is seen that the chemical composition of the third component (ϕ_{a3}) sensitively determines whether there are three critical points (three-phase separation) or one (two phases). In a qualitative sense, this is exactly what Molau reported (36), a 1:1:1 mixture of SAN copolymers in a 25% MEK solution may form two or three layers, depending on the ϕ_a values that do not have to change much for passing from one situation to the other.

Another example of interest was reported by Teramachi et al (79) who studied solutions in toluene of two samples of SAN differing only 2% in AN content. The cloudpoint curves of quasi-ternary mixtures of these two SAN samples in toluene depend greatly on the mixing ratio and show pronounced dents and shoulders.

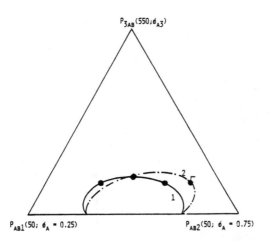

Fig. 22. Spinodal curves and critical points ($\bullet$, $\bullet$) calculated for
ϕ_o=0.5 (solvent concentration), g_{ab}=0.4, g_{oa}=0, g_{ob}=0,
σ_a^o/σ_o=1, σ_b/σ_o=1, ϕ_{a1}=0.25, ϕ_{a2}=0.75, m_1=m_2=50, m_3=550.
Curve 1: ϕ_{a3}^b=0.5; curve 2: ϕ_{a3}=0.475

Mixtures of polymethylmethacrylate and SAN copolymers exhibit
extreme sensitivity to small changes in the chemical composition
of the SAN constituent (14,54). Not only can a flat-bottomed lower
critical miscibility gap with a weak shoulder turn into two distinct
minima but, at the same time, develop a second upper critical
miscibility gap (see fig.23). We have found that the description of
such fine details call for a free volume treatment. For instance,
the MFLG model covers the phenomenon quite easily, as is illustrated
in fig.23. All parameters in the model were kept constant, only the
chemical composition ϕ_a in the copolymer was changed by less than
2%. The spinodals so calculated correspond in principle to the
measured effect. A more quantitative agreement could possibly be
achieved upon introduction of interacting surface areas and chain
length distributions. This is a point of further research.

7. CONCLUSIONS

Phase diagrams of polymer mixtures often show a complex structure,
dents, shoulders or two extremes not being rare occurrences. Such
shapes can be rationalized on the basis of classic thermodynamics
and be demonstrated to relate to polymolecularity and/or concentration dependence of the interaction function. It is a subtle balance
between entropic and energetic terms in the free enthalpy of mixing,
all small, that triggers these peculiarities. Such aspects of phase
behaviour were already a century ago known to and described by
Gibbs (12), Korteweg (38) and Van der Waals and his school (30). The

112

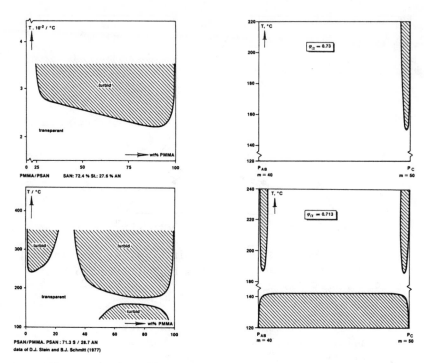

Fig. 23. Influence of a small variation in composition of copolymer
SAN on phase behaviour in the system poly(acrylonitrile/
styrene)/polymethylmethacrylate. Left: Data by Schmitt et
al (14,54). Right: spinodals by MFLG model

facts that polymolecularities (chain length, chemical composition)
are always present, and that the g-function is usually found to
depend on concentration, compels us to make allowance for such
effects if consistent descriptions are to be obtained. In particular,
quantitative confrontations between molecular theories and experiment
would otherwise be meaningless.

As to the molecular origin of the concentration dependence of the
interaction function, a variety of causes can be advanced. Each of
them could be responsible by itself but practice will probably prove
combinations to be operative. At this moment it would seem not to be
possible to decide which molecular effect is predominant, much work
is still needed for which experimental tools like PICS and centrifugal
homogenizer are available. We simply list the molecular parameters
that have presented themselves in the course of this presentation:
1) disparity in numbers of nearest-neighbour contacts between the
   various repeat units and molecules in the system,
2) changes in flexibility of polymer chains by their immediate
   surroundings,
3) nonuniform segment density at both ends of the concentration scale,

4) dependence of the equilibrium free volume on molar mass (distribution) and chemical composition (distribution).

REFERENCES:

1. Stockmayer, W.H., Koningsveld, R., Kleintjens, L.A., Solc, K., unpublished work.
2. Flory, P.J., J. Chem. Phys., 9 (1941), 660; 10 (1942), 51.
3. Huggins, M.L., J. Chem. Phys., 9 (1941) 440; Am. N.Y. Acad. Sci., 43 (1942) 1.
4. Staverman, A.J., Van Santen, J.H., Rec. Trav. Chim., 60 (1941) 76.
5. Tompa, H., Polymer Solutions, Butterworth London, 1956.
6. Flory, P.J., J. Chem. Phys., 12 (1944), 425.
7. Scott, R.L., Magat, M., J. Chem. Phys., 13 (1945) 172.
8. Koningsveld, R., Staverman, A.J., J. Polym. Sci., Part A2, 6, (1968) 325.
9. Koningsveld, R., Adv. Coll. Interface Sci., 2 (1968) 151.
10. Holleman, Th., Physica, 29 (1963) 585.
11. Karasz, F.E., MacKnight, W.J., Pure Appl. Chem. 52 (1980) 409.
12. Gibbs, J.W., Collected Works, Dover Publ. Repr., N.Y., 1961, Vol. I, p-132.
13. Krause, S., J. Macromol. Sci., Rev. Macromol. Chem., C7 (1972) 251.
14. Schmitt, B.J., Angew. Chem., Intern. Ed. Engl., 18 (1979) 273.
15. Casper, R., Morbitzer, L., Angew. Makromol. Chem., 58/59 (1977) 1.
16. Paul, D.R., Newman, S., Eds. Polymer Blends. Acad. Press, New York (1978).
17. Koningsveld, R., Kleintjens, L.A., J. Polymer. Sci., Polymer Symposium Nr 61 (1977) 221.
18. Onclin, M.H., Kleintjens, L.A., Koningsveld, R., Makromol. Chem., Suppl. 3 (1979) 197.
19. Koningsveld, R., Onclin, M.H., Kleintjens, L.A., in Solc, K. (ed): Polymer Compatibility and Incompatibility in MMI Press, 1982, p-25.
20. Koningsveld, R., Kleintjens, L.A., Schoffeleers, H.M., Pure Appl. Chem., 39 (1974) 1.
21. Atkin, E.L., Kleintjens, L.A., Koningsveld, R., Fetters, L.J., Polymer Bull., 8 (1982) 347; Makromol. Chem., 185 (1984) 377.
22. This Volume chapter by M. Gordon.
23. Gordon, M., Ready, B.W., US Pat. 4, 131, 369.
24. Gordon, M., Kleintjens, L.A., Ready, B.W., Torkington, J.A., Brit. Polym. J., 10 (1978) 170.
25. Koningsveld, R., Staverman, A.J., J. Polym. Sci., Part C, Nr 16 (1967) 1775.
26. Rietbeld, B.J., Brit. Polym. J., 6 (1974) 181.

114

27. Allen, G., Fee, G., Nicholson, J.P., Polym. 2 (1961) 8.
28. Powers, P.O., Am. Chem. Soc. Polym. Prepr., 13 (1973) 528; 14 (1973) 1321.
29. Bank, M., J. Leffingwell, Thies, C., J. Polym. Sci., A2 (1972) 1097.
30. VanderWaals, J.D., Kohnstamm, Ph., Lehrbruch der Thermodynamik, Vol.III, Barth, Leipzig, 1912.
31. Schreinemakers, F.A.M., in Bakhuis Roozeboom, H.W., Die heterogenen Gleichgewchte vom Standpunkte der Phasenlehre Vieweg, Braunschweig, 1923, Vol.III, part 2.
32. Tompa, H., Trans. Farad. Soc., 46 (1950) 970.
33. Koningsveld, R., Staverman, A.J., Kolloid Z&Z Polym., 218 (1967) 114.
34. Rehage, G., Möller, D., Ernst, O., Makromol. Chem., 88 (1965) 232.
35. Patat, F., Traxler, G., Makromol. Chem., 33 (1959) 113.
36. Molau, G., J. Polym. Sci., Polym. Lett. Ed., 3 (1965) 1007.
37. Koningsveld, R., Staverman, A.J., Kolloid Z&Z Polym., 220 (1967) 1.
38. Korteweg, D.J., Sitzungsber. Kais. Akad. Wiss. Wien., Math.-Naturwiss. K., Abt. A, 98 (1889) 1154.
39. Tompa, H., Trans. Farad. Soc., 45 (1949) 1142.
40. Solc, K., Macromolecules, 3 (1970) 665.
41. Koningsveld, R., Kleintjens, L.A., Pure Appl. Chem. Macromol. Chem., 8 (1973) 197.
42. Kleintjens, L.A., Schoffeleers, H.M., Domongo, L., Brit. Pol., J., 8 (1976) 29.
43. Solc, K., Kleintjens, L.A., Koningsveld, R., Macromolecules 17 (1984) 573.
44. Nies, E., Koningsveld, R., Kleintjens, L.A., Proceedings IUPAC Symposium on Macromolecules, Bucharest, 1983.
45. Staverman, A.J., Rec. Trav. Chim., 56 (1937) 885.
46. Staverman, R., Kleintjens, L.A., Macromolecules, 4 (1971) 637.
47. Koningsveld, R., Kleintjens, L.A., Macromolecules, 4 (1971) 637.
48. Flory, P.J., J. Am. Chem. Soc., 87 (1965) 1833.
49. Orwoll, R.A., Flory, P.J., J. Am. Chem. Soc., 89 (1967) 6814, 6822.
50. Huggins, M.L., J. Am. Chem. Soc., 86 (1964) 3535.
51. Huggins, M.L., J. Phys. Chem., 74 (1970) 371; 75 (1971) 1258; 80 (1976) 1317.
52. Koningsveld, R., Kleintjens, L.A., J. Polym. Sci., Polymer Symposium, 61 (1977) 221.
53. Silberberg, A., J. Chem. Phys., 48 (1968) 2835.
54. Kruse, W.A., Kirste, R.G., Haas, J., Schmitt, B.J., Stein, D.J., Makromol. Chem., 177 (1976) 1145.
55. Schmitt, B.J., Kirste, R.G., Jelenic, J., Makromol. Chem., 181 (1980).

56. Koningsveld, R., Stepto, R.F.T., Macromolecules, 10 (1977) 1161.
57. Koningsveld, R., Stockmayer, W.H., Kennedy, J.W., Kleintjens, L.A., Macromolecules, 7 (1974) 73.
58. Kleintjens, L.A., Koningsveld, R., Stockmayer, W.H., Brit. Polym. J., 8 (1976) 144.
59. McMaster, L.P., Macromolecules, 6 (1973) 760.
60. Olabisi, O., Macromolecules, 8 (1975) 316.
61. Chai, Z. and Walsh, D.J., Makromol. Chem. 184 (1983) 1459.
62. Rostami, S. and Walsh, D.J., Macromolecules 17 (1984) 315.
63. Sanchez, I.C., Lacombe, R.H., J. Phys. Chem., 80 (1976) 2352, 2568.
64. Sanchez, I.C., Am. Rev. Mater. Sci., 13 (1983) 387.
65. Jain, R.K., Simha, R., Macromolecules, 13 (1980) 1501.
66. Jain, R.K., Simha, R., Zoller, P., J. Polym. Sci., Polym. Phys. Ed., 20 (1982) 1399.
67. Jain, R.K., Simha, R., Macromolecules, 1984 in press.
68. Kleintjens, L.A., Ph.D. Thesis, University of Essex, UK, 1979.
69. Kleintjens, L.A., Koningsveld, R., Colloid & Polym. Sci., 258 (1980) 711.
70. Kleintjens, L.A., Koningsveld, R., Sep. Sci. & Technol., 17 (1982) 215.
71. Trappeniers, N.J., Schouten, J.A., Ten Seldam, C.A., Chem. Phys. Letters, 5 (1970) 541.
72. Schouten, J.A., Ten Seldam, C.A., Trappeniers, N.J., Physica, 73 (1974) 556.
73. Stockmayer, W.H., unpublished.
74. Koningsveld, R., Kleintjens, L.A., Macromolecules (1984) in Press.
75. Scott, R.L., J. Polym. Sci., 9 (1952) 423.
76. Kollinsky, F., Markert, G., Makormol. Chem., 121 (1969) 117.
77. Kollinsky, F., Markert, G., Adv. Chem. Ser., No. 99 (1971) 175.
78. Koningsveld, R., Kleintjens, L.A., Markert, G., Macromolecules 10 (1977) 1105.
79. Teramachi, T., Tomioka, H., Sotokawa, M., J. Macromol. Sci. Chem. A6 (1972) 97.

POLYMER BLEND MODIFICATION OF PVC

George H. Hofmann

E. I. du Pont de Nemours & Co.

1   INTRODUCTION

        Polyvinyl chloride (PVC) resins are a family of polymers and copolymers derived from vinyl chloride monomer ($CH_2 = CHCl$). The vinyl chloride content of such resins is usually greater than 85% by weight but in most commercial applications PVC homopolymer (100% vinyl chloride) is used. PVC resins surpass every class of polymers, in worldwide commercial importance, both in diversity of applications and in total tonnage of finished product. PVC compositions generally contain significant amounts of modifying agents (plasticizers, fillers, etc.). The production of such compounded products was estimated to be 10-15 million metric tons while the PVC resin portion of this tonnage was estimated to be about 8 million metric tons (Western Europe, U.S.A., Canada, Japan) in 1983 (1).

        The factors responsible for PVC's number one position are low cost and the ability to be compounded into various flexible and rigid forms with good physical, chemical, and weathering properties. In addition, PVC has broad processibility including calendering, extrusion, molding (injection, compression, rotational blow), fluid (solution, latex, organosol, plastisol) coating and impregnation. This broad menu of PVC technology has generated an almost endless formulary of PVC compounds. PVC formulations can contain varying amounts of the following ingredients in order to give the polymer the desired processing and end-use characteristics.

Plasticizers
    monomerics
    oligomerics
Polymeric Modifiers
    solid plasticizers
    impact modifiers
    heat distortion improvers
    processing aids
Fillers
Heat Stabilizers
Lubricants
Colorants
Light Stabilizers
Blowing Agents
Flame Retardants
Reodorants
Antistats

This paper will attempt to focus on the area of solid polymeric modifiers for PVC homopolymer. The degree of miscibility of the various polymers used in blends with PVC will be discussed followed by some examples of applications. Finally, the requirements for successfully melt-compounding of typical polymeric modifiers into PVC will be given.

2   POLYMERIC MODIFIERS

This subject may be divided into four categories of polymers, those that; plasticize, increase impact strength, increase heat distortion temperature and improve processibility.

The property modification is highly dependent upon the miscibility of the modifying polymer with PVC. In general, blends of PVC with polymers that show high miscibility will be microscopically homogeneous (single phase) and display a single glass transition temperature (Tg) of intermediate value between the Tg's of the pure component polymers (Figure 1). In cases where the blend Tg is lower than that of PVC plasticization is said to occur. When the blend Tg is increased then heat distortion temperature improvement is obtained. An additional benefit that can be imparted by miscible polymers is an improvement in processibility. These polymers can increase the rate of fusion of the PVC powder leading to a more rapid formation of a homogeneous melt having higher melt strength.

By far the widest range of polymeric additives for PVC fall into a broad category that can be classified as partially miscible. A broad range of properties may be obtained depending on the degree of miscibility. Partially miscible blends display

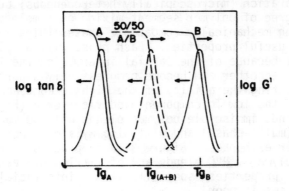

Fig. 1   Generalized behaviour of the dynamic mechnical properties
          of a miscible blend.——— , pure components, ----,
          mixture (2).

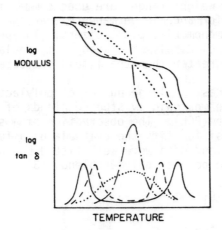

Fig. 2   Generalized mechanical loss (tan δ) and modulus behaviour
          for different types of polymer blends.———·——— ,miscible;
          ----, partial miscibility; ...., microheterogeneous
          (partially miscible); ——— , heterogeneous (2).

some phase separation (microscopically heterogeneous) but a significant degree of polymer segment mixing on a molecular scale occurs producing mechanically compatible phases (high interfacial adhesion) with useful properties. Each phase will display a distinct Tg but because of the partial intermixing the Tg's will be moved closer together and become progressively suppressed in intensity reflecting the ability of the high Tg component to raise the Tg of the low Tg component and vice versa (Figure 2). On the other hand, immiscible polymer blends will be macroscopically heterogeneous (multi-phase) and will display a distinct and unshifted Tg for each phase (Figure 3). In practice, truly incompatible polymer - PVC blends are of little interest since they yield poor properties especially if the interfacial adhesion between the phases is poor.

2.1 Polyesters

Polyester oligomers are well known for their high miscibility with PVC. These liquid polymers, typically in the 1000 to 4000 molecular weight range, are used commercially providing improved performance in permanence over the standard low molecular weight (monomeric) plasticizers. The polyesters to be discussed here are solid polymers having a molecular weight range of 10-100 times greater than the liquid polyesters.

2.1.1 Polylactones. A large number of polylactones are miscible with PVC. The most widely studied blends of this type are of polycaprolactone (PCL). The observation of a single Tg that varies with composition fits the criteria of a fully miscible blend. The blend data were well fitted by the Fox expression for miscible polymer-polymer blends (3).

$$\frac{1}{Tg_{12}} = \frac{W_1}{Tg_1} + \frac{W_2}{Tg_2}$$

Tg_{12} = Tg of the blend
Tg_1 = Tg of Component 1
Tg_2 = Tg of Component 2
W_1 = Weight fraction of component 1
W_2 = Weight fraction of component 2

The solubility parameters (δ) of these polymers are very similar (PVC = 9.6, PCL = 9.4) and the interaction between the carbonyl group of the polylactone and the α-hydrogen of the PVC is also likely to contribute to miscibility (Figure 4).

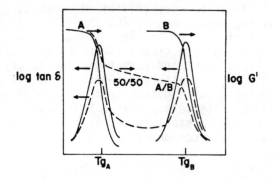

Fig. 3   Generalized behaviour of the dynamic mechanical properties
of a two-phase blend. ——— , pure components; ----,
mixture (2).

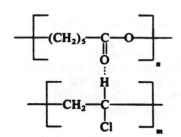

Fig. 4   Hydrogen bonding between PVC and polycaprolactone (2).

Pure PCL is about 50% crystalline (T_m = 60°C, Tg = 71°C). In blends containing less than 40% PCL the PVC effectively destroys the crystallinity of the PCL giving flexible transparent products. Blends containing greater than 40% PCL over a period of time, become translucent and more rigid due to crystallization of the PCL. Small angle scattering (SAXS) and differential scanning calorimetry (DSC) indicate that the two polymers mix in the amorphous phase of these mixtures. Both crystallization rate and induction time for crystallization is critically affected by the concentration of the components (4-7).

Polyvalerolactone (PVL) having a similar structure and Tg to PCL has also been shown to be fully miscible with PVC over the entire composition range as indicated by a single Tg for each blend (8). PVL also crystallizes in blends containing less than 50% PVC.

Other polylactones reported to behave similarly to PCL and PVL are polypivalolactone, poly(α-methyl-α-n-propyl-β-propiol-actone) and poly(α-methyl-α-ethyl--β-propiolactone (9-12). In contrast, poly-β-propiolactone is reported to be completely immiscible (13).

2.1.2 Polyester elastomer (PEEL). A segmented polyether ester prepared by melt transesterification of dimethyl terephthalate with poly (tetramethylene ether) glycol and 1,4-butanediol has been studied extensively in blends with PVC. This random block copolymer contains crystallizable tetramethylene terephthate (4GT) hard segments and poly (tetramethylene ether) glycol terephthate soft segments *).

Blends with PVC exhibit melting behavior by DSC associated with the hard segments that remain segregated and crystallize below the blending temperature (150°C). During blending the soft segments become miscible with the PVC and when quenched the mixtures exhibit, by dynamic mechanical analysis (DMA), a single Tg intermediate between that of the PEEL (-32°C) and the PVC. There is evidence, however, of partial phase separation at high levels (80%) of PEEL.

When quenched blends of PEEL-PVC are isothermally annealed (130°C) they exhibit properties typical of a multiphase system. The phase domains have been estimated to be less than 100 Å and it is suggested they are prevented from growing by the presence of the crystalline hard segments which act as crosslinks in the mixture. There appears to be an optimum condition for heat treatment, probably related to both phase composition and domain size, in attaining optimum physical properties (14-17).

* Hytrel® polyester elastomer

2.1.3 Miscellaneous polyesters. Blends of polybutylene terephthalate (PBT) and PVC, prepared by solution blending because of the high melting point (m.p. 220°, Tg 35°C) of PBT, are clear and exhibit a single Tg intermediate between the values for the unblended constituents. Annealing, above the Tg, results in crystallization of the PBT but single phase behavior is maintained in the amorphous phase. The single Tg of this phase, however, has increased probably due to increased PVC content and crystallinity restricting the amorphorous phase chain mobility (18).

Poly(neopentyl glycol adipate) (PDPA) has a Tg of about -50°C and all blends with PVC exhibit transparency and a single intermediate Tg. Total miscibility is maintained even after annealing (19). Other polyesters reported to be miscible with PVC are poly(butylene adipate) and poly(hexamethylene sebacate). In contrast poly(ethylene adipate) is reported to be completely immiscible (20).

2.2 Polymethylmethacrylate (PMMA)

Unlike most of the polymers miscible with PVC, which are used as plasticizers to lower the Tg of PVC, PMMA and selected copolymers of MMA provide miscible blends having increased Tg's. Tacticity of the PMMA, however, plays an important role in its miscibility. With isotactic PMMA, immiscible blends are observed over the entire composition range as indicated by blend opacity and by two distinct Tg's (∿40°C and 70°C by DSC). The solubility parameter of i-PMMA is 9.28 compared to 9.60 for PVC. Syndiotactic PMMA, however, having a solubility parameter of 9.55 is miscible with PVC up to the 60% (by weight). A single Tg (by DSC) increasing regularly from pure PVC (∿70°C) to ∿90°C is observed over this range of compositions. The 60% composition corresponds to a MMA:VC monomer ratio of 1:1 indicating a strong interaction between PVC and S-PMMA. At higher S-PMMA contents phase separation occurs between the 60:40 S-PMMA:PVC phase and pure S-PMMA. This is demonstrated by turbidity and the measurement of two Tg's of about 90°C and 120°C corresponding to the two phases, respectively (21).

Since atactic PMMA (Tg 115°C) is much more syndiotactic than isotactic it is understandable that blends of α-PMMA with PVC resemble S-PMMA:PVC blends very closely. Microscopic heterogeneity on the order of 1-2μm, however, has been detected by some workers. Tensile strengths are equal to, or somewhat less, than expected by simple additivity (22).

The differences in miscibility of i-PMMA and S-PMMA with PVC may be attributed to different chain conformations. Isotactic PMMA prefers a helical conformation in which the ester groups are rotated inwards and become less accessible to intermolecular H-bonding with the PVC α-hydrogens. The S-PMMA chains have predominantly a planar structure more susceptible to hydrogen bonding interactions (21).

2.3 Butadiene Acrylonitrile (BAN) Copolymers

Miscibility of polymer blends was first observed with BAN/PVC systems. These blends probably have been the most widely studied, having significant commerical interest (2). They have been described as miscible, partially miscible and even immiscible based on the different experimental techniques and copolymer compositions. Miscible blends are widely used as flexible compositions with the BAN acting as a permanent plasticizer.

Partial miscibility was determined at the 20% acrylonitrile (AN) level and high miscibility at 40% AN (23). Microscopic analyses, however, indicated that micro-heterogeneity (<100Å) still remained in the highly miscible blends. Recent work confirmed the excellent compatibility at 31% AN but an upper limit nearing 44% AN was indicated by the presence of two phases; one composed of pure BAN and the other a miscible BAN/PVC phase (24).

In general, the AN content sufficient to yield maximum miscibility appears to fall in the range of 25-40% with gradual immiscibility occuring with either decreasing or increasing AN levels. When high BAN viscosity is combined with a high AN level, low levels of BAN yield impact modified rigid PVC with optimum properties. It is postulated that the high viscosity encourages separate rubbery domains while the AN compatibilizes the domains at the interface giving good mechanical compatibility (25,26).

2.4 Ethylene Copolymers (ECR)

Non-polar polyethylene (δ = 7.9) is immiscible with relatively polar PVC (δ = 9.6). Attempts to compatibilize these polymers by use of a third polymer have been reported (27-29). Increasing the polarity of polyethylene by copolymerizing with polar vinyl acetate (PVA, δ = 9.4), however, leads to a wide range of useful PVC additives ranging from plasticizers to impact modifiers. Miscibility appears optimum at VA contents of 65-70% (Tg approx. -30°C) as evidenced by a single Tg (approx. 30°C) in 50:50 blends (30). No evidence of crystallinity was found by x-ray diffraction. These copolymers can be blended with PVC in any ratio yielding transparent blends. Some workers have found evidence that miscibility is not complete with about one-third of the PVC phase remaining in the 50:50 blend (31,32).

An alternate technique for achieving complete miscibility of ethylene copolymer is to substitute part of the VA with another polar comonomer. Substitution of SO_2 for VA to give E/VA/SO$_2$ (72/19/9) showed the much greater compatibilizing effect of the SO_2 monomer relative to VA. Miscible (single Tg) blends were achieved over the entire range studied (up to 40% copolymer) (33).

Replacing some of the VA with still another polar comonomer, CO, also yields miscible blends over the entire blend ratio range (34). Using ethyl acrylate (EA) in place of VA to make E/EA/CO also yields completely miscible systems. The copolymer E/EA, however, has been reported to be immiscible (35).

Partial miscibility is achieved with copolymers of E/VA containing, typically, 45% VA. These are of commercial significance as impact modifiers for PVC (36).

2.5 Vinyl Chloride Copolymers

Blends of poly(vinyl chloride-co-vinyl acetate) with PVC were shown to be completely miscible over the entire range of composition. Microscopically, no heterogeneities greater than 1μm could be detected and a single Tg falling between the pure component Tg's was obtained. Tensile strength measurements show that the values are at least equal to those expected by simple additivity (22).

PVC blends containing chlorinated PVC (65.2% Cl) also have been shown to display a single Tg over the entire composition range indicating high miscibility. A higher level of chlorine (67.5%), however, in the CPVC resulted in miscibility (single Tg) only at 25% (or less) of CPVC. Higher levels led to two Tg's indicating immiscibility (37).

2.6 Styrenics

The blends of the azeotropic copolymer of styrene-acrylo-nitrile (SAN) contains 27% AN and has been reported to be miscible in PVC based on similar solubility parameters (δ = 9.8 for SAN) and the observation of a single blend Tg (38-41). Others, however, have found two distinct Tg's corresponding to the individual components (42-44). Despite this discrepancy in the literature, SAN (Tg 112°C) copolymers are commercially available as heat distortion temperature (HDT) improvers for PVC.

126

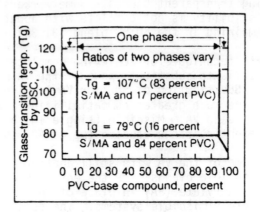

Fig. 5   Miscibility is indicated by shifted single Tg s at low
          levels of alloying. Intermediate levels result in phase
          separation into PVC rich and S/MA rich phases (47).

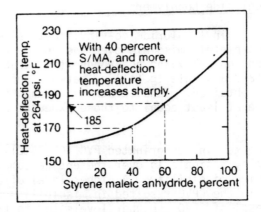

Fig. 6   Deflection temperature under load increases with
          increasing S/MA content (47).

In fact, good compatibility of these copolymers is based on the linearity of the HDT curve vs. the SAN-PVC blend ratio (45).

The azeotropic copolymer of α-methyl sytrene (αMS) and AN (32%) has a Tg of 130°C and is reported to be miscible with PVC over the entire composition range (44). Terpolymers of αMS/S/AN have also been reported to be completely miscible with PVC (46).

Styrene-maleic anhydride (SMA) copolymers also are commerically available as HDT improvers for PVC. They are reported to have limited miscibility with PVC. At low levels of each component (4% SMA and 10% PVC) only one Tg appears. At intermediate levels two constant Tg's are obtained representing a PVC rich phase containing 16% SMA and a SMA rich phase containing 17% PVC (Figures 5 and 6). These phases are well intermixed, having enough miscibility to give good mechanical compatibility and useful physical properties (47).

2.7 Polyurethanes ·

Polyurethanes have shown practical promise in polyblends with PVC. These blends combine the toughness and abrasion resistance of the polyurethane with the stiffness and high modulus of the PVC. Polyurethane chemistry encompasses a wide variety of aliphatic, aromatic, polar and hydrogen-bonding structures. Other parameters, besides chemical structure, are the molar ratio of diisocyanates to the total glycol and polyester or polyether content, and the molar ratio of low MW glycol to polyester or polyether.

Many polyurethanes show partial miscibility with PVC as indicated by separate, but significantly, shifted Tg's of the individual phases. Their useful mechanical properties are due to good adhesion between the phases (48,49). Due to the heat sensitivity of PVC, however, only a limited number of polyurethanes can be considered compatible with respect to processing characteristics (50).

2.8 Acrylonitrile Butadiene Styrene (ABS)

Emulsion polymerization is the standard method of manufacturing ABS and related modifiers including AMBS (acrylonitrile-methacrylate-butadiene-styrene), and MBS (methacrylate-butadiene-styrene). The process comprises two basic stages as follows:

1. rubber (polybutadiene) latex emulsion

2. monomers (e.g., styrene and acrylonitrile) are polymerized in the presence of the latex.

Stage 2 of the process results largely in the formation of a comonomer graft (e.g. SAN) to the polybutadiene (PBD). The resulting ABS particle of about 0.5μm has a "core-shell" morphology consisting of a PBD core surrounded by a shell of SAN (51). When the PVC and ABS are mixed, the polar SAN becomes miscible with the PVC while the PBD phase remains immiscible. Some workers report the detection of two Tg's over the entire blend range. One Tg corresponding to a miscible blend of PVC-SAN while the other corresponds to the PBD (38-41). Other workers, however, in studying similar systems found two distinct Tg's corresponding to the SAN and PVC (in addition to the PBD) indicating immiscibility of these phases (42-44). Further work is needed to clarify this anomaly. Based on the successful impact improvement afforded PVC by the presence of ABS, certainly enough miscibility of SAN must occur to result in good mechanical compatibility of the rubber phase with the PVC matrix.

2.9 Methacrylate Butadiene Styrene (MBS)

As mentioned in the previous section, the MBS modifiers are graft polymers prepared by polymerizing MMA (with or without small amounts of acrylic comonomers) in the presence of poly butadiene (PBD) or PBD-styrene rubber emulsion. Typical emulsion core-shell graft polymerization structures are obtained with the outer acrylic shell being compatible with the PVC matrix while the cross-linked rubber core remains a well-defined separate dispersed phase of uniform particle size. Although these polymers function primarily as impact modifiers they also function to some extent as processing aids (52).

3.0 Methacrylate Butyl Acrylate (MBA)

These modifiers, generally known as acrylics, also incorporate the core-shell graft polymer technology in order to generate uniform particles of cross-linked poly butyl acrylate (PBA) core surrounded by a PMMA (or copolymers) shell. These modifiers offer a combination of high impact strength, low die swell and excellent retention of physical properties on prolonged exposure to the outdoors (52).

3.1 Chlorinated Polyethylene (CPE)

CPE modifiers are prepared by chlorinating high density polyethylene. The compatibility of these modifiers is dependent upon the chlorine content and the distribution of the chlorine atoms on the polyethylene backbone. Polymers containing less than 25% Cl are incompatible with PVC and in general are not used in PVC. Those with 25-40% Cl are the best impact modifiers having partial miscibility (52). At 20% loading of CPE (38% Cl) microscopic heterogeneity consisting of 1-2μm inclusions is observed. Raising the CPE level to 50% increases the size of the inclusions to 10μm. Blends containing 10% CPE display a single Tg near but below that of PVC. Blends containing 20% or more CPE display Tg's near those of the pure components (22).

Most rigid PVC blends are modified with CPE modifiers containing about 36% Cl since these offer the best combination of processing, dispersibility and impact strength. For improved clarity slightly higher Cl levels (42%), giving higher miscibility, are used with some sacrifice in impact strength.

4   APPLICATIONS

Polymeric additives in blends with PVC can range from "process aids" by addition of relatively low levels (<5%) to "alloys" where the additive level approaches or exceeds 50%. Additives used in an intermediate range (5-20%) are considered to be modifiers. PVC impact modifiers, in general, fall into this category.

Most PVC is sold in powder form and is usually modified (or alloyed) by the processor. Since PVC, before it can be processed, must be mixed with stabilizers, lubricants, etc., polymeric additives can be incorporated quite easily at this stage.

Table 1 summarizes polymer blend modification of PVC; describing the major polymeric additives used in PVC, the typical concentration range, the resulting improvement and characteristic applications.

4.1 Plasticizers

As already stated, different structural systems can be produced depending upon the miscibility of the polymeric additive. Completely miscible mixtures will result in completely transparent systems in any mixing ratio (when amorphous). Only one phase can be distinguished with the electron microscope and only one Tg detected. If the Tg of the miscible polymer is significantly below that of PVC it can be considered to be a plasticizer imparting a high degree of flexibility to the blend (Figure 7). Good low temperature flexibility requires a plasticizer with a Tg well below 0°C. Solid high molecular weight polymeric plasticizers exhibit high permanence in PVC as displayed by low extraction in contact with liquids, low volatility under warm conditions and low migration to adjacent surfaces.

TABLE I

POLYMER[1] BLEND MODIFICATION OF PVC

Polymer	Typical level (%)	Category	Properties[2] Achieved	Applications	Miscibility[3]
ECR[4]	up to 50	Alloy	Flexibility, permanence	Sheet, film, shoe soles	M
PVCA	up to 30	Alloy	Flexibility, permanence	Film, sheet	M
BAN[5]	30 to 50	Alloy	Flexibility, permanence	Hose, cable jacket	M-P
PCL	up to 30	Alloy	Flexibility, permanence	——	M
PVL	up to 50	Alloy	Flexibility, permanence	——	M
PEEL	up to 50	Alloy	Flexibility, permanence Impact strength	Hose, cable jacket	P
PU	up to 50	Alloy	Flexibility, permanence	Sheet, film, shoe sholes	P
PMMA	up to 50	Alloy	Heat distortion temp.	Sheet	M
SAN	up to 50	Alloy	Heat distortion temp.	Profiles, film bottles	M-P
αMSAN	up to 50	Alloy	Heat distortion temp.	Profiles, film bottles	M
SMA	up to 50	Alloy	Heat distortion temp.	Molded goods	P
EVA[6]	5-10	Modifier	Impact strength weatherability	Profile, sheet	P
CPE	4-20	Modifier	Impact strength weatherability	Profile, sheet film	P
ABS	5-15	Modifier	Impact strength	Profiles, sheet, film	P
MBS[7]	5-15	Modifier	Impact strength Clarity	Films, bottles	P
MBA	5-15	Modifier	Impact strength weatherability	Profiles, sheet	P
PMMA	1-4	Process aid	Fushion rate, thermoformability	Profiles, bottles	M
MMA-AC[8]	1-4	Process aid	Fushion rate, thermoformability	Profiles, bottles	M
αMSAN	1-4	Process aid	Fushion rate, thermoformability	Film	M

(1) Abbreviations for polymers may be found in the appendix.
(2) In general, an increase in these key properties.
(3) M is highly miscible (Single Tg), P is partially miscible.
(4) With greater than 60% VA or with compatibilizing termonomer (e.g. CO).
(5) Also known as NBR.
(6) Typically 40-60% VA.
(7) Also AMBS.
(8) Acrylate either BA or EA.

131

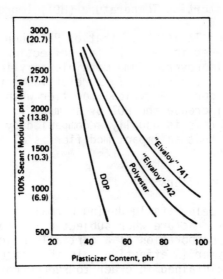

Fig. 7   Effect of plasticizer resins on vinyl compound modulus
         (60)

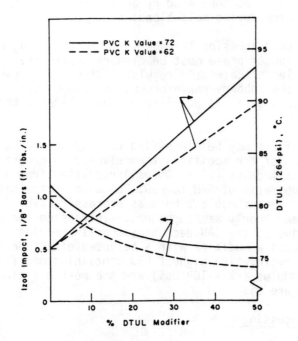

Fig. 8   Effect of DTUL modifier concentration on impact strength
         and DTUL (52).

4.2 Heat Distortion Temperature (HDT) Improvers

When the blend Tg is above that of PVC then the polymeric additive is considered to be a HDT (or deflection temperature under load, DTUL) improver. Some miscible acrylic and styrenic polymers, because of their high Tg's relative to that of PVC, fall into this category. A typical modifier when added to PVC at the 50% level can increase the HDT by as much 15°C (Figure 8). In actual practice this is usually accompanied by a decrease in impact strength so that an impact modifier is also incorporated into the formula (52).

4.3 Impact Modifiers

Rigid PVC is below its Tg during normal use and therefore may undergo brittle fracture when subjected to impact. For good impact strength, incorporation of a partially miscible glassy/rubbery polymer is effective. In this case a discrete rubbery phase is maintained in order to reduce notch sensitivity by the ability of the rubber to deform producing a combination of crazing and/or shear banding. For the rubber toughening to be effective, stresses must be transmitted across the rubber-plastic interface. This is accomplished by achieving at least some interfacial mixing during melt blending.

Besides good adhesion through interfacial mixing the heterogeneous rubber phase must be dispersed effectively to produce particles of carefully regulated size. When insufficient dispersion of the rubbery phase occurs the toughness, as measured by notched izod impact, is equal to or only slightly better than unmodified PVC.

ABS modifiers may be classified roughly according to the tensile modulus of the modifiers themselves. In general, they range from high modulus ($>3 \times 10^5$ psi) resulting from a relatively high ratio of SAN to PBD. Because of the high level of compatible phase these act best as processing aids (see Section 4.4) and poorly as tougheners. Increasing the PBD content relative to the SAN decreases tensile modulus ($2\text{-}3 \times 10^5$ psi) and produces a modifier with a balanced effect on processing and impact strength. The highest PBD containing modifiers with low tensile modulus ($<2 \times 10^5$ psi) are the most efficient impact modifiers (Figure 9).

4.4 Processing Aids

Miscible polymers are typically used in small quantities (below 5%) in PVC to give substantial improvement in processing behavior without any significant reduction in other properties. The improvements are in glass, surface quality, hot elongation and hot tear strength (Table 2).

The miscible polymers most widely used are composed of the monomer combinations consisting of acrylates, methacrylates and acrylonitrile (53-55). Typical compositions are listed in Table 1.

5   MELT COMPOUNDING

In order to take full advantage of the beneficial properties of polymeric modifiers it is essential that the PVC blends are made homogeneous through effective melt compounding. The equipment used for this purpose consists of three basic types of relatively high-shear, high-intensity mixers; roll mills, internal mixers and extruders.

The requirements for melt compounding (or alloying) a typical solid polymeric plasticizer into PVC will be discussed in detail. This will be followed by a brief description of the melt compounding characteristics of processing aids, HDT modifiers and impact modifiers.

5.1 Plasticizers

Conventional liquid plasticizer is completely absorbed by the PVC powder, in the dry-blending step, prior to melt compounding (56-58). With the solid polymeric plasticizers this cannot occur; therefore, a salt and pepper blend of plasticizer pellets in PVC powder is usually prepared prior to compounding. Typical plasticizers of this type are ethylene copolymer resins (ECR) (59,60).

During the melt compounding step, the low melting ECR (mp 66 C) melts first forming a relatively low viscosity phase in which the higher melting (mp $>170\,°C$) and more viscous PVC powder grains are suspended. Under these conditions it is difficult to get enough shear energy into the system to completely break down and disperse the PVC grains. As a result, many PVC grains can pass through the melt compounding step intact. These unfluxed PVC grains, referred to as gel, have been positively identified as the heterogeneities in PVC/ECR blends that can produce rough, grainy extrudates with reduced physical properties.

The goal is to develop techniques for melt compounding PVC/ECR into blends equivalent to PVC/DOP (liquid plasticized) blends in homogeneity. Using a specific gel test, a gel count of 50 or less indicated a blend homogeneity suitable for most applications (acceptable surface quality and physical properties). Blends plasticized with DOP (only) fell into this category under most melt compounding conditions. The formulations evaluated are found in Table 3.

5.1.1 Morphology of melt compounding unplasticized PVC. Previous workers (61,62) characterized unplasticized PVC melt

134

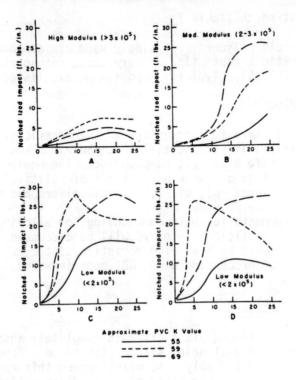

Fig. 9   Impact efficiency of various ABS polymers in PVC (52).

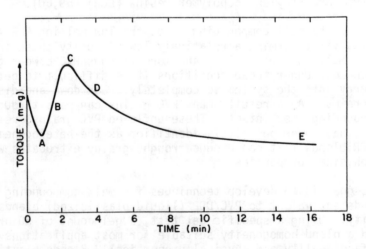

Fig. 10   Brabender plastogram - PVC (73).

Table 2

Hot Elongation of PVC Containing a Typical
Acrylic Processing Aid (52)

% Processing Aid	% Elongation at 100°C
0	185
5	270

Table 3

FORMULATIONS USED IN MELT
COMPOUNDING STUDY
(PHR)

	I	II	III	IV
PVC[1]	100	100	100	100
ECR[2]	VARIED	100	70	70
LIQUID PLASTICIZER	–	30	–	30
CaCO$_3$	25	25	25	25
Ba/Cd SOAP	2	2	3	3
PHOSPHITE CHELATOR	1	1	1	1
EPOXY SOYA OIL	10	10	10	10
WAX	–	–	4	4

(1) K VALUE OF 67 UNLESS NOTED OTHERWISE
(2) ELVALOY® 741 RESIN MODIFIER

compounding morphology by interpretation of characteristic Brabender plastograms (Figure 10). Curves of essentially the same shape are produced by PVC/ECR blends and should, therefore, be instructive in the interpretation of PVC/ECR melt compounding morphology. Photomicrographs of the unplasticized samples, taken at points corresponding to the letters on the curve, showed the PVC morphology changes during the melting process. They showed that as the torque approaches minimum A, the 150μm PVC grains are torn apart and most of them are broken into 1μm primary particles. The temperature is low at this point in the process and little fusion interaction occurs between the primary particles as indicated by the low torque value. As the temperature increases, due to shear heating, the primary particles begin to melt and fuse together, increasing the torque (B). At the maximum torque, C (fusion point), most of the primary particles are fused together, but are still distinguishable. After the fusion point (C) the melt temperature increases significantly, causing a reduction in the melt viscosity and torque (D). At this point the particulate structure starts to disappear. The temperature continues to rise and the viscosity decreases until they level out to relatively constant values. At this equilibrium point (E) all primary particles have disappared and a continuous melt is formed completing the process.

5.1.2 Liquid plasticized PVC. Similar studies on DOP plasticized PVC (63) show that the same melt flow behavior pattern occurs. The only differences are a reduction in the temperature and torque at which these morphological transitions occur. The liquid plasticizer which produces these improvements in processing characteristics was shown to function by being absorbed into the 1μm primary particles in the dry blending step, prior to melt compounding, causing a reduction in Tg and viscosity (56,57).

5.1.3 ECR plasticized PVC. Such an absorption mechanism is impossible with a solid high MW plasticizer such as ECR. When the PVC powder and ECR pellets are added to the melt compounding equipment they are, in fact, mixed but still two separate phases. Early in the melt compounding cycle before minimum torque A is reached, two processes are occuring. The 150μm PVC grains are eroding into unmelted 1μm primary particles and at the same time the ECR pellets are melting into a low viscosity melt phase. The low viscosity ECR melt encapsulates many of the 150μm grains before they are eroded into primary particles. The primary particles that are formed, however, blend with the ECR and begin the fusion process (B) as well-dispersed phase. At the fusion point (C), however, many of the encapsulated 150μm grains still remain intact. During further heating and melting (D-E) the dispersed 1μm grains disappear, forming a continuous PVC/ECR

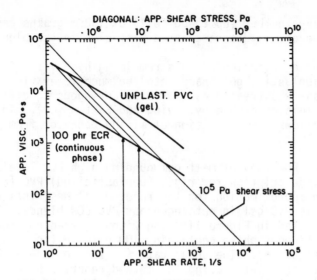

Fig. 11   Rheology PVC/ECR (190°C) Brabender mixing (73).

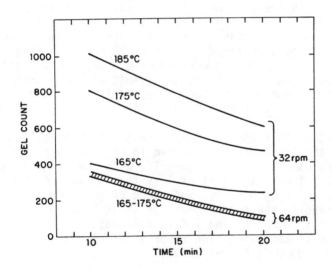

Fig. 12   Homogeneity vs. time, Brabender (73).

melt. A large number (up to 50 vol%) of 150μm grains can remain, however, as discrete large particles in the blend.

The homogenization of this remaining high viscosity, but still thermoplastic, gel phase into the much lower viscosity PVC/ECR phase was treated as a problem analogous to difficult mixing problems described by Irving and Saxton (64). This treatment allows the prediction of mixing behavior from rheological data.

5.1.4 Rheology of melt compounding high ECR levels. A rheological description for rigid (unplasticized) PVC (65) was taken as an approximation for the rheological description of the thermoplastic PVC gel in heterogeneous PVC/ECR blends. These data are plotted in Figure 11 along with the rheology for the PVC/ECR continuous phase containing 100 phr ECR (ignoring effect of minor gel phase). The shear rate ($37 s^{-1}$) on the Brabender (32 rpm) was calculated using the method reported by Goodrich and Porter (66). At this shear rate a shear stress of approximately 40 kPa is imparted to the major component (PVC/ECR). Because shear is the same in both the major and minor (PVC gel) components, the shear stress isobar is followed until it intersects the line describing the minor component. As seen in Figure 11, the shear stress isobar intersects the PVC gel curve at a very low shear rate ($2s^{-1}$), indicating that only a low rate of mixing will occur. This, in fact, was the case as seen in Figure 12 and Table 4. A very high gel count remained even after 20 min of mixing at 185°C. As the temperature settings were lowered, however, the gel level decreased accordingly. This is the result of the increasing shear stresses generated by the increasingly viscous major component as its melt temperature decreased.

A further increase in shear stress was obtained by increasing the Brabender rotor speed to 64 rpm (shear rate of $73 s^{-1}$). As can be seen in Figure 12, a further reduction in gel count was obtained but a goal of less than 50 was not achieved. Temperature settings had less influence on gel count at this higher rpm due to the higher level of shear heating. The shear heating at the 185°C setting led to degradation, pointing out the desirability of achieving high shear stresses without generating excessively high melt temperature.

Shear rates in the 35-75 s^{-1} range of these experiments are representative of mixing devices such as roll mills and single screw extruders. Even at the fusion point, relatively low (<75 kPa) shear stresses were developed (with high ECR levels), making high homogeneity difficult to achieve.

Table 4

COMPOUNDING OF ECR WITH STANDARD SUSPENSION PVC[1]

VARIABLES	ECR[2] phr	RPM	TEMP[3] (°C)	SHEAR[4] RATE (sec^{-1})	FUSION POINT[5] (sec)	FUSION POINT[5] TORQUE (m-g)	FUSION POINT[5] SHEAR STRESS[4] (Pa×10^{-3})	EQUILIBRIUM POINT[6] TORQUE (m-g)	EQUILIBRIUM POINT[6] SHEAR STRESS[4] (Pa×10^{-3})	EQUILIBRIUM POINT[6] MELT TEMP(°C)	EQUILIBRIUM POINT[6] GEL COUNT
TEMP AND SHEAR RATE	100	32	185	37	120	950	39	750	31	191	800
	100	32	175	37	150	1210	49	920	38	182	600
	100	32	165	37	180	1430	58	1040	42	175	300
	100	64	175	73	75	1590	65	1085	44	192	250
	100	64	165	73	90	1805	74	1155	47	186	200
ECR LEVEL	100	32	175	37	150	1210	49	920	38	182	600
	70	32	175	37	90	1660	68	1105	45	184	250
	40	32	175	37	75	2230	91	1340	55	186	0

(1) BRABENDER PLASTI-CORDER
(2) FORMULATION I, TABLE I
(3) MIXING CHAMBER TEMPERATURE AT START
(4) APPROXIMATION USING THE METHOD OF REFERENCE II
(5) POINT C IN FIGURE 1
(6) 15 MINUTES FROM START

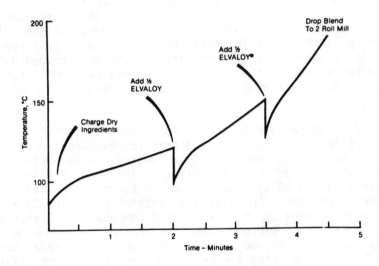

Fig. 13   Typical Banbury sequential addition cycle (67).

5.1.5 Reduced ECR levels. A technique found useful for minimizing the effect of encapsulating PVC grains with ECR at the beginning of the mixing cycle was to initially withhold a major portion of the ECR. The blend containing 40 phr ECR has a much higher initial viscosity and fusion torque than blends with 70 or 100 phr ECR. This higher torque effectively breaks down the 150µm grains. After 10 min, the gel count was 10. At this time the additional 60 phr was added to produce a homogeneous blend with slight additional mixing. This sequential addition technique (Figure 13) is very effective for batch compounding of PVC with high levels of ECR and is currently in use with Banbury mixers (67).

5.1.6 Optimum PVC. Table 5 summarizes data obtained with various types and grades of PVC. Emulsion polymerized PVC, as expected, produces gel-free blends because the 150µm grains, produced by suspension and bulk polymerization, are not present. The more economical suspension and bulk polymerized resins, however, show a wide variation in compounding performance with ECR.

The best choice of resin, as indicated by the very low gel counts, is the large particle grade of suspension PVC. This material is polymerized under conditions that produce an average powder grain size 2-3 times larger than standard PVC. Additionally, these grains are less dense, having a more open porous structure and not having a surrounding membrane (68,69). This produces a more friable, shear sensitive powder that readily erodes into primary 1µm particles early in the melt compounding process despite the presence of ECR. The advantages of large grain PVC with conventional liquid plasticizers have also been reported (70).

5.1.7 High rate, high shear continuous melt compounding. Using Figure 14, it was anticipated that the high shear rate Farrel Continuous mixer would produce low gel blends based on its high shear stress (>300 kPa) capability. The short residence time (approximately 30 s), however, was a factor that could negate the higher shear rate. Formulations containing 70 phr ECR were compared over a range of shear rates and temperatures (Table 6) on a No. 4, Farrel Continuous mixer. As can be seen, low gel levels (5-20) were achieved on a composition containing a standard grade of PVC provided that the melt temperature was kept relatively low (160°C). As on the Brabender, raising the melt temperature (180°C) reduced the viscosity and shearing action on the PVC grains, causing the gel level to increase (60-110). The same composition, modified by the inclusion of DOP in the recipe, behaved in a similar manner and had comparable gel levels.

Table 5

HOMOGENEITY VS PVC TYPES[1]

PVC TYPES	K VALUE	FUSION POINT			EQUILIBRIUM POINT		GEL COUNT
		sec.	TORQUE (m-g)	SHEAR STRESS (Pa X 10⁻³)	TORQUE (m-g)	SHEAR STRESS (Pa X 10⁻³)	
EMULSION							
A	68	105	1330	54	960	39	0
B	76	160	1310	53	1200	49	0
STANDARD SUSPENSION [2]							
C	54	45	1080	44	440	18	>2000
D	67	150	1210	49	920	38	600
E	67	195	1010	41	900	37	>2000
F	80	250	1575	64	1195	49	1300
BULK [2]							
G	55	15	1400	57	495	20	>2000
H	67	15	1400	57	850	35	>2000
LARGE PARTICLE SUSPENSION [3]							
I	69	95	1720	70	940	38	40
J	82	80	2400	100	1320	54	0

(1) 100 phr ECR, 175°C CHAMBER, 32 rpm, 15 min.

(2) <5% RETAINED ON 60 mesh.

(3) >90% RETAINED ON 60 mesh

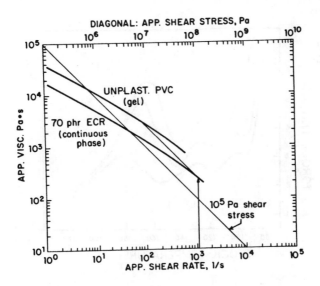

Fig. 14   Rheology PVC/ECR (190°C) FCM mixing (73).

142

Table 6

HIGH INTENSITY CONTINUOUS MIXER[1]

FORMULATION[2]	MELT. TEMP. °C	SHEAR STRESS[3] (Pa × 10⁻³)	GEL COUNT[4]
III	160	500	5-20
(STANDARD PVC)	180	400	60-110
IV	160	300	20-40
(STANDARD PVC + DOP)	180	200	40-130
III	160	500	0
(LARGE PARTICLE PVC)	180	400	<5

(1) FOUR INCH FARREL CONTINUOUS MIXER,
   500 rpm (1125 sec⁻¹ SHEAR RATE), 544 kg/hr (1200 pph)

(2) EACH CONTAINS 70 phr ECR

(3) CONSTANT RATE CAPILLARY RHEOMETER

(4) CONTROL CONTAINING 60 phr DOP HAD A GEL COUNT RANGE OF 20-40

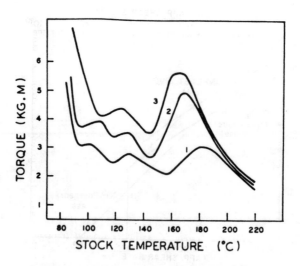

Fig. 15   Torque - temperature profiles for compounds with
processing aid. 1 = 0 phr; 2 = 2 phr; 3 = 5 phr
processing aid (55).

Substitution of the standard PVC by the large grain PVC also confirmed the Brabender results by producing an essentially gel-free blend at low temperature with only very slight residual gel remaining in the blends prepared at high temperatures.

5.2 Processing Aids and HDT Improvers

Milling experiments clearly demonstrate the value of processing aids in PVC. Untreated PVC will produce a crumbly and nonuniformly banded stock on the rolls. Hot elongation and tear strength are poor. Production of a uniform sheet of good surface quality and appearance is difficult at best.

The addition of a miscible processing aid results in a drastic change in appearance of the material on the mill. The PVC becomes clear, glossy, smooth and uniformly banded. Because good elongation and tear strength are now present, an attractive uniform sheet of good surface quality can be cut and pulled off the mill. Processing aids also can improve the processibility of plasticized and impact modified PVC.

Rheology of mixing studies indicate that the processing aid increases the rate of fusion of the PVC powder blend. In addition, at the time of fusion, PVC blends containing processing aids show considerably higher torque than unmodified PVC (Figure 15). When the torque and temperature reach equilibrium essentially no difference in these parameters are detected between the modified and unmodified blends. This observation indicates that the presence of a processing aid does not affect the melt viscosity of the PVC.

Presumably due to an inter-particle (and particle metal) increase in adhesion and friction, attributed to the miscible processing aid, significant increases in torque and work input occur. This, in turn, increases the rate and extent of resin particle (grains) break down into primary particles. The processing aids are thought to be distributed between the primary (micro) particles, wet their surface and act as a binder. Later, at higher temperatures, their presence at the primary particle interface can modify the rate of formation of a molecular network across the microparticle boundaries and the development of interparticle strength (55).

The mechanism of the fusion process for miscible HDT improvers is believed to be identical to that described for the miscible processing aids and, in fact, some HDT modifiers can serve a dual purpose by also acting as processing aids (Figure 16).

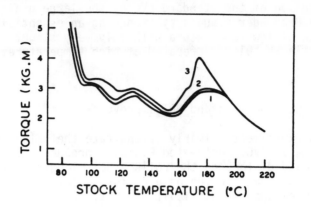

Fig. 16   Torque - temperature profiles for compounds with DTUL
           modifier.   1 = 0 phr; 2 = 2 phr; 3 = 5 phr modifier (55).

Table 7

EXTRUSION CONDITIONS, PROPERTIES AND WEATHERABILITY OF IMPACT MODIFIED PVC (71)

COMPOUND NUMBER	O	S	P	R	T
INHERENT VISCOSITY	0·92	0·92	1·12	1·12	0·92
SCREW	TWIN	TWIN	SINGLE	SINGLE	SINGLE
SCREW SHEAR RATE, SEC.$^{-1}$	Low	Low	13	20	28
DIE SHEAR RATE, SEC.$^{-1}$	1030	1150	170	220	180
MELT TEMPERATURE FROM THE DIE °C	182	200	178	193	204
VARIABLE HEIGHT IMPACT TEST JOULES/CM	110	129	110	160	137
MODE OF FAILURE	BRITTLE	DUCTILE	BRITTLE	BRITTLE/ DUCTILE	DUCTILE
PERCENT IMPACT RETENTION					
ONE YEAR*	-	-	18	-	80
TWO YEARS*	-	-	7	-	86

*AVERAGE OF ARIZONA, FLORIDA AND OHIO

5.3 Impact Modifiers

As discussed in Section 5.1.1, unplasticized PVC undergoes a series of morphology changes during the melting process. After fusion many primary particles, though fused together, are still distinguishable. If excessive particulate structure remains in the final product, brittle failure of the rigid composition is likely to occur due to poor adhesion between the particles. It has been shown that melt compounding temperature is a key factor in obtaining rigid blends with minimum particulate structure (55,71,72). At a melt temperature of below 188°C a product with distinguishable particles, was prepared which failed in a brittle mode. At about 190°C agglomeration and better adhesion between particles occur. Above 195°C a continuous melt with loss of particle structure occurs resulting in a product with superior impact properties (Table 7).

In impact modified PVC, as previously mentioned, good mechanical compatibility between the rubbery domain and the continuous PVC phase is necessary for the rubber to deform and absorb energy through crazing and/or shear banding. This effect, however, will not be realized if the continuous phase is weakened by the presence of an excessive number of primary particles which will allow the crack to propagate along the particle boundaries.

REFERENCES

1.	Modern Plastics International, January 1984 (McGraw-Hill Publishers), pp. 21-35.
2.	O. Olabisi, L. Robeson and M. T. Shaw, Polymer-Polymer Miscibility (Academic Press, New York, 1979).
3.	J. Koleske and R. Lundberg, J. Polym. Sci. Part A-2. 7 (1969) p. 795.
4.	M. Coleman and J. Zarian, J. Polym. Sci. Polym. Phys. Ed. 17 (1979) p. 837.
5.	F. Khambatta, et. al., J. Polym. Sci. Polym. Phys. Ed. 14 (1976) p. 1391.
6.	D. Hubbell and S. Cooper, J. Appl. Polym. Sci., 21, (11) (1977) p. 3035.
7.	L. Robeson, J. Appl. Polym. Sci., 17 (1973), p. 3607.
8.	M. Aubin and R. Prud'homme, J. Polym. Sci., Polym. Phys. Ed. 19 (8), (1981), p. 1245.
9.	D. Varnell and M. Coleman, Polymer 22 (10) (1981), p. 1324
10.	M. Aubin, et. al., MMI Press Symp. Ser. 2 (Polym. Compat. Incompat.), (1982), p. 223.

146

11. R. Prud'homme, Polym. Eng. Sci. 22 (2) (1982), p. 90.
12. M. Aubin and R. Prud'homme, Macromolecules, 13 (2) (1980), p. 365.
13. M. Coleman and D. Varnell, J. Polym. Sci. Polym. Phys. Ed. 18 (6), (1980), p. 1403.
14. T. Nishi, et. al., J. Appl. Phys. 46 (1975), p. 4157.
15. T. Nishi and T. Kwei, J. Appl. Polym. Sci. 20, (1976), p. 1331.
16. D. Hourston and I. Hughes, J. Appl. Polym. Sci. 21, (1977), p. 3099.
17. D. Hourston and I. Hughes, Polymer 20, (1979), p. 823.
18. L. Robeson, J. Polym. Sci. Polym. Lett. Ed. 16 (1978), p. 261.
19. S. Goh, et. al., J. Appl. Polym. Sci. 27 (3) (1982), p. 1091.
20. R. Prud'homme, Polym. Eng. Sci. 22 (17) (1982), p. 1138.
21. J. Schurer, et. al. Polymer 16 (1975), p. 201.
22. M. Dimitrov and E. Foldes, Angew. Markromol. Chem. 106 (1982), p. 91.
23. M. Matsuo, et. al., Polym. Eng. Sci. 9 (1969), p. 197.
24. C. Wang and S. Cooper, J. Polym. Sci. Polym. Phys. Ed. 21 (1), (1983), p. 11.
25. M. Woods and D. Frazer, SPE Ann. Tech. Conf. 32nd 20, (May 1974) p. 426.
26. R. Deanin and K. Sheth, Org. Coat. Plast. Chem. 43 (1980), p. 23.
27. C. Lin and N. Hsieh, Bulletin of the College of Engineering, National Taiwan Univ., No. 20 (July 1976), p. 72.
28. P. Bataille et. al., J. Vinyl Tech. 2 (4), (1980), p. 218.
29. D. Paul and C. Locke, Polym. Eng. Sci, 13 (1973), p. 308.
30. C. Hammer, Macromolecules, 4 (1971), p. 69.
31. Y. Shur and B. Ranby, J. Appl. Polym. Sci. 19 (1975), p. 1337.
32. B. Ranby, J. Polym. Sci. Polym. Symp. 51 (1975), p. 89.
33. J. Hickman and R. Ikeda, J. Polym. Sci. Polym, Phys. Ed. 11 (1973), p. 1713.
34. E. Anderseon, et. al., Adv. Chem. No. 176 (ACS), (1979), p. 413.
35. L. Robeson and J. McGrath, Polym. Eng. Sci. 17 (1977), p. 300.
36. R. Deanin and N. Shah, Org. Coat. Plast. Chem. 45 (1981), p. 290.
37. B. Carmoin, et. al., J. Macromol. Sci. Phys. 14 (1977), p. 307.
38. H. Breuer, et. al., J. Macromol. Sci. Phys. B14 (1977), p. 387.
39. F. Haaf, et. al., Angew. Makromol. Chem. 58-59 (1977), p. 95.

40. R. Deanin and C. Moshar, Polym. Prepr. A.C.S. Div. Polym. Chem. 15 (1), (1974), p. 403.
41. Y. Shur and B. Ranby, J. Appl. Polym. Sci. 20 (1976), p. 3121.
42. A. Pavan, et. al., Mater. Sci. Eng. 48 (1)(1981), p. 9.
43. W. Congdon et. al., Polymer Alloys, Vol. II (Plenum, New York, 1980). D. Klempner and K. Frisch, ed.
44. M. Rink et. al., Plast. Rubber Process Appl. 3 (2) (1983), p. 145.
45. W. Hornibrooke, Pure Appl. Chem. 53 (2) (1981), p. 501.
46. D. Leisz, et. al., Thermochim. Acta. 35 (1) (1980), p. 51.
47. L. Bourland and A. Wambach, Plast. Eng. 39 (5) (1983) p. 23.
48. N. Kalfoglou, J. Appl. Polym. Sci. 26 (3) (1981) p. 823.
49. J. Piglowski et. al., Angew. Makromol. Chem. 85 (1980) p. 129.
50. H. Bonk et. al., J. Elastoplast. 3, (1971), p. 157.
51. C. Bucknall, Toughened Plastics, (Applied Science Publishers, London, 1977).
52. L. Nass, Encyclopedia of PVC, Vol. 2, (Marcel Dekker, New York, 1977).
53. R. Gould and J. Player, Kunstoffe 69 (1979), p. 10.
54. F. Ide and K. Okano, Pure Appl. Chem. 53 (2) (1981), p. 489.
55. R. Krzewki and E. Collins, J. Vinyl Tech. 3 (2) (1981), p. 116.
56. J. Wingrave, J. Vinyl Tech. 2 (1980), p. 204.
57. P. McKenney, J. Appl. Polym. Sci. 9 (1965), p. 3359.
58. J. Defiefe, J. Vinyl Tech. 2 (1980), p. 95.
59. C. Hammer, ACS, Coat. Plast. Prepr., 37 (1977), p. 234.
60. J. Tordella, Mod. Plast., January 1976 (McGraw-Hill Pub.), p. 64.
61. E. Rabinovitch and J. Summers, J. Vinyl Tech. 2 (1980), p. 165.
62. J. Wingrave and M. Peden, J. Vinyl Tech. 1 (1979), p. 107.
63. T. Chapman, et. al., J. Vinyl Tech. 1 (1979), p. 131.
64. V. Uhl and J. Gray, Mixing Theory and Practice, Vol. II. (Academic Press, New York, 1967).
65. J. Chauffoureaux et. al., J. Rheol. 23 (1979), p. 1
66. J. Goordich and R. Porter, Polym. Eng. Sci. 7 (1967), p. 45.
67. Technical Information Bulletin, Using a Banbury to compound PVC blends containing Elvaloy Resin Modifiers, Polymer Products Dept., E. I. Du Pont Co., Wilmington, Delaware 19898.
68. Technical Information Bulletin, Vygen 300 Series Resins, Chemical Plastics Division, General Tire and Rubber Co., Akron, Ohio.

148

69.   Technical Service Report, Geon 90 Series Resins, TSR
        71-04, B. F. Goodrich C., Cleveland, Ohio.
70.   P. Shah and V. Allen, SPE J. 26 (1970), p. 56.
71.   J. Summers et. al., SPE Ann. Techn. Conf. 36th, (1978),
        p. 757.
72.   T. Hattori et. al., Poly,. Eng. Sci. 12 (1972), p. 199.
73.   G. Hofmann, Polymer Alloys, Vol. III (Plenum, New York,
        1981). D. Klempner and K. Frisch, ed.

APPENDIX

COPOLYMER ABBREVIATIONS

αMSAN   αMethylstyrene acrylonitrile
ABS   Acrylonitrile butadiene styrene
AMBS   Acrylonitrile methacrylate butadiene styrene
BAN   Butadiene acrylonitrile
CPE   Chlorinated polyethylene
ECR   Ethylene copolymer resin
EVA   Ethylene vinylacetate
MBA   Methacrylate butyl acrylate (core shell)
MBS   Methacrylate butadiene styrene
MMA-AC Methacrylate acrylate
NBR   Acrylonitrile butadiene rubber
PBT   Polybutylene terephthalate
PCL   Polycaprolactone
PEEL   Polyether ester elastomer
PMMA   Polymethylmethacrylate
PU   Polyurethane
PVCA   Vinyl chloride vinylacetate
PVL   Polyvalerolactone
SAN   Styrene acrylonitrile
SMA   Styrene maleic anhydride

SYNTHESIS OF BLOCK AND GRAFT COPOLYMERS

G. Hurtrez, D.J. Wilson* and G. Riess

Ecole Nationale Supérieure de Chimie de Mulhouse
3, rue Alfred Werner, 68093 Mulhouse Cedex (France)

Block and graft copolymers are gaining considerable importance both in scientific research and in their application to industry; a large number of publications, patents, review articles and symposia exist in this field. Although these copolymers may vary considerably in structure, two general groups exist:

<u>Block Copolymers</u>

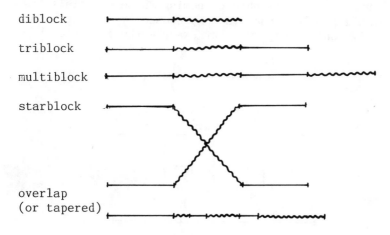

diblock

triblock

multiblock

starblock

overlap
(or tapered)

---

*on leave from the University of Sussex, England.

Graft Copolymers

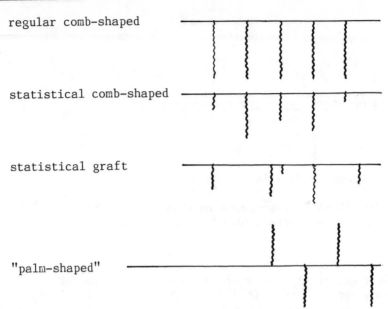

regular comb-shaped

statistical comb-shaped

statistical graft

"palm-shaped"

Although a concise method of naming these materials presents a considerable problem, the system proposed by IUPAC (1) has all but been universally adopted. For example the copolymers,

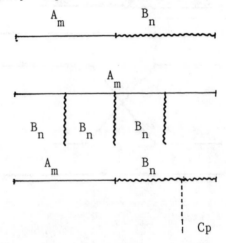

are named respectively:

poly A - <u>block</u> - poly B
poly A - <u>graft</u> - poly B
poly A - <u>block</u> - (poly B - <u>graft</u> - poly C)

There exist two general methods for the synthesis of block and graft copolymers. In the first one, active sites on a macromolecular chain are created which subsequently initiate the polymerization of a suitable monomer. The active sites may be either at the ends (leading to block copolymers) or situated along the length of the chain (leading to graft copolymers). Such a polymerization can be free-radical, anionic or cationic and is schematically represented as follows:

<u>Block</u>

$$\underset{A_m}{\rule{7em}{0.4pt}} \ \circledast \ + \ nB \quad \longrightarrow \quad \underset{A_m}{\rule{5em}{0.4pt}}\ \underset{B_n}{\sim\!\sim\!\sim}$$

<u>Graft</u>

$$\underset{A_m}{\rule{7em}{0.4pt}} \ + \ 3\ nB \quad \longrightarrow \quad$$

The second method is essentially a condensation reaction involving end-groups, be they cationic, anionic or of a functional group nature. In the latter case this may be illustrated:

<u>Block</u>

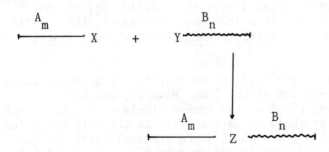

Graft

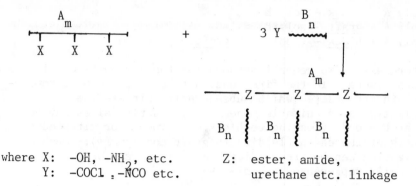

where X: -OH, -NH$_2$, etc.   Z: ester, amide,
      Y: -COCl , -NCO etc.   urethane etc. linkage

In this article we have outlined the main synthetic
techniques which are presently in use; more detailed information
may be obtained from the numerous reviews which exist on the
subject (2), (3), (4), (5).

ACTIVE-SITE CREATION

This encompasses a wide range of techniques the most
important of which are:

        radical polymerization
        anionic polymerization
        cationic polymerization
        active-centre transformations

Radical Polymerization

The synthesis of block copolymers by radical means was first
performed by Melville (6) using a photochemical set-up as shown in
Figure 1. Block copolymer A-B is formed by irradiation of monomer
A as it passes through a capillary; the lifetime (ζ) of the
growing chain is greater than its residence time in the capillary
with the result that the still growing chains of poly A arrive in
monomer B.

The active sites produced in the reaction were not isolated
and the product yield was rather low. However, another
possibility of using non-isolated free-radicals exists i.e. that
of polymerization by chain transfer. This is the polymerization
of a monomer A in the presence of a free-radical initiator (or
polymeric radical) and a preformed polymer B, possessing labile
groups (e.g. allylic or tertiary hydrogen).

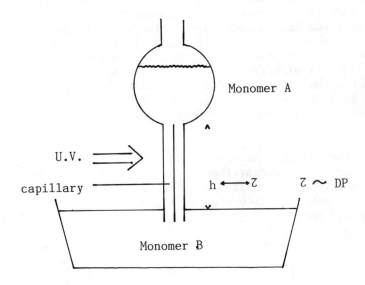

Fig. 1. Photochemical apparatus of Melville

A macroradical is produced on polymer B which subsequently reacts
with monomer A as illustrated:

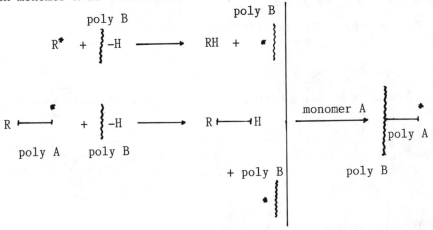

    Although this technique is relatively simple and industrially
useful, a complex mixture such as:

- poly A (by direct polymerization of monomer A)
- poly B (non-grafted backbone)
- graft copolymer A–B

may result, and alternative end products may be formed. For example if chain termination occurs by recombination, a crosslinked polymer may result. The commercially important products high-impact polystyrene (HIPS) and poly(acrylonitrile/butadiene/styrene) (ABS) may be manufactured in this manner i.e. by incorporation of the appropriate monomer (s) onto a polybutadiene chain (styrene and acrylonitrile/styrene respectively). Similarly a grafting reaction by chain transfer on a stereoblock polybutadiene containing 1,4 (cis or trans) and 1,2 vinyl blocks results in the "palm-shaped" copolymer mentioned previously; grafting occurs preferentially at the 1,2 sites.

The use of molecules which generate free-radicals represents another route whereby the synthesis of block and graft copolymers may be attained. Molecules of this type possess initiating groups such as azo, peroxide etc. in a polymer chain. Radicals are produced thus:

$$P-O-O-P' \qquad\qquad P-N=N-P'$$

heat, U.V. etc.

$$P-\overset{\cdot}{O}\ \overset{\cdot}{O}-P' \qquad\qquad \overset{\cdot}{P}\ \overset{\cdot}{P'} + N_2$$

where P, P' = polymer chain

For example polymeric initiators were prepared by Simionescu (7) using azo-peroxide derivatives as shown:

$$R-OO-\overset{\overset{O}{\|}}{C}-(CH_2)_2-\overset{\overset{CH_3}{|}}{\underset{\underset{CN}{|}}{C}}-N=N-\overset{\overset{CH_3}{|}}{\underset{\underset{CN}{|}}{C}}-(CH_2)_2-\overset{\overset{O}{\|}}{C}-OOR$$

$$R = NO_2 -\text{[benzene ring]}- \overset{\overset{O}{\|}}{C}-$$

or

$$\text{[benzene ring with Cl]}- \overset{\overset{O}{\|}}{C}-$$

If methylmethacrylate is used as monomer, this leads to a polymeric peroxide on a poly(methylmethacrylate) chain thus:

$$R-OO-\overset{\overset{O}{\|}}{C}-(CH_2)_2-\overset{\overset{CH_3}{|}}{\underset{\underset{CN}{|}}{C}}\overset{PMM}{\sim\sim\sim\sim\sim}\overset{\overset{CH_3}{|}}{\underset{\underset{CN}{|}}{C}}-(CH_2)_2-\overset{\overset{O}{\|}}{C}-OOR$$

Subsequent degradation of the peroxide in the presence of a second monomer leads to the desired block copolymer. In the same way, tetraethylthiouram disulphide (8) and polyazo derivatives (9), (10) have been used to synthesise block copolymers.

Graft copolymers can be synthesised by analogous techniques. For example a polymeric chain can be peroxidised and thus used as a radical initiator thus:

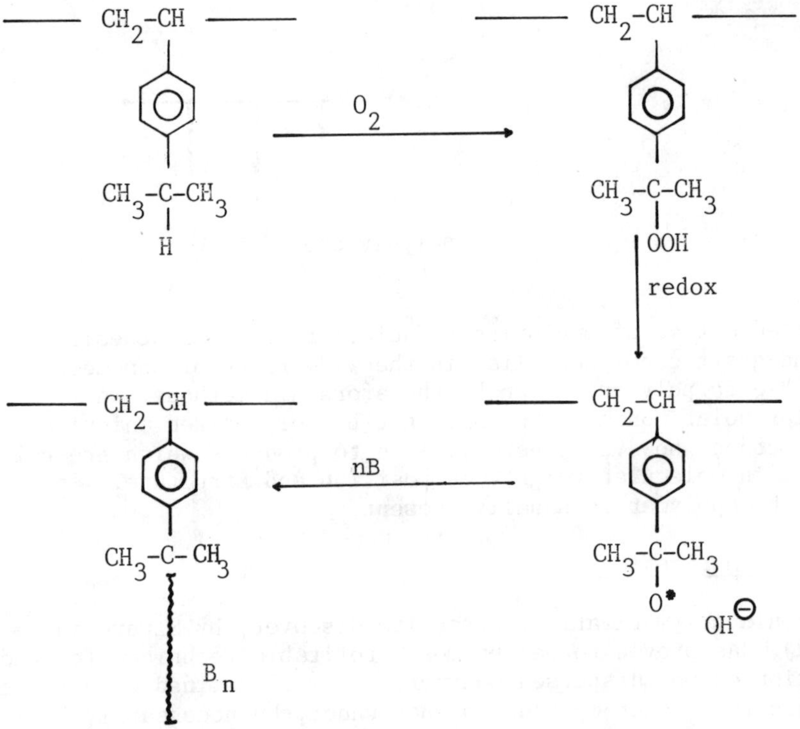

Graft copolymers may also be obtained from what is termed a macromonomer; Milkovich (11) has used such a method to prepare graft copolymers from styrene and vinylic (or acrylic) monomers as illustrated below:

1st. step: deactivation of the "living" anionic polymer:

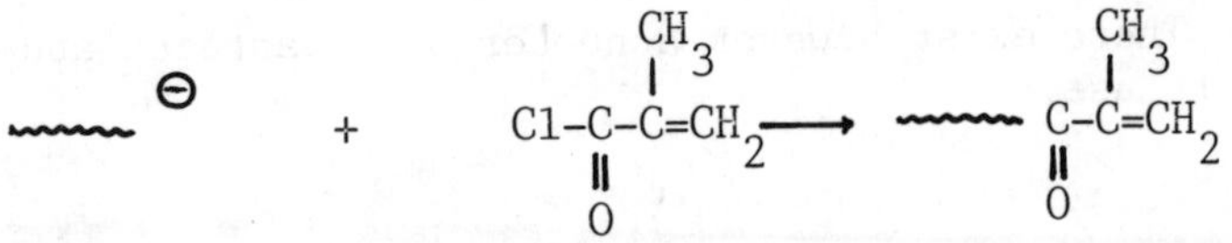

2nd. step. free-radical copolymerization with suitable monomer (e.g. methylmethacrylate):

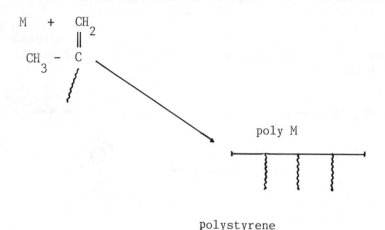

poly M

polystyrene

The advantage of employing a radical route to synthesize block and graft copolymers lies in the wide range of monomers susceptible to polymerization by the aforesaid methods, in particular polar monomers which cannot be polymerized anionically. Radical methods however generally lead to products which are poly-dispersed in molecular weight, composition and structure. In addition homopolymer is usually present.

Anionic Polymerization

Anionic polymerization, since its discovery by Szwarc in 1956 (12), has proved to be the most profitable technique for the preparation of monodispersed polymers and well-defined copolymers. It differs from other methods in that under the necessarily stringent conditions required for polymerization, a termination or transfer step is absent; for this reason it is termed "living anionic polymerization". The possibility of preparing almost pure copolymer of a specific molecular weight, composition and structure becomes realised by anionic methods and monomers of a diene, vinylic (non-polar) and cyclic ether nature may be polymerized by this technique.

There exist however a number of advantages and disadvantages in its use.

advantages:

  (a) by successive polymerization of monomers via mono- or
       difunctional initiators or by coupling, block and graft
       copolymers of well-defined structure etc. may be
       obtained;
  (b) functionalised block copolymers with specific end-groups
       may be obtained by means of suitable reagents.

disadvantages:
      Aside from the limited number of monomers which can be used,
a number of conditions must be fulfilled in order to obtain the
required product, namely:
  (a) low temperature;
  (b) high vacuum or inert blanket gas;
  (c) ultra-pure reagents and solvents.

      The mono- and difunctional initiators used are most frequently
lithium derivatives; the monomers currently in use in anionic
polymerization are styrene, butadiene, isoprene and methylstyrene.

Block Copolymers

monofunctional initiation:

$$RLi \;+\; mA \longrightarrow R \vdash\!\!-\!\!-\!\!-\!\!\dashv A^{\ominus} Li^{\oplus}$$

$$\text{e.g. } CH_3\!-\!\underset{\underset{CH_3}{|}}{\overset{\overset{CH_3}{|}}{C}}{}^{\ominus} Li^{\oplus}, nBu^{\ominus} Li^{\oplus},$$

$$\underset{\underset{CH_3}{|}}{\overset{\overset{CH_3}{|}}{\langle\!\bigcirc\!\rangle\!-\!C}}{}^{\ominus} Li^{\oplus}, \text{ etc.}$$

nB

$$R \vdash\!\!-\!\!-\!\!-\!\!\dashv A \,\rightsquigarrow\, B^{\ominus} Li^{\oplus}$$

pC

$$R \vdash\!\!-\!\!-\!\!-\!\!\dashv A \,\rightsquigarrow\, B \vdash\!-\!\!-\!\dashv C^{\ominus} Li^{\oplus}$$

bifunctional initiation:

$$Li\!-\!R\!-\!Li \;+\; mA \longrightarrow {}^{\oplus}Li^{\ominus}A \vdash\!\!-\!\!-\!\dashv R \vdash\!\!-\!\!-\!\dashv A^{\ominus} Li^{\oplus}$$

nB

$$Li^{\oplus} B^{\ominus} \rightsquigarrow A \vdash\!\!-\!\!-\!\dashv R \vdash\!\!-\!\!-\!\dashv A \rightsquigarrow B^{\ominus} Li^{\oplus}$$

pC

$$Li^{\oplus} C^{\ominus} \vdash\!-\!-\!-\!\dashv B \rightsquigarrow A \vdash\!\!-\!\!-\!\dashv R \vdash\!\!-\!\!-\!\dashv A \rightsquigarrow B \vdash\!-\!-\!-\!\dashv C^{\ominus} Li^{\oplus}$$

158

e.g.

functionalisation: Block copolymers may be functionalised to give a variety of end-groups (15). For example:

coupling: This technique proves useful particularly in the synthesis of star block copolymers.

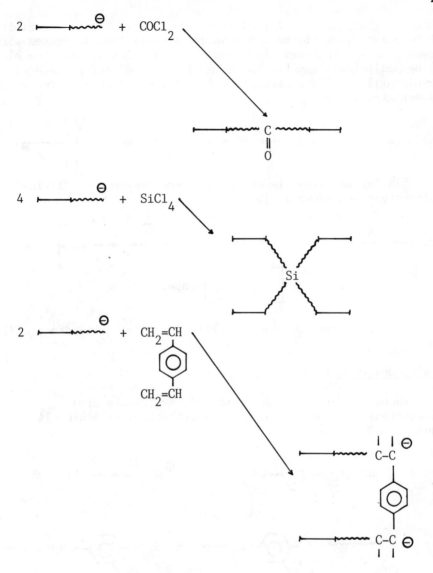

It should be noted that the final structure of the block copolymer depends on a number of factors, namely:

- reaction conditions (type of solvent, temperature etc.).
- type of initiator used.
- composition of initial reaction mixture.

A case in point as regards the latter is that of the preparation of an overlap block copolymer possessing a tapered sequence. If, for example in the preparation of a polystyrene/ polybutadiene block copolymer, non-polar solvents are used, pure block polybutadiene is formed first (due to the higher reactivity of butadiene). As butadiene monomer concentration decreases with respect to styrene there is a tendency for short polystyrene blocks to be built in; these blocks become longer and longer (tapered part) until finally a pure block polystyrene block is formed. This is depicted in (a):

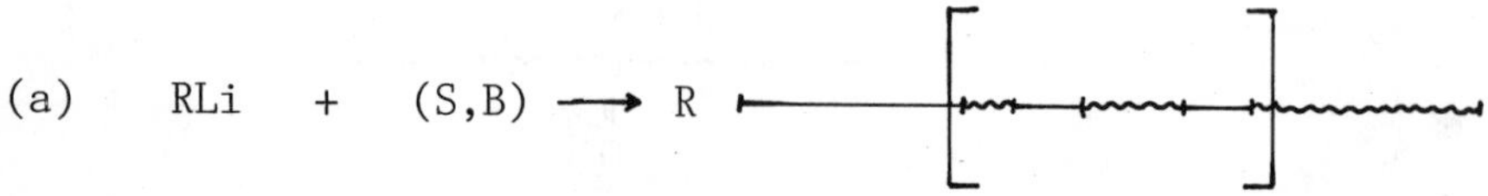

(a)   RLi   +   (S,B) ⟶ R

Similar behaviour is observed in the presence of "living" polystyrene, as shown in (b):

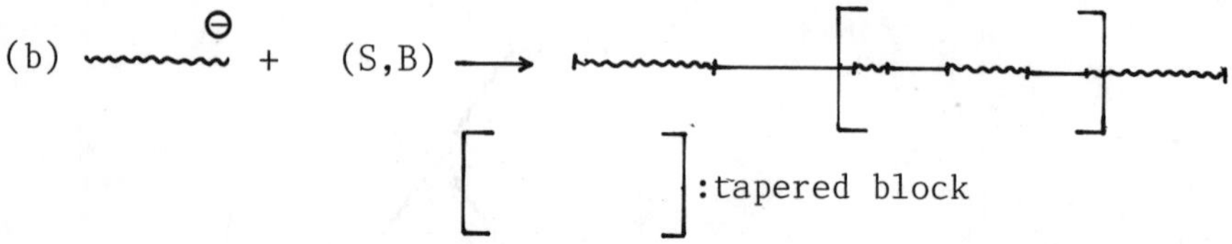

(b)   ⊖   +   (S,B) ⟶

[   ] :tapered block

Mixtures of styrene and isoprene exhibit a similar tendency (13).

Graft Copolymers

These may be obtained for example by the use of a difunctional initiator plus a small connecting molecule (14) thus:

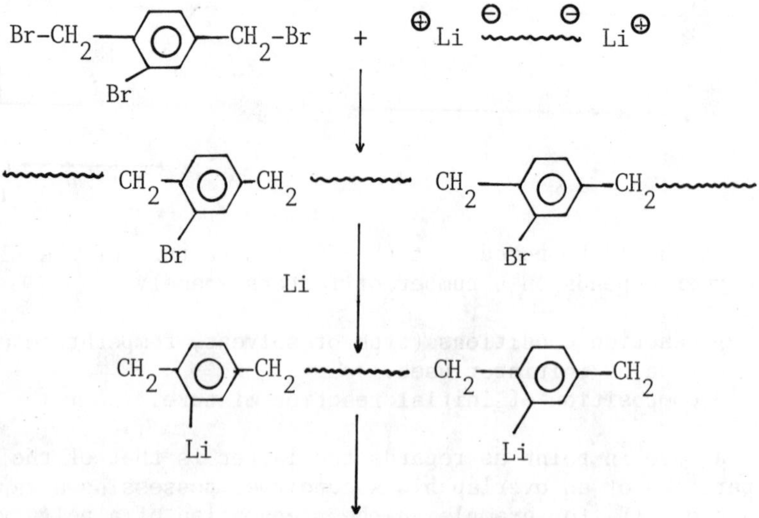

grafting of appropriate monomers at Li sites.

or by the deactivation of a "living" chain on an existing polymer
backbone thus:

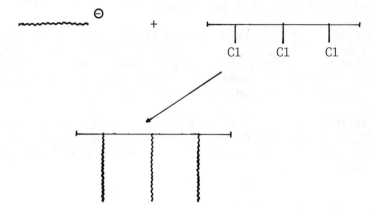

Cationic Polymerization

    Although attempts at block and graft copolymer synthesis by
cationic routes up until the 1960s was largely unproductive, work
done in recent years by Kennedy (16) and others. in the sphere of
controlling the various steps involved (initiation, propagation and
termination) has done much to rectify this situation.

controlled initiation: By means of alkyl aluminium compounds
(e.g. Et_3Al, Et_2AlCl) in conjunction with active cationogenic
species (e.g. tertiary, benzylic or allylic halides), polymerization
of very reactive cationic monomers via the formation of a specific
head group may be induced. Thus:

$$RCl \quad + \quad Et_2AlCl \longrightarrow R^{\oplus}Et_2AlCl_2^{\ominus}$$

$$R^{\oplus} \quad + \quad CH_2{=}C \longrightarrow R{-}CH_2{-}C^{\oplus}$$

    The use of halogen-containing macromolecules (R=polymeric
species) led to the discovery of a new family of block and graft
copolymers (17), e.g. polyvinylchloride-graft-polystyrene.

controlled propagation: Different techniques may be employed in
order to prevent chain transfer to monomer.

(a) "quasi-living" polymerization: This consists of feeding the
monomer at a given rate to the active initiating system such that
the rate of monomer addition and monomer consumption are equal.

Recently triblock copolymers of poly αmethylstyrene-block-polyisobutylvinyl ether-block-poly αmethylstyrene have been prepared by means of this technique (18).

(b) proton traps: Certain sterically-hindered amines, for example 2,6-di-tert-butyl pyridine, can suppress chain transfer by trapping the offending protons. An efficient process for block and graft copolymer preparation makes use of this type of molecule in conjunction with a conventional Lewis acid (19).

(c) "inifer" system: This is defined as a initiating and transfer system and leads to head and tail functionalised polymers. The inifer may be a mono-, bi-, tri- or multifunctional species. For example,

(mono)inifer

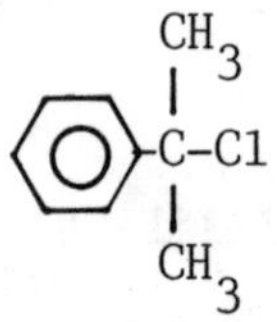

$$\text{C}_6\text{H}_5-\overset{\overset{\displaystyle CH_3}{|}}{\underset{\underset{\displaystyle CH_3}{|}}{C}}-Cl$$

(bi)inifer

$$Cl-\overset{\overset{\displaystyle CH_3}{|}}{\underset{\underset{\displaystyle CH_3}{|}}{C}}-\text{C}_6\text{H}_4-\overset{\overset{\displaystyle CH_3}{|}}{\underset{\underset{\displaystyle CH_3}{|}}{C}}-Cl$$

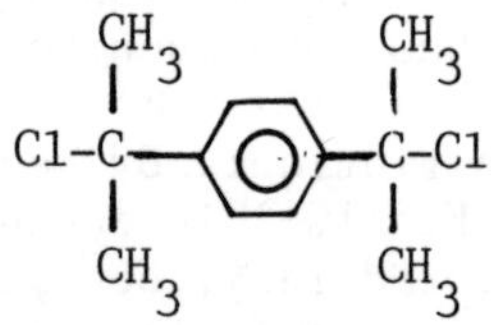

(tri)inifer

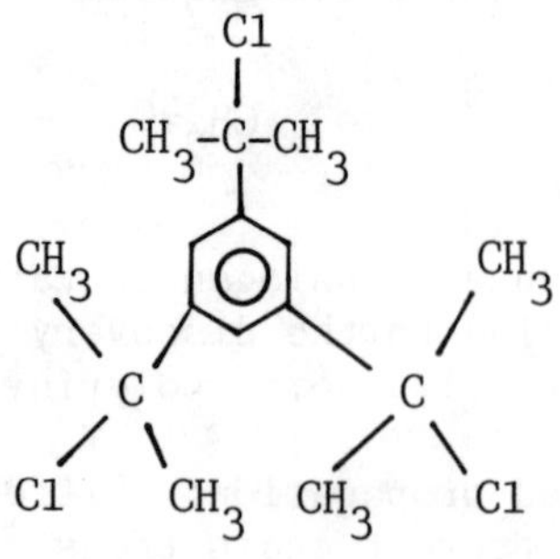

A simplified mechanism may be described:

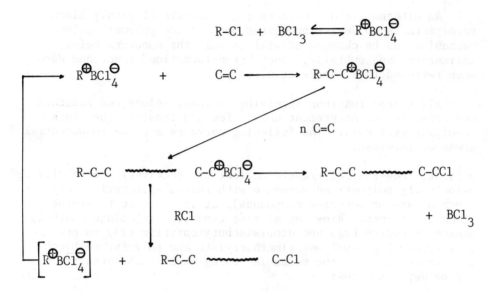

where R–Cl is a monofunctional inifer such as cumyl chloride. For the inifer to be useful, the requirement must be met that the rate of chain transfer to inifer is faster than that of chain transfer to monomer.

Such telechelic polymers prepared by this method are extremely useful for the block copolymerization of olefins and cyclic ethers (20) as illustrated:

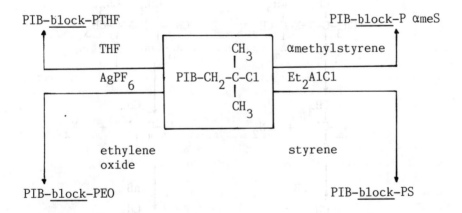

In addition, the chlorinated terminal groups of such polymers may be converted into hydroxyl, olefin, epoxide, acid, silane etc. for further applications in copolymer synthesis.

Active-Centre Transformation

An alternative approach to the synthesis of mainly block copolymers is to devise processes whereby the polymerization mechanism can be changed at will to suit the monomers being polymerised sequentially. Such "transformation" reactions have been reviewed by Richards et al (21).

All transformations involving cations, anions and radicals are possible but at present only a few are feasible for block copolymer synthesis. The following serve as a guide to potential areas of interest.

anion to free-radical transformation: By deactivation of a "living" anionically polymerized sequence with radical-generating reagents (such as azo or peroxide compounds), it is possible to synthesise block copolymers. Riess et al have used p,p bis(halogenomethyl) benzoyl peroxide (22) and azobisisobutyronitrile (23) to prepare polystyrene-block-polymethylmethacrylate and polystyrene-block-polyvinylchloride. The mechanism using azobisisobutyronitrile may be depicted thus:

$$
A_m^{\ominus} + \underset{\substack{|\\CN}}{\overset{\substack{CH_3\\|}}{CH_3-C}}-N=N-\underset{\substack{|\\CN}}{\overset{\substack{CH_3\\|}}{C-CH_3}} \longrightarrow \underset{\substack{|\\N=C\\|\\A_m}}{\overset{\substack{CH_3\\|}}{CH_3-C}}-N=N-\underset{\substack{|\\C=N^{\ominus}\\|\\A_m}}{\overset{\substack{CH_3\\|}}{C-CH_3}}
$$

(left branch)

$$
\underset{\substack{|\\A_m}}{\overset{\substack{CH_3\\|}}{CH_3-C}}-N=N-\underset{\substack{|\\A_m}}{\overset{\substack{CH_3\\|}}{C-CH_3}}
$$

$$
\underset{\substack{|\\CH_3}}{\overset{\substack{CH_3\\|}}{A_m-C^{\bullet}}} + N_2
$$

$$
\downarrow nB
$$

$$
\underset{\substack{|\\CH_3}}{\overset{\substack{CH_3\\|}}{A_m-C-B_n^{\bullet}}}
$$

(right branch)

$$
\downarrow H^+/H_2O
$$

$$
\underset{\substack{|\\A_m-C=O}}{\overset{\substack{CH_3\\|}}{CH_3-C}}-N=N-\underset{\substack{|\\C=O\\|\\A_m}}{\overset{\substack{CH_3\\|}}{C-CH_3}}
$$

$$
\underset{\substack{\|\\O\\|\\CH_3}}{\overset{\substack{CH_3\\|}}{A_m-C-C^{\bullet}}} + N_2
$$

$$
\downarrow nB
$$

$$
\underset{\substack{\|\\O\\|\\CH_3}}{\overset{\substack{CH_3\\|}}{A_m-C-C-B_n^{\bullet}}}
$$

Metal salts may also be used to synthesise block copolymers of, for example, styrene and methylmethacrylate (24) thus:

$$A_m^{\ominus} Na^{\oplus} \; + \; Et_3PbCl \longrightarrow A_m-PbEt_3 \; + \; NaCl$$

$$\searrow AgClO_4$$

$$A_m^{\bullet} \; + \; Ag \longleftarrow A_m-Ag \; + \; Et_3PbClO_4$$

$$\searrow \; nB$$

$$Am-B_n^{\bullet}$$

Another possible route is via metal carbonyls; dimanganese decacarbonyl ($Mn_2(CO)_{10}$) in conjunction with benzylic bromine terminated polystyrene produces radicals capable of polymerizing methylmethacrylate (25).

anion to cation transformation: Of the range of transformations possible, anion to cation is perhaps the most studied. A general scheme may be illustrated thus:

$$A_m^{\ominus} Li^{\oplus} \; + \; \begin{matrix} X_2 \\ \text{or} \\ X-R-X \end{matrix} \longrightarrow \begin{matrix} A_m X \\ \text{or} \\ A_m RX \end{matrix} \; + \; LiX$$

$$\begin{matrix} A_m X \\ \text{or} \\ A_m RX \end{matrix} \; + \; Ag^{\oplus} Y^{\ominus} \longrightarrow \begin{matrix} A_m^{\oplus} Y^{\ominus} \\ \text{or} \\ A_m R^{\oplus} Y^{\ominus} \end{matrix} \; + \; AgX$$

where $A_m^{\ominus}$ is the "living" anionic polymer, $Y^{\ominus}$ an anion such as $PF_6^{\ominus}$, $SbF_6^{\ominus}$, $ClO_4^{\ominus}$ etc. and X a halogen. The first step, which is in fact the functionalisation of the "living" species, is best achieved by using α, α-dibromoxylene. The yield of this first reaction can be greatly increased by using $MgBr_2$ in conjunction with a halogenated species; block copolymers such as polystyrene-block-polybutadiene-block-polytetrahydrofuran have been prepared in this way (26).

Block copolymers of a similar nature have been prepared using phosgene as deactivating agent (27) thus:

$$A_m^{\ominus}Li^{\oplus} + COCl_2 \longrightarrow A_m\text{-}COCl + LiCl \xrightarrow{\;AgSbF_6\;} A_m\text{-}\overset{\oplus}{\underset{\underset{O}{\parallel}}{C}}SbF_6^{\ominus} + AgCl$$

$$\Big\downarrow\, nB$$

$$A_m\text{-}\overset{\oplus}{\underset{\underset{O}{\parallel}}{C}}\text{-}B_n$$

cation to anion transformation: This may be approached in two ways. In the first method, we functionalise a "living" cationic species with a reagent bearing a styryl unit (e.g. the lithium alcoholate of cinnamyl alcohol) and subsequently anionically polymerize a second monomer in conjunction with this species. If the original species is a polytetrahydrofuryl cation (28) then:

$$A_m\text{-}O(CH_2)_4\text{-}\overset{\oplus}{O}\!\!\bigcirc \quad PF_6^{\ominus} \;+\; LiOCH_2\text{-}CH\!=\!\underset{\underset{\phi}{\big|}}{CH}$$

$$\phi = \text{phenyl group}$$

$$\Big\downarrow$$

$$A_m\text{-}O(CH_2)_4\text{-}OCH_2\text{-}CH\!=\!\underset{\underset{\phi}{\big|}}{CH} \;+\; LiPF_6$$

$$\Big\downarrow\, BuLi$$

$$A_m\text{-}O(CH_2)_4\text{-}OCH_2\underset{\underset{Bu}{\big|}}{CH}\!\text{-}\underset{\underset{\phi}{\big|}}{CH}^{\ominus} Li^{\oplus}$$

$$\Big\downarrow\, nB$$

$$A_m\text{-}O(CH_2)_4\text{-}OCH_2\text{-}\underset{\underset{Bu}{\big|}}{CH}\!\text{-}\underset{\underset{\phi}{\big|}}{CH}\text{-}B_n^{\ominus}$$

The second method involves functionalisation by a primary amine followed by metallation and anionic polymerization (29) thus:

$$A_m\text{-}O(CH_2)_4\text{-}\overset{\oplus}{O}\!\!\bigcirc \quad PF_6^{\ominus} \;+\; R\text{-}NH_2 \longrightarrow A_m\text{-}O(CH_2)_4\text{-}NHR$$

$$\Big\downarrow\, \text{metallation}$$

$$A_m\text{-}O(CH_2)_4\text{-}\underset{\underset{R}{\big|}}{N}\text{-}B_n^{\ominus} \xleftarrow{\;nB\;} A_m\text{-}O(CH_2)_4\text{-}\underset{\underset{R}{\big|}}{N}^{\ominus} ME^{\oplus}$$

cation to radical transformation: This transformation can be achieved by using, as deactivating agent for the cationic polymer, a radical initiator bearing an acid or alcohol function. For example (30):

$$2\ A_m^{\oplus}Y^{\ominus} \quad + \quad Na\text{-}O\text{-}\overset{O}{\overset{\|}{C}}\text{-}(CH_2)_2\text{-}\overset{O}{\overset{\|}{C}}\text{-}O\text{-}O\text{-}\overset{O}{\overset{\|}{C}}\text{-}(CH_2)_2\text{-}\overset{O}{\overset{\|}{C}}\text{-}O\text{-}Na$$

$$\downarrow$$

$$A_m\text{-}O\text{-}\overset{O}{\overset{\|}{C}}\text{-}(CH_2)_2\text{-}\overset{O}{\overset{\|}{C}}\text{-}O\text{-}O\text{-}\overset{O}{\overset{\|}{C}}\text{-}(CH_2)_2\text{-}\overset{O}{\overset{\|}{C}}\text{-}O\text{-}A_m$$

$$2\ A_m\text{-}O\text{-}\overset{O}{\overset{\|}{C}}\text{-}(CH_2)_2^{\cdot} \ + \ CO_2 \xrightarrow{\quad 2nB \quad} 2\ A_m\text{-}O\text{-}\overset{O}{\overset{\|}{C}}\text{-}(CH_2)_2\text{-}B_n^{\cdot}$$

Another possibility involves using a transfer agent such as mercaptan as deactivator (30) thus:

$$A_m^{\oplus}\ Y^{\ominus} \quad + \quad SH_2 \longrightarrow A_m\text{-}SH \quad + \quad YH$$

$$A_m\text{-}SH \quad + \quad \sim\!\!\sim\!\!\sim M^{\cdot} \longrightarrow \sim\!\!\sim\!\!\sim MH \quad + \quad A_m\text{-}S^{\cdot}$$

$$A_m\text{-}S^{\cdot} \xrightarrow{\quad nB \quad} A_m\text{-}S\text{-}B_n^{\cdot}$$

radical to ion transformation: The difficulty involved in a radical to ion transformation is the short lifetime of a radical. This has been overcome however, by Mendjal (31), who has synthesised a brominated polystyrene by radical means and subsequently reacted it with $AgClO_4$ or $AgSbF_6$. The resultant species is able to polymerize, cationically, a second monomer such as tetrahydrofuran, indene etc.

Another possible method (32) involves using a solvent in which the polymer is insoluble; generated macroradicals are trapped in the precipitate matrix. Subsequent swelling in the presence of a suitable reducing agent (e.g. sodium napthalene), produces the desired transformation, i.e.

$$A_m^{\cdot} \quad + \quad [\text{naphthalene}]^{\ominus}\ Na^{\oplus} \longrightarrow A_m^{\ominus}Na^{\oplus} \quad + \quad [\text{naphthalene}]$$

POLYCONDENSATION

This method, though useful for the preparation of diblock, triblock and graft copolymers, has particular importance in the synthesis of multiblock copolymers; Goodmann (33) has recently reviewed block copolymer preparation by this method. In essence the technique involves the reaction of end-groups which may be anionic, cationic, or consist of a functional group such as hydroxyl, amino etc. The following methods illustrate the breadth of this type of process (X=-OH, -NH$_2$, etc.; Y=-COCl, -NCO, etc.):

(a)   condensation of a "living" anionic and/or cationic polymeric sequence;

(b)   deactivation of a "living" anionic or cationic polymeric sequence via a functionalised polymer;

(c)   condensation (either in the presence or absence of a coupling agent) of different preformed polymeric sequences possessing functional end-groups.

(a)   The anionic or cationic groups may interact directly,

$$A_m^{\ominus} \quad + \quad B_n^{\oplus} \longrightarrow A_m B_n \qquad\qquad (\longrightarrow \text{block})$$

or may involve an additional molecule thus:

$$A_m^{\ominus} \quad + \quad X-R-X \quad + \quad B_n^{\ominus} \longrightarrow A_m -R-B_n \qquad (\longrightarrow \text{block})$$

(b)   If poly B is functionalised:

$$A_m^{\ominus} \quad + \quad X-B_n \longrightarrow A_m -B_n \qquad\qquad (\longrightarrow \text{block})$$

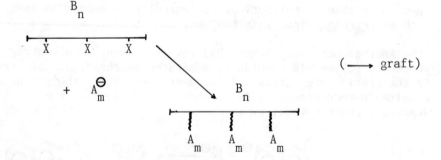

$(\longrightarrow \text{graft})$

(c)   If both poly A and poly B are functionalised:

$$A_m -X \quad + \quad Y-B_n \longrightarrow A_m -Z-B_n \qquad (\longrightarrow \text{block})$$

or in the presence of a coupling agent:

$$A_m-X \quad + \quad Y-R-Y \quad + \quad X-B_n \longrightarrow A_m-Z-R-Z-B_n \; (\longrightarrow \text{block})$$

For a multifunctional sequence:

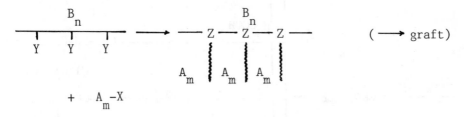

$$(\longrightarrow \text{graft})$$

In addition, if one uses difunctionalised preformed polymeric species:

$$X-A_m-X \quad + \quad Y-B_n-Y \longrightarrow A_m-Z-B_n-Z-A_m-Z-B_n \quad (\longrightarrow \text{multiblock})$$

$$- \text{alternating structure}$$

in the presence of a low molecular weight reagent:

$$X-A_m-X \quad + \quad Y-R-Y \quad + \quad X-B_n-X \qquad (\longrightarrow \text{multiblock})$$

$$- \text{random structure}$$

$$B_n-Z-A_m-Z-B_n-Z-B_n-Z-A_m$$

Numerous examples of these methods may be found in the Encyclopedia of Polymer Science and Technology (2). Synthesis of block and graft copolymers by condensation however, is not without its disadvantages. The principal one is that when one increases the molecular weight of the precursor polymer sequences the concentration of end-groups decreases. This may be accompanied by incompatibility of the sequences leading to a two-phase system. In both cases a drastic reduction in reaction rate is observed.

CHEMICAL MODIFICATION

The chemical modification of a block or graft copolymer is essentially the transformation of one or more differing sequences to lead to a new copolymer. Modification of polybutadiene-based copolymers have attracted a great deal of interest (34); a number of modifications are possible:

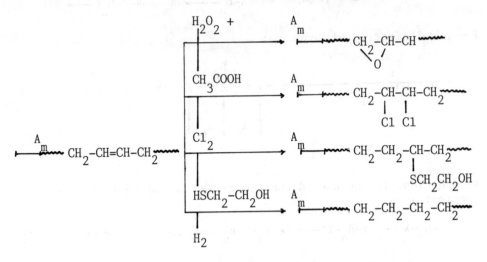

Sequences other than polybutadiene can be modified; Reeb (35) has described the esterification of the maleic anhydride units of a polyisoprene-<u>block</u>-poly(styrene-<u>alt</u>-maleic anhydride) copolymer.

Chemical modification of a copolymer is a potentially powerful technique whereby new properties may be created or existing ones enhanced. By way of an example, if a styrene-(4-vinylbenzyl dimethylamine)-isoprene block copolymer is quaternised with ethylbromide and sulphonated by sulphuric acid, the resultant product is capable of acting as a amphoteric ion exchanger (36).

CONCLUSION

This short review serves to illustrate the large range of techniques existing at present, which allow us to obtain "tailor-made" block or graft copolymers for which we can control the structure, molecular mass and composition.

Of the variety of methods mentioned, anionic and to a lesser extent cationic polymerization, are the most suitable leading to well-defined copolymers in good yield. Other techniques such as radically-initiated graft polymerization via chain transfer, although generally having a relatively low yield, are of industrial interest. The reaction product mixture is composed of, in addition to graft copolymer, a significant proportion of the corresponding homopolymers. Yet for certain applications, e.g. HIPS and ABS materials, only a small amount of graft copolymer is required to achieve the desired properties of the final product.

At present block copolymers available commercially are those obtained principally by anionic or polycondensation routes. In the first category, the most important are di-, tri- and star-block copolymers of diene (butadiene, isoprene)/styrene composition, i.e. S-B, S-B-S and $(SB)_x$ respectively (S: polystyrene; B: poly butadiene). The condensation method proves most useful for the synthesis of multiblock copolymers formed by functional oligomers (telechelic polymers). For example:

segmented	polyester-polyurethane
"	polyester-polyether
"	polyamide-polyether

Block copolymers from both of these methods may be used in a variety of applications, e.g.

thermoplastic elastomers
moulding products
plastic modifiers
adhesives   etc.

A special class is that of block or graft copolymers composed of hydrophilic and hydrophobic sequences (e.g. polyoxyethylene and polydimethylsiloxane respectively). The presence of groups of this type impart a surface-active nature to the copolymer, resulting in what are termed non-ionic surfactants (2).

REFERENCES

1. IUPAC, Commission on Macromolecular Nomenclature, Report "Source-Based Nomenclature for Copolymers", Amherst Meeting, 1982.
2. Riess, G., P. Bahadur and G. Hurtrez, Block Copolymers (to be published in The Encyclopedia of Polymer Science and Technology, 1984).
3. Gallot, Y. Colloques Nationaux du CNRS, Ed. CNRS, Paris, N°938 (1979) 149.
4. McGrath, J.E. J. Chem. Ed. 58 (11) (1981) 914.
5. McGrath, J.E. Pure and Appl. Chem. 55 (10) (1983) 1573.
6. Hicks, J.A. and H.W. Melville. Nature (London) 171 (1953) 300.
7. Dimitriu, S., A.S. Smith, E. Comanita and C. Simionescu. Eur. Polym. J. 19 (1983) 263.
8. Otsu, T. and M. Yoshida. Makromol. Chem., Rapid Commun. 3 (1982) 127.
9. Anand, P.S., H.G. Stahl, W. Heitz, G. Weber and L. Bottenbruch, Makromol. Chem. 183 (1982) 1685.
10. Laverty, J.L. and Z.G. Gardlung, J. Polym. Sci., Polym. Chem. Ed. 15 (1977) 2001.

11. Milkovich, R. Polym. Prepr. ACS Div. Polym. Chem. 21 (1980) 40.
12. Szwarc, M., M. Levy and R. Milkovich, J. Am. Chem. Soc. 78 (1956) 2656.
13. Corbin, N. and J. Prud'homme, J. Polym. Sci., Polym. Chem. Ed. 14 (1976) 1645.
14. Hahnel, H. and K. Gehrke. Plaste-Kautschuk 31 (4) (1984) 123.
15. Hurtrez, G. Thesis to be submitted, University of Haute-Alsace, Mulhouse (France).
16. Kennedy, J.P. Makromol. Chem. Suppl. 7 (1984) 171.
17. Kennedy, J.P. and E. Maréchal, Carbocationic Polymerization (John Wiley Interscience Pub., New York, 1982).
18. Sawamoto, M. and J.P. Kennedy, J. Macromol. Sci.,Chem. A18(8) (1982) 1293.
19. Kennedy, J.P., S.C. Guhaniyogi and L.R. Ross, J. Macromol. Sci., Chem. A18 (1982) 119.
20. Kennedy, J.P. Rubber Chem. Technol. 56(3) (1983) 677. Makromol. Chem. Suppl. 7 (1984) 171.
21. Abadie, M.S.M. and D.H. Richards. Inf. Chimie 208 (1980) 135.
22. Nicolova-Nankova, Z., F. Palacin, R. Raviola and G. Riess. Eur. Polym. J. 11 (1975) 301.
23. Riess, G. and R. Reeb, ACS Symp. Ser. 166 (1981) 477.
24. Abadie, M.J.M., F. Schue, T. Souel and D.H. Richards. Polymer 22 (1981) 1076.
25. Bamford, C.H., G.C. Eastmond, J. Woo and D.H. Richards, Polymer 23 (1982) 643.
26. Burgess, F.G., A.V. Cunliffe, J.R. McCallum and D.H. Richards. Polymer 18 (1977) 726.
27. Franta, E., L. Reibel, J. Lehmann and S. Penczek. J. Polym. Sci., Polym. Symp. 56 (1976) 139.
28. Abadie, M.J.M., F. Schue, T. Souel, D.B. Hartley and D.H. Richards. Polymer 23 (1982) 445.
29. Cohen, P., M.J.M. Abadie, F. Schue and D.H. Richards, Polymer 23 (1982) 1105.
30. Burgess, F.J. Master of Science, University of St. Andrews (1976).
31. Mendjel, H. Magister, Alger (1979).
32. Richards, D.H. Br. Polym. J. 12 (1980) 89.
33. Goodman, I. Development in Block Copolymers-1, Ed. I. Goodman (Appl. Science Publ. London 1982, p.127).
34. Merton, M., N.C. Lee and E.R. Terrill, ACS, Symp. Ser. 193 (1982) 101.
35. Reeb, R. Thesis to be submitted, University of Haute-Alsace, Mulhouse (France).
36. Fujimoto, T., M. Nagasawa and S. Ohno (to Toyo Soda Mfg. Co.), Fr. Demande 2,470,139 (1981).

BLOCK COPOLYMERS
MORPHOLOGICAL AND PHYSICAL PROPERTIES

Dale J. Meier

Michigan Molecular Institute, Midland, Michigan, U.S.A.

1 INTRODUCTION

It is generally recognized that two dissimilar polymers will
be thermodynamically incompatible with one another providing that
specific interactions between them do not lead to a negative heat
of mixing. This follows from the negligible entropy of mixing of
high molecular weight species coupled with a typical positive heat
of mixing. It is then not surprising that when such incompatible
polymers are joined together to form a block copolymer that this
incompatibility should still be manifested – as it is. However,
the union of incompatible species in a block copolymer has profound
effects on the phase behaviour and morphology of the resulting
phase-separated system. There are fundamental differences between
a system consisting of a simple mixture of two polymeric species
and a system consisting of the same species joined together to form
a block copolymer. Whereas the thermodynamically stable state of a
simple mixture is the gross separation of the two species to a
completely segregated state – with a minimum of interfacial contact
between the two separated phases, such gross phase separation
cannot occur with block copolymers since the incompatible species
are joined together. This factor leads to phase separation that
is restricted to molecular dimensions – "microphase separation" –
with the microphases existing in a domain morphology. Such systems
are unique in that the dispersed domain structures are thermodynamically stable in the dispersed state.

That microphase separation and domain formation occurred in
block copolymers was first clearly shown by Sadron and associates
in Strasbourg (1) more than twenty years ago. However, this
phenomenon remained an interesting curiosity until it was found (2)

that such microphase separation, when it occurred in block copolymers
of an appropriate molecular architecture, could lead to materials
possessing unique and technologically-useful physical properties,
e.g. thermoplastic elastomers from tri-block copolymers. That
demonstration of the profound influence of molecular architecture
on physical properties, and the attendant recognition that block
polymers could be designed to have specific end-use properties, has
lead to the explosive interest in block copolymers in following
years.

In this over-view of the morphology and properties of block
copolymers, the molecular and thermodynamic factors that control
microphase separation will be considered first. This will then be
followed by a discussion of the domain morphologies that are
developed by microphase separation, and then by a discussion of the
influence of domain formation and morphology on the mechanical and
rheological properties of microphase separated systems. The various
topics will be restricted to phenomena involving the "simple" block
copolymers, i.e. di- and tri-block amorphous polymers. This
restriction is made since it is with this class of polymers that
one can clearly show the relationships between structure and properties of block copolymers, and at the same time show that such
relationships have a theoretical background. In contrast, those
block copolymers having a multi-block architecture or having (quasi)
crystallizable components, e.g. segmented polyurethanes or polyether-
ester block copolymers, have morphologies that are much more complex,
and the theoretical relationships between molecular structure,
morphology and properties are poorly established at the present
time.

2 MICROPHASE SEPARATION AND DOMAIN MORPHOLOGY

It was mentioned in the introduction that phase separation in
block copolymers is restricted to molecular dimensions as a
consequence of the incompatible block components being joined
together, thus preventing gross physical separation of the two
components as would occur with their simple mixture. As a result
of many experimental and theoretical investigations, it is recognized
that domain structures which are formed by microphase separation
in block copolymers have dimensions and shapes that are directly
related to molecular dimensions and to thermodynamic interaction
of the block components.

In the simple block copolymers, there are three equilibrium
shape morphologies that can form, i.e. (1) alternating lamellae of
the two components, and (2) cylinders or (3) spheres of one
component imbedded in a matrix of the other component. To a first
approximation, the size of such domain structures is governed by
the molecular weights or molecular volumes of the block components,

while their shape is governed by the relative volume fraction of the components. These relationships can be put on a much more rigorous basis, as will be shown below.

The various molecular and thermodynamic factors which are important in governing the morphology of the microphases in block copolymers were first recognized by Meier (3), and used by him to develop a statistical mechanical theory of microphase separation. The following overview of the theory of microphase separation will follow - in modified form - his formulism.

The overriding thermodynamic factor which controls domain sizes is the requirement for the uniform filling of space by chain segments. This is a result of the usual very low compressibility of matter. For example, compressional moduli of most materials (including polymers) are higher than 10^9 Pa, and to create a density change of only 10% in such a material would require a stress of at least 10^8 Pa, or 1000 Atm. Thus we may assume that regardless of whatever chain perturbations must occur in forming domains, the equilibrium system will be one of uniform density. Fig. 1 shows cross-sections through domains formed from di-block polymers, in which the top diagram shows chains of each block type originating from the block junctions located in the boundary or interfacial region of the domains, with the remaining segments of the chains occupying the appropriate domain spaces. The block junctions must lie in the interfacial region where segments of the two block types are still intermixed. The middle diagram of Fig. 1 shows the relative segment densities of chains which originate in the inter-facial region and are constrained to stay within the domain boun-daries. The overall density of segments within the domain results from chains which originate on both sides of the domain, as shown. The final diagram in Fig. 1 shows the equilibrium domain size as one where the segmental densities are uniform throughout. This diagram also shows that although the major portion of the different block species are segregated into their different domain regions, there will be a finite (interfacial) region where mixing of the components still occurs.

Consider the thermodynamic changes that accompany the trans-formation of a di-block copolymer system from a randomly mixed state to a microphase-separated state, as shown in Fig. 2. In the initial randomly mixed state, the molecules have no spatial constraints since they can occupy any portion of the total available space. Also, the interaction of the different components giving an attendant heat of mixing occurs throughout the system. However, in the microphase-separated state (shown in Fig. 2 as having a spherical morphology), there are now constraints on the placement of molecules and segments thereof in space, and the interaction of the components now occurs only in the interfacial region of the domains. Contributions to the free energy change ΔG accompanying microphase

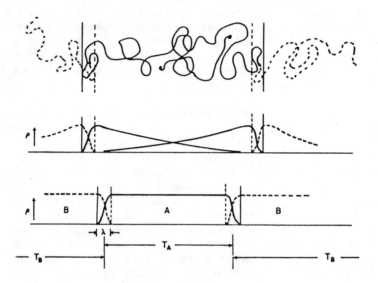

Fig. 1 Chains and Segment Densities in Domains

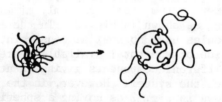

Fig. 2 Transformation to Domain Morphology

separation can be identified as:

ΔH(mixing). The interaction between the block components changes as the initial randomly mixed state transforms to a microphase separated state. It is this change which provides the driving energy for microphase separation.

ΔS(placement). The segregation of the components from one another in the microphase separated state requires that the junction between the components be constrained to be in the interface region. The resulting reduced volume available for the placement of molecules is responsible for a loss of entropy.

ΔS(segment exclusion). The formation of a domain state requires that each block segment be restricted into its appropriate domain space. This exclusion of some of total volume for the placement of chain segments also is responsible for a loss of entropy in the microphase separated system.

These various contributions to the free energy of domain formation can be formulated in the following manner (3):

ΔH(mixing). In the initial randomly mixed state, the heat of mixing H_o can be expressed in terms of Flory–Huggins theory (4),

$$H_o(\text{mixing}) = V\phi_A\phi_B\chi_{AB}kT \tag{1}$$

where V is the volume of the system, the ϕs are volume fractions of the two components, χ_{AB} is the interaction parameter between the components (expressed on a segmental volume basis), k is the Boltzmann constant and T·is the temperature. In the domain state, the residual interaction between the two components occurs in the interfacial region. The heat of mixing in a volume element can be given in the same form as in Eq. (1), but since the local composition will change across the interface region, the total heat of mixing in the interfacial region $H_i(\text{mixing})$ will be obtained by integrating over the total interfacial volume. In addition, a corrective term (Cahn–Hilliard, 5) must be introduced if the interfacial thickness is comparable to the range of inter-molecular forces ("non–local" interactions). Thus the interfacial heat of mixing becomes

$$H_i(\text{mixing}) = \chi_{AB}kT\int[\phi_A\phi_B + \frac{t^2}{6}(\frac{d\phi}{dx})^2]dV \tag{2}$$

where the ϕs are the composition with the volume element dV, and t is a measure of the range of intermolecular forces (approximately equal to the size of a segment). The second term within the brackets is the Cahn–Hilliard corrective term, but it will not be retained in the following presentation since its influence is very

small for the block copolymer interfacial thicknesses to be
discussed. In order to use Eq. (2), it is necessary to specify the
spatial dependence of the composition ϕ_A and ϕ_B of the components
in the interface. Although in principle the composition can be
given by theory (6), this has only been possible for infinite
molecular polymers where the origin of chains can be ignored.
However, the final results from present theory are not sensitive
to the particular spatial dependence assumed for the interfacial
composition, so that from the theory (6) for infinite molecular
weight polymers will be assumed, i.e. for a planar interface
(lamellar domains),

$$\phi_A(x) = (1 - \exp a'x)^{-1} \quad \text{with} \quad \phi_B(x) = 1 - \phi_A(x) \tag{3}$$

where a' is related to the thickness of the interface region.
With this composition in the interface, Eq. (2) gives

$$H_i = NX_{AB}kT\lambda/2T_A \tag{4}$$

where N is the number of molecules, and λ is the "linear" interfacial thickness

$$(\lambda = (\delta\phi/\delta x)_{x=0} = 4/a' \tag{5}$$

and T_A is the thickness of the A-domain. The change in the heat
of mixing of the components then becomes

$$\Delta H(\text{mixing}) = NX_{AB}kT\lambda/2T_A - V\phi_A\phi_B X_{AB}kT \tag{6}$$

$\underline{\Delta S(\text{placement})}$. The entropy change associated with the restricted
volume available for the placement of molecules in space (their
block junctions) is directly related to the ratio of the volume
available in the domain state to that of the random state,

$$\Delta S(\text{placement}) = Nk \ln(V_i/V_o) \tag{7}$$

where V_i is the volume of the interface. This assumes that the
total interfacial volume is available to each of the block
junctions. If, as is likely, the block junctions are restricted
to "cells", with each cell excluding neighbouring junctions, then
an additional term $-Nk$ is required in Eq. (7).

$\underline{\Delta S(\text{constraints})}$. The entropy change associated with the restriction
of segments to their domain regions is evaluated from the probabil-
ity that the A- and B-chains starting from an initial point in the
interface will have all of their respective segments within their
own domain space. Each chain is prohibited from being in the
primary domain space of the other species other than in the mixed
interfacial region. The required probability functions can be
evaluated by obtaining solutions to the diffusion equation (here

in one-dimensional form) for the statistics of random flight chains,

$$\delta P_n(x,x_o)/\delta n = (1^2/6)\delta^2 P_n(x,x_o)/\delta x^2 \qquad (8)$$

where $P_n(x,x_o)$ is the probability of finding the second end of a chain of n statistical segments at x (within dx) with the first end at x_o, and 1 is the length of a statistical segment. Since the chains are constrained to stay out of the region occupied by the other component, "absorbing" boundary conditions are appropriate (7), i.e. $P_n(x'_k,x_o) = 0$, where x'_k is a domain boundary (k = A,B).

With this boundary condition, the solution to Eq. (8) for a k-th component having n_k statistical segments is

$$P_k(x,x_o)=(2/T_k)\sum_m \sin(m\pi x_o/T_k)\sin(m\pi x/T_k)\exp-(m^2\pi^2 n_k 1^2/6T_k^2)dx$$

$$(9)$$

where absorbing boundaries are placed at x = 0 and x = T_k.

We have taken λ as a measure of the thickness of the interfacial region, i.e. the "linear" thickness. Although the block junctions are restricted to the interfacial region, a restriction to a thickness λ is too restrictive, so the interfacial thickness that will be allowed for the placement of the junctions will be expressed as $z\lambda$, where z is a parameter (of the order of 1) to be established. In order to obtain ΔS(constraints) with Eq. (9), we must obtain the joint probability $P_A P_B$ that (1) the two chain types can have a common origin anywhere within the interface, (2) satisfy the boundary conditions, and (3) satisfy the condition that space be uniformly filled with segments (constant density). If the chain origin for one species is taken as x_o, the origin for the other species relative to its coordinate system will be at $z\lambda - x_o$. Since the chain origins (block junctions) can be anywhere within the interfacial region, we allow for this by integrating over a product of chain origin terms from Eq. (9) to obtain the probability that the chain origins can be anywhere within z,

$$P_{AB}(z\lambda) = (z\lambda)^{-1}\int_o^{z\lambda} \sum_{m,p} \sin(m\pi x_{Ao}/T_A)\sin(p\pi[z\lambda-x_{Ao}]/T_B)dx_{Ao}$$

$$= \sum_{m,p} mp(6T_A T_B)^{-1}(\pi z\lambda)^2 \qquad (10)$$

where for simplicity the assumption has been made that the interface is thin with respect to domain dimensions.

In order to introduce the constraint that the domain space be of uniform segment density, it is necessary to establish the relationship between chain and domain dimensions that gives the required uniform density. The number density $\Omega_\sigma(x,x_o)$ of segments at x of a chain of σ total statistical segments wit one end at x_o is given by

$$\Omega_\sigma(x,x_o) = \sum P_n(x,x_o) P_{\sigma-n}(V,x) \qquad (11)$$

where V is the volume available to the chains. For lamellar domains, the first term within the summation is Eq. (9), while the second term is obtained from Eq. (9) by integrating the position of the free end of the chain over V. Eq. (11) can be evaluated for various ratios of chain dimension to domain dimensions, i.e. $\sigma l^2/T^2$, thus allowing the evaluation of that value which gives the desired uniform segment density. For the lamellar system, this value is $\sigma l^2/T^2 = 2$. It is now necessary to introduce this value as a constraint which will guarantee that the uniform density requirement is obeyed, even though chain perturbations occur in the system as a result of e.g. differences in block molecular volumes, addition of solvents into the domain structure, etc. This is done by specifying that the free end of the chain must be found (on the average) at that position where it is when segmental densities are uniform. Thus, when $\sigma l^2/T^2 = 2$, the most probable position $\langle x \rangle$ of the free end of a chain in a uniform density lamellar system is at $\langle x \rangle/T = 1/2$, i.e. at the centre of the domain. With this value in Eq. (9), and with the results of Eq. (11), the desired probability function for the constraints on segmental placement becomes

$$P_A P_B = (2/3)(\pi z \lambda)^2 (T_A T_B)^{-2} \sum_{\substack{m,p \\ \text{odd}}} (-1)^{(m+p+2)/2} mp\exp{-[(\pi^2/6)(m^2\sigma_A l^2/T_A^2 +}$$
$$p^2\sigma_B l^2/T_B^2)] \, dx_A dx_B \qquad (12)$$

from which

$$\Delta S(\text{constraints}) = Nk \ln(P_A P_B) \qquad (13)$$

For lamellar domains, $\Delta S(\text{placement})$ (=Nk $\ln(V_i/V_o$ from Eq. (7)) becomes

$$\Delta S(\text{placement}) = Nk \ln(2z\lambda/(T_A + T_B)) \qquad (14)$$

<u>$\Delta G(\text{domain formation})$</u>. The free energy change for the transformation of the a di-block copolymer system from a randomly mixed state to the domain state:
$\Delta G(\text{domain formation}) = \Delta H(\text{mixing}) - T[\Delta S(\text{placement}) + \Delta S(\text{constraints})]$,
and can be evaluated from Eqs. (6), (13) and (14). The resulting equation is a function of two independent domain variables, the thickness λ of the interface region and the thickness T_k of the domains (the thicknesses of the two domain regions are directly related to one another by their relative molecular volumes, i.e.

$T_A/T_B=V_A/V_B$ for di-block copolymers or $T_A/T_B=2V_A/V_B$ for A-B-A tri-block copolymers). The equilibrium interfacial and domain thicknesses are obtained by setting the partial derivatives of the free energy with respect to λ and T_A (or T_B) equal to zero. The resulting two equations can then be solved simultaneously to obtain equilibrium properties. Fig. 3 shows a comparison of results calculated from this theory with experimental results for S-1 (polystyrene-polyisoprene) block copolymers.

The molecular and thermodynamic parameters used to obtain the theoretical results are: $\chi_{s-1} = .150\ N_s$ (8), where N_s is the degree of polymerization of the polystyrene block, $K_s=4.49\times10^{-17}M_s cm^2$ and $K_1 = 5.13\times10^{-17}M_1 cm^2$, where $K_k= q_k l^2/M_k$, the M's are molecular

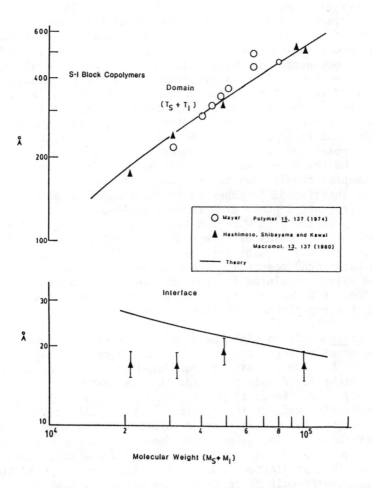

Fig. 3 Comparison of theory and experiment for domain and interfacial dimensions vs. molecular weight

weights, and the polymer densities used are $\rho_s=1.05$ gm cm^{-3} and $\rho_1=0.92$ gm cm^{-3}. Also the parameter z was taken to be 2 (results are not particularly sensitive to reasonable values for this parameter).

Fig. 3 shows a remarkably good agreement between theory and experiment for the domain repeat thickness $(T_A + T_B)$ as a function of molecular weight (probably better agreement than should be expected given the problems associated with obtaining accurate values for interaction parameters χ_{AB}). The agreement between theory and experiment for the interfacial thickness does not appear as good, particularly at low polymer molecular weights. Theory predicts that the interfacial thickness will be a function of molecular weight, the thickness decreasing with increasing molecular weight. However, experimental evidence (SAXS (9) and SANS (10)) appear to show that the interfacial thickness is independent of molecular weight. This fact presents a dilemma, since any theory that assumes a Flory–Huggins pair–wise interaction between the components (as here) will find the heat of mixing for a lamellar system to be directly proportional to the interfacial thickness (and to the total interfacial volume). Also, in any realistic theory, the interfacial thickness will appear in logarithmic form in the placement entropy and probably (as here) also in the constraint entropy terms. These considerations guarantee a molecular weight dependence for the interfacial thickness. On the other hand, the dilemma may not be real since the experimental results may be in error. The experimental determination of interfacial thicknesses by either SAXS or SANS techniques presents a grave experimental problem, in that data must be taken in the high Q region (Q is the scattering vector), with an attendant problem of background scattering. As yet, there is no definitive way to subtract the background scattering, so the empirical background scattering function used for subtraction will affect the value calculated for the interfacial thickness. At the present time, it is not clear whether the predictions of theory are correct or the experimental results.

<u>Domain Formation</u>. By setting the above free energy of formation of domains equal to zero, criteria for their formation may be established. For the S-1 system considered above, the critical molecular weight for domain formation in this system is approximately 5000. An early prediction from theory (3) comparing critical molecular weights for domain formation with critical molecular weights for simple phase separation of the same components stated that the molecular weights for domain formation would have to be 3-5 times higher than would be required for simple phase separation. This prediction seems to be established experimentally, although it is difficult to establish the exact critical molecular weights. It should be remembered that the theory is for a "fully-developed" domain system, whereas criteria for domain formation

could also be established for the "onset conditions" where aggregation of the block components just commences but the domain structure has not formed. The latter problem has been treated by Leibler (11).

<u>Domain Shapes</u>. It has been mentioned that there are three equilibrium shapes that domains from simple block copolymers can adopt, i.e. 1) lamellar, 2) cylindrical, and 3) spherical. The above theory for lamellar domains can be extended to include the cylindrical and spherical forms by appropriate modifications to take account of the change in geometry. However, the resulting somewhat more complex expressions for the free energy of formation of each of the domain types will not be reproduced here. By comparing the free energy of formation for each of the domain shapes for various molecular parameters, e.g. molecular weights, the equilibrium morphology for those parameters can be established. The first theoretical calculations for the relative stability of the various domain morphologies as a function of block molecular weights was by Meier (12). His results showed that each morphological form would be stable only over a restricted composition range. From experimental observations, Molau (13) had also reached this conclusion. Although that early theory has undergone a number of modifications in recent years, the general conclusions from it have not changed, nor have the experimental observations. Fig. 4 shows the results of recent calculations. As the volume fraction of the minor component ("A" in the Figure) increases, the morphology with the lowest free energy (and hence being thermodynamically most stable) changes from spheres to cylinders to lamellae. Of course, above 50% A, there will be an inversion in the role of the two components.

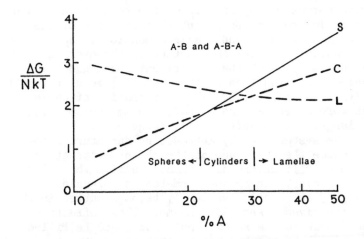

Fig. 4 Domain morphology vs. composition

It is of interest to inquire of the molecular basis for such morphological transformations. The overriding factor responsible is that due to chain perturbations brought about by differences in molecular volumes. Consider first a system where the molecular volumes of the components of a di-block copolymer are equal. From symmetry considerations alone, a lamellar morphology must be the equilibrium morphology since it is the only morphology which is symmetrical across the microphase boundaries – as is required by the equal molecular properties. However, when the molecular volumes begin to differ, perturbations of chain dimensions will occur. Lamellar domain thicknesses are directly related to the molar volumes of the components

$$T_A/T_B = V_A/V_B \tag{15}$$

At the same time, theory (and confirmed by the excellent agreement between theory and experiment) indicates the thickness is related to chain dimensions by $T_A = K_A (\overline{\sigma l^2})_A{}^\beta$, where K_A is a constant and β is approximately 2/3. The chain dimension $(\overline{\sigma l^2})_k$ of the k-th component as it exists in the domain system can be written in terms of the unperturbed dimensions $(\overline{\sigma l^2})_{ko}$ as $(\overline{\sigma l^2})_k = \mu_k (\overline{\mu l^2})_{ko}$, where μ_k is a perturbation parameter relating perturbed and unperturbed dimensions. Since $(\overline{\sigma l^2})_{ko} = K'M_k$, the domain thicknesses T_A and T_B for components A and B can then be written as

$$T_A = K_A{}''(\mu_A M_A)^\beta \quad \text{and} \quad T_B = K_B{}''(\mu_B M_B)^\beta \tag{16}$$

Assuming $K_A{}'' = K_B{}''$ and equal densities for the two polymers, combining Eqs. (15) and (16) gives, $\mu_A/\mu_B = (M_A/M_B)^{1/2}$, with β taken as 2/3. Thus, if molecular weight ratios differ from 1.0, perturbations of chain dimensions will occur – with chains of the higher molecular weight species being expanded and the lower molecular species being contracted, and with corresponding effects on the respective domain dimensions. It is of interest to point out that in spite of the perturbations of individual chain and domain dimensions, the opposing effects essentially cancel one another in the overall domain repeat spacing $T_A + T_B$, such that the repeat spacing becomes essentially a single-valued function of the overall molecular weight of the block copolymer, as in Fig. 3. The perturbation of chain dimensions becomes more severe as the mismatch in molecular weights (volumes) becomes larger. At a sufficiently large mismatch in molecular weights, the system can, by introducing curvature into the domain shape, reduce the magnitude of the chain perturbation energy. The curvature is first one-dimensional to give the cylindrical shape to the domains, but as the mismatch in molecular weights becomes larger, the system eventually becomes curved in the second direction which gives a spherical shape. These effects can ultimately be traced back to the requirement that space be filled uniformly. Chains will be perturbed rather than the system have regions of non-uniform density, but the chain perturbations are affected by the curvature of the constraining boundaries.

3   BLOCK COPOLYMERS AND SOLVENTS

3.1   Non-Selective Solvents

The addition of a compatible but non-selective solvent to a
block copolymer system will have the effect of <u>reducing</u> equilibrium
domain sizes.   Although a reduction in size with the addition of a
solvent may appear somewhat puzzling, it is observed experimentally
(14) and is predicted by theory (15).   The effect is easily
rationalized by noting that the size of domains is a function of
the interfacial enthalpy, with the system attempting to minimize
this positive contribution to the free energy of the system.   The
interfacial enthalpy is proportional to the interaction parameter
χ_{AB} and the volume of the interface (the product of the interfacial
area and the interfacial thickness).   Hence the interfacial
enthalpy can be reduced if the interfacial area is decreased, as
will occur if the domain thickness is increased, and this increase
in thickness will be proportional to the interaction parameter, i.e.
the larger χ_{AB} is, the larger will be the domain thickness.   In
turn, if χ_{AB} is reduced, the domain thickness will decrease.   This
reduction in χ_{AB} will occur with the addition of a non-selective
solvent, since the A-B contacts will be reduced in proportion to
the amount of added solvent, e.g. the effective χ'_{AB} becomes
$\chi'_{AB}=\phi_p\chi_{AB}$, where ϕ_p is the volume faction of polymer.   Fig. 5
shows a comparison of theory and experiment for the effects on
domain dimensions produced by adding a non-selective solvent
(toluene) to a styrene-isoprene block copolymer system.   The

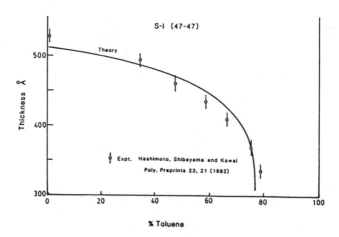

Fig. 5   Domain size vs. concentration of non-selective solvent

theoretical results were obtained using the theory outlined in the earlier sections, using χ'_{AB} in place of χ_{AB}. The agreement of theory and experiment is excellent.

3.2 Selective Solvents

In the earlier sections, it was noted that the shape of domains, e.g. lamellae, cylinders or spheres, was governed primarily by the relative volume fraction of the two block components. It is not surprising then that the addition of a selective solvent to one or the other components of a block copolymer can affect the morphology merely by changing the effective volume fractions of the components. The results of a theory (16) for the effect of selective solvents on morphology of a block copolymer is shown in Fig. 6, for a polymer having equal block molecular weights (results for unequal block molecular weights have been published elsewhere (17)). In this Figure the relative free energies of each domain shape is shown as a function of the amount of selective solvent added wholly to the B-component to the left and wholly to the A-component to the right of centre (the results are not symmetric about the centre line since even on the A-rich side it is assumed that the A-component is the dispersed component. The results for B-dispersed on the A-rich side would appear as the mirror image of the B-rich side). The equilibrium morphology for the pure polymer is, of course, lamellar. But now consider the morphological changes produced by the addition of solvent to the B-component. The results shown in Fig. 6 indicate that when the polymer concentration is reduced to

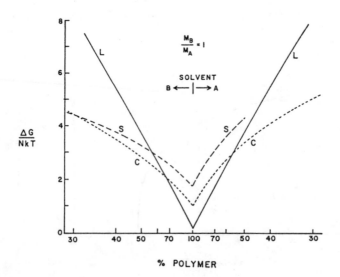

Fig. 6 Morphology vs. concentration of selective solvent

about 65%, the equilibrium morphology becomes cylindrical, and at polymer concentrations below about 30% the equilibrium morphology is spherical. This general trend is to be expected, but the results do provide an explanation for the observed (18) profound effects of casting solvents on the mechanical properties of cast block copolymer films. Assume that the block copolymer of Fig. 6 is dissolved in a selective solvent is for the B-component. As the solvent is removed to form a cast film, the polymer concentration will eventually become high enough that microphase separation will occur. The concentration at which microphase separation occurs will naturally depend on the particular interaction of the solvent with the polymer components. If the interaction is such that microphase separation occurs at polymer concentrations below about 30%, Fig. 6 shows that the equilibrium morphology that would form would be spherical. On the other hand, if microphase separation occurred with polymer concentrations in the range 30-65%, the morphology would be cylindrical, and above 65% lamellar. After a particular domain morphology has formed, and solvent continues to be removed, the transformation to another perhaps more stable morphology is essentially prohibited because of a large kinetic barrier that must be overcome for morphological transformations to occur. Thus, the transformation from a spherical morphology to a cylindrical morphology would require the diffusional transport of one block species through the other — with which it is incompatible. This diffusional barrier will prevent the transformation, and the system will be "locked" into the non-equilibrium spherical morphology even as solvent is completely removed. On the other hand, it is possible that at the concentration where microphase separation occurred the cylindrical morphology would be the equilibrium morphology. If so, upon the complete removal of solvent, this now non-equilibrium morphology would also be "locked in". Also possible is the formation of the equilibrium (in the absence of solvent) lamellar morphology if microphase separation occurred at a high polymer concentration in the selective solvent. Thus, depending on the casting solvents, the recovered polymer films could have completely different morphologies and concomitantly completely different mechanical properties. The formation of non-equilibrium states is a ubiquitous property of block copolymers. This has lead to many misconceptions in the literature since these states appear to be stable, e.g. annealing has little if any effect in bringing about a transformation from one morphological form to another, more stable, form.

4 MOLECULAR ARCHITECTURE, DOMAIN MORPHOLOGY AND MECHANICAL PROPERTIES

Early investigations (1) of block copolymers concentrated exclusively on di-block copolymers, and thus missed the discovery of the enormous influence of molecular architecture on mechanical properties of microphase separated systems. This influence can be demonstrated by comparing the mechanical properties of a group of polymers whose overall composition is the same, but whose molecular architecture differs. We shall assume for this comparison that the copolymer components are styrene (25%) and butadiene (75%), and the copolymers each have an overall molecular weight of 10^5.

Influence of Molecular Architecture on Mechanical Properties

25% Styrene
75% Butadiene
MW = 10^5

Molecular Architecture	Tensile Strength ($N\ m^{-2} \times 10^{-5}$)
Random SB	~ 0
	50 (when vulcanized)
Di-block S-B	~ 0
Tri-block S-B-S	300
Tri-block B-S-B	~ 0
Multi-block $(S-B)_n$	~ 0 (n>3)

The enormous differences between the tri-block S-B-S polymers and the di-block S-B or tri-block B-S-B polymers is due to the existence in the S-B-S system of a virtual network of the polybutadiene chains, with the polybutadiene chains being terminated on each end by the domains of rigid polystyrene. The system then resembles a filled, crosslinked elastomer. However, such networks cannot be formed by the di-block S-B or the "inverted" tri-block B-S-B copolymers, and as a result these materials have little strength. Multi-block polymers $(S-B)_n$ with n>3 also have little strength since the molecular weights of the polystyrene blocks are then below the critical molecular weight for domain formation. These polymers would resemble the random copolymer.

Morphology and Mechanical Properties. It was recognized from the earliest investigations (19-21) of the physical properties of tri-block copolymers that composition played a dominate role in governing properties. This is shown in Fig. 7, which shows tensile data (of Holden, et. al (19)) of S-B-S copolymers as a function of composition (at constant overall copolymer molecular weight). At the lowest concentration (13%) of the polystyrene component, the material is essentially of zero strength because the molecular weight of the polystyrene blocks is too low for domains to form. At 28% polystyrene, the material is a high-strength elastomer, with a domain morphology that undoubtedly is spherical. The deformation of this material is concentrated in the rubbery polybutadiene matrix, with little effect on the rigid polystyrene spheres (until fracture occurs). When the concentration of the polystyrene is increased to 39%, the tensile behaviour becomes much different, in

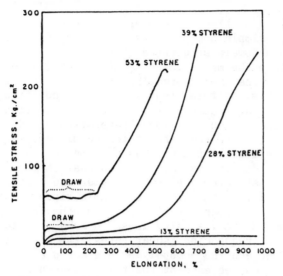

Fig. 7   Tensile properties of S-B-S polymers vs. composition.
(Adapted from Ref. 19)

that the initial modulus is much greater and "drawing" occurs.
The change in mechanical behaviour is the result of a morphological
transformation of the polystyrene domains from a spherical to a
cylindrical morphology. The macroscopic deformation of a system
with polystyrene cylinders must involve the deformation and drawing
of the (rigid) polystyrene component, with effects due to these
properties thus appearing in the overall tensile behaviour. The
same behaviour is noted in the sample with 53% polystyrene, with
the same explanation, except that the morphology is lamellar.

An S-B-S system having a cylindrical polystyrene morphology
resembles a fibre-reinforced elastomer. It has been shown by
Folkes and Keller (22) that such materials can be oriented by
extrusion techniques, to yield "macroscopic crystals" in which the
domain order is essentially perfect over macroscopic dimensions (mm).
They found that the mechanical properties of such ordered materials
were highly anisotropic (moduli in the axial and transverse directions differing by two orders of magnitude), as would be expected
for a system consisting of oriented polystyrene "fibres" imbedded
in a rubbery matrix. The mechanical properties also obeyed the
theoretical properties for a fibre-reinforced composite, using the
tensile moduli of macroscopic polystyrene and polybutadiene. Thus
it appears that the macroscopic mechanical properties of polystyrene
and polybutadiene are retained even on the microscopic scale of the
domains, i.e. a few hundred Å.

190

5 BLOCK COPOLYMER RHEOLOGY

The melt rheology of block copolymers is profoundly different
from the rheology of simple homopolymers or copolymers if the block
copolymer exists in a domain–structured state (23). The reason
is quite simple, since the macroscopic deformation of a domain–
structured state requires the deformation or destruction of the
structured state. This, in turn, requires the forced mixing of
the components (whose incompatibility lead to the formation of the
domain state) and hence to an additional rate of energy dissipation
that will be proportional to the rate of forced mixing. The rheological behaviour can thus be pictured as a formation–destruction
process involving the domain system. The contrast in the rheological
behaviour of a domain–structured block copolymer and a simple homopolymer is shown in Fig. 8, in which the dynamic viscosities are
shown as a function of frequency (the viscosity behaviour appears
similar if the steady state viscosity is shown as a function of
shear rate). At high frequencies (shear rates) the viscosities
of the block copolymer and the homopolymer are not dramatically
different. This is a consequence of the time–dependence for
reformation of the domain state, i.e. at the higher rates of
deformation, the system does not have time to reform the domain
state and so the behaviour is similar to that of a randomly mixed
(co)polymer. However, at low frequencies (shear rates), the homopolymer shows the typical Newtonian behaviour, while the viscosity
of the block copolymer appears to increase without limit as the

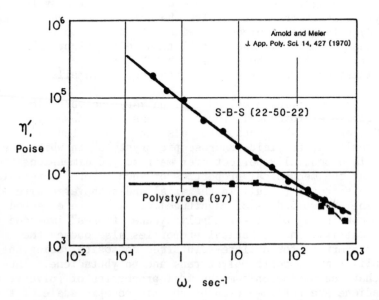

Fig. 8 Comparison of dynamic viscosities vs. frequency of a
homopolymer and a block copolymer. (MW x 10^{-3})

frequency (shear rate) decreases. That this behaviour is due to a domain structure is clearly shown by the date of Chung and Gale (24) and Gouinlock and Porter (25), who independently showed with a block copolymer having a critical temperature for domain formation within an experimentally accessible range that the rheological behaviour was typical of an ordinary homopolymer above the critical temperature (no domains), and typical (as in Fig. 8) of a block copolymer with domains below the critical temperature. The rheological behaviour of the domain state can be rationalized as follows. As mentioned above, the macroscopic deformation of the domain system requires the forced mixing of the two incompatible components. The rate of energy dissipation $\dot{E}_1$ accompanying this mixing will be proportional to the incompatibility (χ_{AB}) and to the rate of mixing, i.e. $\dot{E}_1 \sim \chi_{AB} \dot{\gamma}$ since the rate of mixing will be proportional to the shear rate $\dot{\gamma}$. The dissipation of energy due to ordinary viscoelastic processes $\dot{E}_2$ can be expressed as $\dot{E}_2 = f\dot{\gamma}^2$, where f is a segmental frictional factor (which at low shear rates can be taken as independent of shear rate). The total energy dissipation rate $\dot{E}$ for a block copolymer can then to a first approximation be taken as the sum of these two processes, one $\dot{E}_1$ due to thermodynamics and the other $\dot{E}_2$ due to the usual viscoelastic processes for polymers. Since the viscosity of the system is defined by $\eta = \dot{E}\dot{\gamma}^{-2}$, the above contributions to the rate of energy dissipation of the block copolymer system show that the viscosity will have a constant term ($\sim$f) and a term $\chi_{AB}\dot{\gamma}^{-1}$. The second term arises from the incompatibility of the components and the disruption of the domain structure and will at low shear rates outweigh the first term because of the inverse dependence of this dissipation term on shear rate. Thus, this simple analysis indicates that at low shear rates a block copolymer should have a viscosity that increases continuously with decreasing shear rate – as is experimentally observed (see Fig. 8). Also, the viscosity is predicted to increase with the degree of incompatibility χ_{AB} of the components. This can also be demonstrated experimentally in the following manner (26). The hydrogenation of the centre block of an S-1-S block copolymer (1=polyisoprene) converts it to an S-EP-S copolymer (EP=an alternating copolymer of ethylene and propylene), while further hydrogenation of the polystyrene component will convert the polymer to a cH-EP-cH block copolymer (cH=polyvinylcyclohexane). The incompatibility of the block components is in the order $\chi_{S-EP} > \chi_{S-1} > \chi_{cH-EP}$, which is exactly in the same order as the viscosity of these polymers.

The viscosity of the domain-structured polymers is predicted to be inversely proportional to the shear rate at low shear rates, $\eta \sim \dot{\gamma}^{-1}$. This is equivalent to the existence of a yield stress in the system.

6   INTERFACIAL AND SURFACTANT PROPERTIES

Block copolymers are the polymeric analogues of low molecular weight surfactant molecules. However, it has only been in very recent years that a general appreciation of the efficacy of block copolymers as interfacially active materials has developed as a result of both theoretical and experimental investigations. These investigations have shown: (a) interfacial activity in polymeric systems (27-30), (b) surface activity (31-34), (c) "compatibilizing" activity in polymer blends (35-37), and (d) dispersant activity (38-39) of block copolymers. These areas are under active investigation in many laboratories at the present time, and many new developments may be expected in the future. Unfortunately, space limitations preclude a more complete discussion here of theoretical and experimental phenomena involving the interfacial activity of block copolymers.

7   SUMMARY

The discovery and following investigations of block copolymers over the past few decades has revolutionized polymer science and technology. Such polymers that were laboratory curiosities twenty years ago are now produced commercially at rates of hundreds of million pounds per year. But equally importantly, such polymers have made apparent the tremendous importance of molecular architecture on physical properties, and have shown that it is possible to create polymers with unique properties by combining composition and architecture. These polymers offer the opportunity that has never existed before to design, a priori, polymers for desired end-use properties.

In this overview of the simple di- and tri-block copolymers, an attempt has been made to describe some of the unique morphological and physical properties of such copolymers and then to indicate that these unique properties can be understood or rationalized by theory.

REFERENCES

1. a. Sadron, C. Angew. Chem. Internat. Edition 2 (1963) 248.
    b. Skoulios, A., G. Finaz and J. Parrod. Comp. rend. 251 (1960) 739.
    c. Gallot, B., R. Mayer and C. Sadron. Comp. rend. 263 (1966) 42.
2. Holden, G. and R. Milkovich. Belgian Patent 627,652; C.A., 60 (1964) 14713f.
3. Meier, D.J. J. Poly. Sci.: Part C 26 (1969) 81.
4. Huggins, M.L. J. Phys. Chem. 46 (1942) 151.
    Flory, P.J. J. Am. Chem. Soc. 10 (1942) 51.
5. Cahn, J.W. and J.E. Hilliard. J. Chem. Phys. 28 (1958) 258.
6. a. Helfand, E. and Y. Tagami. J. Poly. Sci. B 9 (1971) 741; ibid. 56 (1971) 3592; ibid. 57 (1972) 1812.
    b. Helfand, E. Macromol. 8 (1975) 552.
7. DiMarzio, E.A. J. Chem. Phys. 42 (1965) 2101.
8. Rounds, N.A. Doctoral Dissertation, Univ. of Akron (1970).
9. Hashimoto, T., M. Shibayama and H. Kawai. Macromol. 13 (1980) 1237.
10. Richards, R.W. and J.L. Thomason. Polymer 22 (1981) 581.
11. Leibler, L. Macromol. 13 (1980) 1602.
12. Meier, D.J. ACS Poly. Preprints 11 (1970) 400.
13. Molau, G.E. In Block Copolymers, ed. by S.L. Aggarwal (New York, Plenum, 1970) p. 79.
14. Hashimoto, T., M. Shibayama and H. Kawai. ACS Poly. Preprints 23 (1982) 21.
15. Meier, D.J. Unpublished results.
16. Meier, D.J. In Block and Graft Copolymers, ed. by J.J. Burke and V. Weiss (Syracuse, NY, Syracuse University Press, 1973, with later unpublished modifications) p. 105.
17. Cowie, J.M.G. In Developments in Block Copolymers-1, ed. by I. Goodman (London, Applied Science Publishers, 1982), p.17.
18. Brunwin, D.M., E. Fischer and J.F. Henderson. J. Poly. Sci: Part C 26 (1969) 135.
19. Holden, J., E.T. Bishop and N.R. Legge. ibid. 26 (1969) 37.
20. Morton, M., J.E. McGrath and P.C. Juliano. ibid. 26 (1969) 99.
21. Beecher, J.F.,L. Marker, R.D. Bradford and S.L. Aggarwal. ibid. 26 (1969) 117.
22. Folkes, M.J. and A. Keller. Polymer 12 (1971) 222.
23. Arnold, K.R. and D.J. Meier. J. App. Poly. Sci., 14 (1970) 427.
24. Chung. C.I. and J.C. Gale. J. Poly. Sci., Poly. Phys. Ed. 14 (1976) 1149.
25. Gouinlock, E.V. and R.S. Porter. Poly. Eng. Sci. 17 (1977) 535.
26. Holden, J. Personal communication.
27. Noolandi, J. and K.M. Hong. Macromol. 15 (1982) 482.
28. Inoue, T. and D.J. Meier. Unpublished work.

29. Cantor, R. Macromol. 14 (1981) 1186.
30. Riess, G., J. Nervo and D. Rogez. Poly. Eng. Sci. 17 (1977) 634.
31. Gia, H.B., R. Jerome and Ph. Teyssie. J. Poly. Sci.: Poly. Phys. Ed. 18 (1980) 2391.
32. Owen, M.J. and T.C. Kendrick. Macromol. 3 (1970) 458.
33. Gaines, G.L., Jr. and G.W. Bender. ibid. 5 (1972) 82.
34. O'Malley, J.J., H. Ronald Thomas and G.M. Lee. ibid. 12 (1979) 996.
35. Ramos, A.R. and R.E. Cohen. Poly. Eng. Sci. 17 (1977) 639.
36. Riess, G., J. Kohler, C. Tournut and A. Banderet. Makromol. Chem. 101 (1967) 58; Rubber Chem. Tech. 42 (1969) 447.
37. Ikada, Y., F. Horii and I. Sakurada. J. Poly. Sci.: Poly. Chem. Ed., 11 (1973) 27.
38. Meier, D.J. J. Phys. Chem. 71 (1967) 1861.
39. Napper, D.H. Polymeric Stabilization of Colloidal Dispersions (New York, Academic Press, 1983).

COLLOIDAL BEHAVIOUR AND SURFACE ACTIVITY OF BLOCK COPOLYMERS

D.J. Wilson*, G. Hurtrez and G. Riess

Ecole Nationale Supérieure de Chimie de Mulhouse
3, rue Alfred Werner, 68093 Mulhouse Cedex (France)

Block copolymers whose methods of synthesis have been covered in a preceeding article (1) represent one of the fastest growing areas of polymer science, interest stemming principally from their unique properties in a solvated, colloidal or solid state.

This article attempts to outline the behaviour of these materials, essentially in the colloidal state, both from a fundamental point of view and in relation to their applications ; we shall deal here primarily with bi- and triblock copolymers.

1 MICELLES

Dilute solutions of block copolymers in selective solvents (i.e. good solvent for one block but a precipitant for the other) behave like typical amphiphiles where the copolymer molecules aggregate reversibily to form "micelles" ; this can be considered analogous to classical surface-active agents. Both monomolecular and multimolecular micelles may form and these species consist generally of a swollen core of the insoluble block surrounded by a flexible fringe of soluble block(s) as shown in Fig. 1. The latter class believed to form by a closed association process, are generally spherical in shape with a narrow size distribution.

Work done prior to 1976 in this field has been covered extensively by TUZAR and KRATOCHVIL (2) with additional reviewing by PRICE (3).

*on leave from the University of Sussex (England)

196

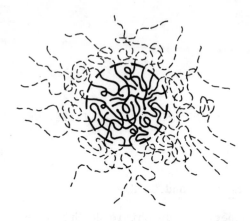

<u>Fig. 1</u> : Sketch of a multimolecular block copolymer micelle in selective solvent (the insoluble block is surrounded by the solvated block).

1.1 Micelle Formation and Surface Activity

Critical micelle concentration : The critical micelle concentration (c.m.c.) of a surfactant is defined as the concentration at which micelles begin to form. Although it is the best method for the characterisation of surface-active agents, values obtained for c.m.c. are somewhat dependant on the technique employed. Block copolymers present an additional number of difficulties. Firstly the c.m.c. for these compounds is often too low to be measured accurately and secondly the copolymer in solution has both unimer and micelles over a wide range of concentration. As a result it is often the concentration at which a copolymer exists mainly as micelles which is measured.

Although TUZAR and KRATOCHVIL (4) have reported a critical concentration similar to c.m.c. for polystyrene-<u>b</u>-polybutadiene-<u>b</u>-polystyrene block copolymers by light scattering, most studies use surface tension to determine c.m.c. As shown in Fig. 2 the surface tension of the block copolymer solution below c.m.c. decreases linearly with increasing concentration and attains an almost constant value when the concentration exceeds the c.m.c. Such plots have been observed for polystyrene-<u>b</u>-poly(ethylene oxide) (5, 6) and star-shaped block copolymers based on polyisoprene and poly(ethylene oxide) (7).

Other studies of interest include AVERBUKH (8) who has shown the existence of a c.m.c. for styrene-butadiene block copolymers in heptane and FLORENCE et al (9, 10) who reexamined

the association behaviour of block copolymers using surface
tension and photon correlation spectroscopy.

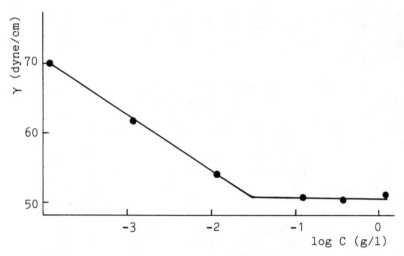

Fig. 2 : Plot of equilibrium surface tension vs log C for an
aqueous solution of polystyrene-b-poly(ethylene oxide) (5)
$\overline{M}_n$ = 3,300   PEO (mole) / PS (mole) : 11.9

Aggregation number :   Aggregation number is defined as the number
of copolymer unimers in a micelle and may be calculated by measurement of the micellar hydrodynamic radius in association with
viscosity data. To this end ORANLI et al (11) have utilised
photon correlation spectroscopy to study a number of styrene-butadiene block copolymer micelles in selective solvents and related
core characteristics to aggregation number. These authors have
also observed the independance of aggregation number with temperature and its dependance on the percentage insoluble block of
the parent copolymer.

Usually large particles may be observed in some cases.
Large particles causing milky opalescence at the onset of micelli-
sation have been observed for example for the system : polystyrene
-b-polybutadiene-b-polystyrene in ethyl acetate (12). This behaviour has been attributed to the transient formation of large
spherical particles (unlike compact micelles) somewhat unstable
with time (13). The same authors (14) discuss the possibility
of a multilayer spherical micelle with alternating concentric
layers of the two blocks, which impart a poor stability to these
particles in solution. Other unusual observations have been
made by PRICE et al (15) who noted the existence of "wormlike"
micelles.

Note that as yet there is no evidence of block copolymer micelles possessing a crystaline or glassy core.

Surface activity : The two incompatible blocks make these copolymers interfacially active. The block copolymer molecules tend to adsorb at various interfaces-solid/liquid, liquid/liquid, gas/liquid etc...

The adsorption characteristics and surface activity of various water-soluble block copolymers has been studied. Generally, surface tension and the pressure - area isotherms for monolayer adsorption of block copolymers at the interfaces have been the methods of study. Monolayer studies on block copolypeptides at the air-water interface have been carried out by PHILLIPS ét al (16). The interfacial properties (surface pressure and surface potential) of a series of synthetic block copolypeptides were investigated. IKADA et al (17) have examined the monolayer formation of block and graft copolymers at the air/water interface by means of pressure-area isotherms, for polyvinylalcohol-b-polystyrene copolymers. They observed that the molecule was orientated such that the dissolved polyvinylalcohol chains (in water) were supported at the interface by the compact, monomolecular particles of polystyrene on the surface.

1.2 Micelle Characterisation

A number of techniques may be employed in order to obtain information as to the size, shape, size distribution, etc. of a particular micellar system ; Table 1 gives a summary of these methods.

New techniques such as small angle neutron scattering may prove useful, since in contrast to SAXS the possibility exists of producing large contrasts by deuteration (3). CANDAU et al (35) have utilised this technique to investigate the styrene-ethylene oxide graft copolymer/water system and HIGGINS et al (36) have used this method to investigate the dimensions of styrene-dimethylsiloxane block copolymers adsorbed on dispersed polymer particles.

In situations where both unimer and micelle are present techniques such as light scattering may give erroneous results ; to circumvent this problem TUZAR et al (37) have used sedimentation data to determine the relative proportion of micelle to unimer and have applied their sedimentation results in conjunction with light scattering to study micellar systems.

TABLE I

Technique of Study	Block Copolymer	Solvent System	Ref.
Light Scattering	PS-PBut	n-alkanes	18
	PS-PIso	n-decane	19
	PS-PMM	acetone	20
Electron Microscopy	PS-PIso	white oil	21
	PS-PMM	acetonitrile	22
Small Angle X-ray Scattering	PS-PBut	butanone, n-heptane	23
	PS-PBut	n-tetradecane	24
Sedimentation	PS-PBut	THF + methoxyethanol	25
	PS-PIso	DMF, n-hexane	26
Ultracentrifugation	PS-PBut-PS	dioxane/ethanol	2,27
Ultramicroscopy	PS-PIso	DMA, n-decane	28
Viscosity	PS-PBut	butanone, n-heptane	29
	PProp-PEO	water	30
Nuclear Magnetic Resonance	PS-PBut	n-heptane	31
	PS-PDS	n-alkanes	32
Gel Permeation Chromatography	PS-PIso	DMA	33
	PS-P(E/P)	base lubricating oil	34
Photon Correlation Spectroscopy	PS-PBut	DMF,DMA,EMK,n-alkanes	11
	PS-PIso	DMA, N-decane	28
Turbidity	PS-PEO	water	6

PS : polystyrene, PBut : polybutadiene, PIso : polyisoprene, PEO : poly(ethylene oxide), PDS : polydimethylsiloxane, P(E/P) : poly(ethylene/propylene), PProp : polypropylene, PMM : polymethyl methacrylate.

1.3 Influence of Temperature, Solvent etc.

The influence of temperature, solvent and the molecular characteristics of the parent copolymer on micellar species has been subjected to some scrutiny utilising primarily light scattering (2, 3, 4, 8, 38).

At temperatures of about 20°C a block copolymer micellar solution is composed mainly of micelles and the particle weight is approximately equal to the micellar weight ; at high temperatures (> 80°C) the particle weight corresponds to the molecular weight of the copolymer. The ratio of micelle to unimer (at intermediate temperatures) is dependant both on temperature and concentration, as shown by TUZAR et al (4), who studied polystyrene-b-polybutadiene-b-polystyrene micelles in dioxane/ethanol Fig. 3.

Gel permeation chromatography studies (33, 34, 39) on polystyrene-b-polyisoprene block copolymer micelles in N,N-dimethylacetamide (selectively bad solvent for polyisoprene block) also

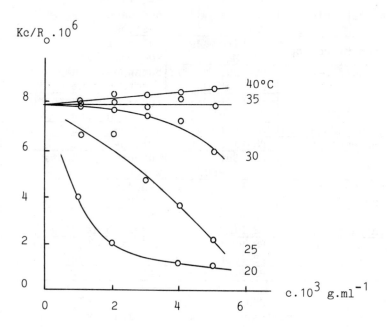

<u>Fig. 3</u> : Dependance of Kc/R_o on the concentration of the copolymer SBS in the mixture dioxan/25 vol. % ethanol at different temperature (4)

SBS : styrene - <u>block</u> - butadiene - <u>block</u> - styrene

$\overline{M}_w$ = 140,000   weight % styrene = 52

K_c : optical constant at copolymer concentration c

R_o : reduced intensity of scattered light at zero angle of measurement

show the micelle/unimer ratio dependance on temperature. These systems, studied in the temperature range 26-110°C, show the presence of both unimer and micelle at intermediate temperatures. A similar study on polystyrene-<u>b</u>-poly(ethylene oxide) micelles in a base lubricating oil have been carried out (40).

Viscosity studies have also contributed to an understanding of the temperature effects on block copolymer micelles (4, 8, 41).

1.4 Further Studies

Solubilization : A block copolymer micelle may solubilize insoluble substances within its core. Such a phenomenon has been reported by TUZAR and KRATOCHVIL (2) and TUZAR et al (42, 43) has solubilized insoluble polybutadiene by means of polystyrene-b-polybutadiene-b-polystyrene micelles possessing a polybutadiene core. This system however has several constraints. The molecular mass of the homopolymer to be solubilized must be lower than the molecular mass of the copolymer polybutadiene block and the amount of homopolymer solubilized has an upper limit which is dependant on its molecular mass, the molecular characteristics of the block copolymer, the solvent selectivity and temperature. Insoluble block copolymer could also be solubilized by the soluble block copolymer ; a mixed micellar system was proposed. In addition the miscibility of diblock copolymer with the parent homopolymer has been reported by GALLOT (44).

Stabilization : TUZAR et al (25, 45) has succeeded in stabilizing or cross-linking the polybutadiene cores of triblock copolymer micelles in selective solvents, i.e. polystyrene-b-polybutadiene--b-polystyrene in THF/alcohol, by means of UV radiation or fast electrons. The stabilised micelles proved to be resistant to heat and remained intact in the presence of a good solvent for both blocks. WILSON et al (46) has extended this technique to styrene-butadiene diblock copolymers.

A-B-C block copolymers : Some studies have been made on triblock copolymer micelles (47) possessing three different blocks ; this author has shown that polystyrene-b-polyisoprene-b-polymethylmethacrylate micelles in acetonitrile form core-shell micelles.

1.5 Applications

Dispersion polymerization : The surface-active properties of block copolymers render them suitable for use as stabilizers for a variety of colloidal dispersions. One application of this phenomenon is the stabilization of dispersed polymer particles formed as a result of dispersion polymerization either in organic or aqueous media (48).

The rate of dissociation of micelles plays an important role in governing the effectiveness of the stabilizer. In organic media especially, the conditions must be carefully chosen to ensure that the block copolymer molecules have enough mobility to maintain adequate coverage of the reaction product ; dispersed polymer particles in the range 10^{-2} - 10^{-4} nm with narrow size distribution can be obtained by carefully controlling the experimental conditions.

EVERETT and STAGEMAN (49) obtained non-aqueous colloidal dispersions of polymethylmethacrylate and polyacrylonitrile in hexane using polydimethylsiloxane-b-polystyrene-b-polydimethylsiloxane copolymers as stabilizers in a dispersion polymerization process.

TADROS and VINCENT (50) have utilized polystyrene-b-poly(ethylene oxide) copolymers to study the temperature and electrolyte effect of their absorption onto polystyrene latex and investigated the stability of the coated particles. Supplementary studies on the stabilization of latices by block copolymers have been mentioned under 2.1.2.

Several authors have used block copolymers in an anionic or cationic dispersion process. DAWKINS and TAYLOR (51-54) report the use of polystyrene-b-polydimethylsiloxane copolymers in the anionic polymerization and radical dispersion polymerization of styrene and methylmethacrylate in aliphatic hydrocarbons and SCHWAB has described the application of tert butylstyrene-styrene block copolymers as dispersant in the anionic synthesis of divinylbenzene-styrene copolymer (55). The use of lyophobic-lyophilic block copolymers as dispersants in the cationic polymerization of isobutylene-isoprene (butyl rubber) is reported by POWERS and SCHATZ (56).

The formation of polyisoprene latex in dimethylformamide using styrene-isoprene block copolymers has been studied by RIESS et al (57).

Viscosity improvers : Block copolymers of styrene and diene or hydrogenated dienes have been used as viscosity improvers for base lubricating oil (58-63). For example CROSSLAND and ST. CLAIR (61) have used a polystyrene-b-polyisoprene copolymer (mol. wt. $\sim$ 32,000-55,000) to improve the viscosity index of multirange lubricants.

Heterogenous catalysis : Micellar species may have potential in heterogenous catalysis since certain block copolymers form stable complexes with a number of transition metal ions. An example of this is the formation of a polyvinylpyridine-b-poly (ethylene oxide) complex with Cu(ii) in benzene (64).

2 EMULSIONS

An emulsion, defined as composed of two immiscible liquids, is generally of two types namely oil/water (both normal and

inverse) and oil/oil and block copolymers may act as emulsifiers
in both cases ; the stability and type of emulsion formed is
related to the structure, molecular characteristics etc. of the
block copolymer in addition to the other constituents of the
emulsion. The block copolymers used are necessarily hydrophobic-
hydrophilic in nature and may possess a number of alternative
groups :

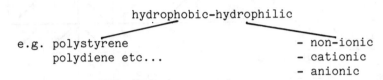

hydrophobic-hydrophilic

e.g. polystyrene   - non-ionic
     polydiene etc...   - cationic
                                          - anionic

2.1 Oil/Water Emulsions

Both non-ionic and cationic based block copolymers are
frequently employed with this type of emulsion.

Non ionic : Here the hydrophobic block may be poly(propylene
oxide) (65), poly(α-butylene oxide) (66) polystyrene (67, 68, 69)
etc. ; the non-ionic hydrophilic block is generally poly(ethylene
oxide). Star-shaped block copolymers of the type $A(B)_2$ where
B is a poly(ethylene oxide) block and A a polydiene block may
also be included in this class and have been used in both oil/water
and water/oil emulsions (70) ; these authors discuss the influence
of molecular characteristics and structure of the block copolymer
in relation to the stability and type of emulsion formed and
conclude that both types of emulsions may be formed using a
block copolymer of low molecular weight (1,000-10,000) containing
~50 % poly(ethylene oxide).

Cationic : Block copolymers with a cationic hydrophilic group
based on vinyl pyridine have been employed by several authors.
GALLOT et al (71, 72) have studied the emulsifying behaviour
of polystyrene-b-poly(2-vinylpyridinium chloride) and polyisopre-
ne-b-poly(2-vinylpyridinium chloride) with oil/water and water/oil
emulsions and CLAYFIELD and WHARTON (73) have obtained stable
water/oil type emulsions with uniform droplet size using block
copolymers with a similar hydrophilic group and polybutadiene
or 2-t-octyl styrene as hydrophobic group.

2.1.1. *Characteristics and influence of constituants* : Viscosity,
resistivity, particle size and emulsion stability studies have
been carried out by RIESS et al (74) who have chosen to study
polystyrene-b-poly(ethylene oxide) copolymers. Their observations
of the effect of parameter variation (e.g. block copolymer molecular weight, concentration etc.) on the resultant emulsion are
shown in Table 2, and the authors have been able to select the
copolymer resulting in the most stable emulsion. The best results

TABLE 2

Variable sense of variation	Domain of emulsion		Particle size	Viscosity		Stability	
	O/W	W/O	O/W	O/W	W/O	O/W	W/O
Concentration ↗			↘	↗	↗	↗	↗
Composition PEO content ↗	↗	↘	↘	↘	↗	↗	↘
Molecular weight ↗	↘	↘	↗	↗	↗	↘	↘
Structure bi → tri — high PS content →	→	↘	↗	↗	-	
high PEO content →	→	↘	↗	↗	→	→

Variation of emulsion characteristics with block copolymer characteristics (74).

for stable inverse emulsions (W/O) were obtained using a di- or triblock block copolymer of polystyrene content 60-80 %. Phase inversion behaviour as determined by viscosity measurements for one such system is depicted in Fig. 4.

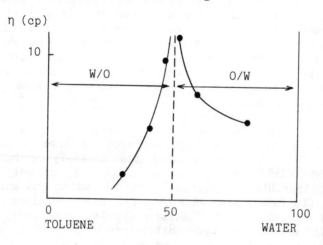

Fig. 4 : Phase inversion of toluene-water emulsions stabilized by styrene-ethylene oxide block copolymers (74).

● : polystyrene - block - poly(ethylene oxide) $\bar{M}_n$ = 44,000

weight % poly(ethylene oxide) = 35

2.1.2. Applications.

Emulsion polymerization/latex : For practical purposes an emulsion polymerization may be defined as "an addition polymerization process which proceeds by a micellar mechanism" (75). Two broad types may be discussed - oil/water (both normal and inverse) and oil/oil and in most cases the final product is what is termed a "latex", i.e. "a stable dispersion of a polymeric substance in an essentially aqueous medium", although oil/oil emulsions are a departure from this definition.

Block copolymers by virtue of their distinctive surface-active properties are invaluable participants in both the polymerization itself and in the stabilization of the final product.

Various aspects of the use of block copolymers in the stabilization of dispersed polymer particles has been covered in section 1.5. A number of other notable studies exist however in particular the work of RIESS et al (76) and ROGEZ (77) who have stabilized polystyrene latex in aqueous media using polystyrene-b-poly(ethylene oxide) copolymers. Similar block copolymers have been used to stabilize the inverse emulsion polymerization of styrene (78).

Foams : Most studies on the foam stabilizing behaviour of block copolymers are restricted to polyurethane foam formation, particularly in "one shot" processes using polysiloxane-b-polyether copolymers. The block copolymer is generally branched and dimethylsiloxane is usually the "silicone" block. The requirements of high surface activity for nucleation and stabilization of the cells combined with good emulsifying abilities for the raw material and blowing agent (79) may be met by designing the block copolymer structure to meet the particular requirements. ROSSMY et al (79) have deduced that one of the principal factors is adequate dispersion of the insoluble polyurea which forms at the onset of polymerization. PLUMB and ATHERTON (80) and OWEN (81, 82) have reviewed the surfactant properties and applications of these block copolymers.

2.2 Oil/Oil Emulsions

Block copolymers act as efficient emulsifiers for stabilizing emulsions composed of two immiscible organic liquids, generally when each of them is a selective solvent of one of the blocks of the copolymer. RIESS et al (83-85) have investigated the emulsifying effect of polystyrene-b-polyisoprene copolymers for dimethylformamide-hexane emulsions and polystyrene-b-polymethylmethacrylate for cyclohexane-acetonitrile emulsions.

Fig. 5 : Microphotograph of an oil-in-oil emulsion.
System : Dimethylformamide-in-hexane emulsion stabilized by
polystyrene-block-polyisoprene.
M_n = 33,000   weight % polystyrene = 44

From viscosity and particle size determination for these
oil/oil emulsions as a function of phase ratio and type of emulsion for block copolymers with different molecular characteristics, these authors have concluded that the solubility of
the blocks in oils is an important parameter and the block copolymers with the lowest degree of association, in both hexane and
dimethylformamide, were generally the most efficient emulsifiers
for oil/oil emulsions. A microphotograph of one oil/oil emulsion
studied by the authors is shown in Fig. 5. It was also concluded
that the block copolymer of molecular weight of about 30,000-
50,000 and a composition of about 50:50 had the best emulsifying
efficiency. Formation of oil/oil emulsions is a very specific
property of block copolymers and such systems are difficult,
if not impossible, to obtain using classical surfactants.

2.3 Microemulsions

A microemulsion is defined as a stable dispersion of one
liquid in another, in the form of spherical droplets whose diameter
is less than 1,000 Å and which confer a translucency to the
system ; it is the small size and thermodynamic stability of
the particles which distinguish it from a classical emulsion.

RIESS et al (74) have reported the formation of microemulsions
using polystyrene-b-poly(ethylene oxide) block copolymers instead
of common surfactants ; microemulsions were formed by successive
addition of isopropanol to oil/water emulsions stabilised by
the block copolymer. A number of other studies in the same vein

followed (86-92). For example MARIE and GALLOT (86, 87) obtained oil/water and water/oil microemulsions using polystyrene-b-poly-vinylpyridinium bromide copolymeric surfactant ; interfacial properties showed that the copolymer molecules were preferentially adsorbed at the interface. As an adjunct to these studies, various authors have used block copolymer/microemulsion systems to polyme-rize water-soluble monomers such as acrylamide (91, 92).

3 POLYMERIC EMULSIONS

These may be defined as systems which consist of two incompa-tible polymers poly A and poly B in a single mutual solvent S forming a two-phase system. The presence of a block copolymer A-B acts as an emulsifier for both types of emulsion which exist, namely water/water and oil/oil.

Oil/oil : A typical example of such an emulsion is the system consisting of polystyrene and polybutadiene in a common solvent such as toluene or styrene. By adjusting the concentration of both homopolymers, phase separation occurs due to the incompatibi-lity of the polymeric species ; one has polybutadiene as its major polymeric component and the other polystyrene. Emulsifica-tion of this system can be achieved by the addition of styrene-butadiene block copolymers (93).

Water/water : RIESS et al (94) has studied systems of this type, namely poly(ethylene oxide) and poly(vinylpyridine hydrochloride) in water, in the presence of poly(ethylene oxide)-b-poly(vinylpy-ridine hydrochloride).

In a systematic study of both types of "polymeric" emulsions, RIESS et al (94, 95) have correlated, for the examples mentioned, the characteristics of such emulsions, e.g. stability, particle size of the dispersed phase, phase inversion etc. to the charac-teristics of the ternary system poly A - poly B - solvent and to those of the A-B block copolymer.

Thus in a typical phase diagram shown in Fig. 6, the following characteristics of the ternary system poly A - poly B - solvent(s) have been extablished :

- the limit of phase separation ; this is the main require-ment for the existence of polymeric emulsions

- the composition and the viscosity of the phases in equili-brium
- the interfacial tension Y_i, between the coexisting phases

208

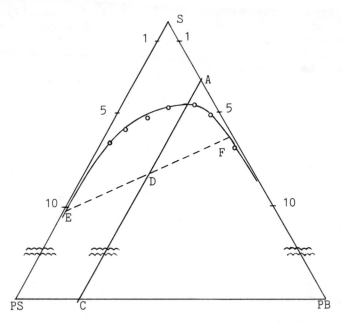

Fig. 6 : Phase diagram for the ternary system poly A - poly B - solvent at 25°C (93).

  AC - reaction path for the polymerization of styrene in the presence of 3 % PB

  B - phase separation limit

  EF - tie line (conjugated diameter)

poly A = polystyrene $\overline{M}_n$ = 130,000

poly B = polybutadiene $\overline{M}_n$ = 120,000

 S = styrene

   It has been further observed that the concentration of block copolymer necessary for the stabilization of "polymeric" emulsions diminishes with the increase in overall polymer concentration of the system. At concentrations near the critical point, it becomes practically impossible to stabilize a polymeric emulsion.

   The stabilizing action of the block copolymer was dependant upon ·their molecular characteristics, e.g. molecular weight, structuré, composition etc. and upon the characteristics of the coexisting phases. Generally copolymers with high molecular weight and molar compositions of about 50:50 were found to be good emulsifiers ; this high molecular weight criterion was fundamental in the stabilization of these "polymeric emulsions".

   Based on FLORY-HUGGINS (96) and VRIJ's (97, 98) theories, OSSENBACH-SAUTER (64) has described the characteristic features

of the ternary system poly A-poly B-solvent(s) such as the phase
separation limit, the critical point, the composition of the
phases in equilibrium, the interfacial tension between these
phases and the thickness of the interfacial layer as a function
of polymerization degree of the homopolymers and the various
interaction parameters. This author has completed his theoretical
approach by calculating the free energy conditions required
to place a diblock copolymer A-B at the centre of the interfacial
zone. His calculations are in agreement with experimental observations of the "stabilization" of these emulsion systems.

3.1 Applications

The oil/oil polymeric emulsion system has specific application in the preparation of "High Impact Polystyrene" (HIPS),
the system being essentially one of polybutadiene (dispersed
phase) in a polystyrene matrix (continuous phase). New types
of HIPS demonstrating special two-phase morphology may be obtained
by the polymerization, without additional grafting, of styrene

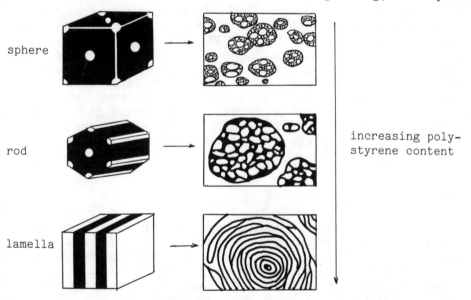

Fig. 7 : High impact polystyrene based on polybutadiene-b-polystyrene-b-polybutadiene block copolymers.
Schematic representation of the different morphologies.
(polybutadiene : black
   polystyrene : white)
fig. drawn according to ECHTE et al (100, 101).

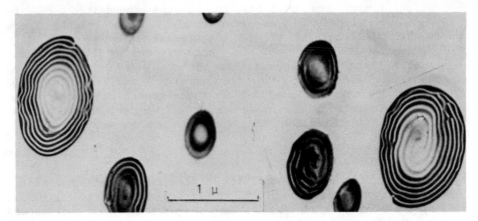

Fig. 8 : Electron micrograph of a polymer blend displaying onion type morphology. Polybutadiene is stained with osmium tetroxide.

in the presence of block copolymers of type polystyrene-b-polybutadiene-b-polystyrene and polybutadiene-b-polystyrene-b-polybutadiene of varying composition ; RIESS et al (93, 99), ECHTE et al (100, 101) and EASTMOND et al (102) have made notable contributions in this field. In the solid state these display mesomorphic structures such as spheres, rods, lamella etc. Fig. 7 illustrates both the mesomorphic structures and the resultant morphologies of the blends obtained by polymerization of styrene in the presence of block copolymers of type polybutadiene-b-polystyrene-b-polybutadiene. Fig. 8 is an electron micrograph of such a blend displaying onion type morphology.

CONCLUSION

The unique surface-active properties of block copolymers have made them a valuable asset in circumventing a number of problems associated with two-phase and multi-phase polymeric systems and these special characteristics have rendered them useful as dispersants, emulsifiers, wetting agents, foam stabilizers, solubilizers, flocculants, demulsifiers, etc. in many industrial and pharmaceutical preparations. The most widely used block copolymeric surfactants are those of ethylene oxide and propylene oxide. Such block copolymers behave like typical hydrophobic-hydrophilic agents where the surface activity may be tailored simply by adjusting the molecular characteristics. For this reason and due to their non-toxicity, ethylene oxide - propylene oxide block copolymers (commercialised by BASF Wyandotte as Pluronics and Tetronics) have a wide range of applications in the cosmetic and paper industries, in textiles as antistatic agents, in agricultural formulations (e.g. in the dispersion of pesticides), as detergents, in the petroleum industry, etc. The numerous applications of these block copolymers have been reviewed extensively by LUNSTED and SCHMOLKA (103) and SCHMOLKA (30).

Many parallels may be drawn between block copolymers and classical surfactants e.g. surface activity, micelle formation etc., but the former possess several important advantages

(a) "tailor-made" block copolymers of a well-defined molecular weight, structure etc. may be prepared to suit a particular application ;

(b) they are capable of solubilizing previously insoluble substances ;

(c) they are able to stabilize a wide range of latexes, dispersions etc. particularily in non-aqueous media.

REFERENCES

1. Hurtrez, G., D.J. Wilson and G. Riess. (in press)
2. Tuzar Z. and P. Kratochvil. Advan. Colloid Interface Sci. 6 (1976) 201
3. Price, C. Development of Block Copolymers I. (Ed. I. Goodman, Applied Science Publishers, England, 1982), pp. 39-80
4. Tuzar, Z. and P. Kratochvil. Makromol. Chem. 160 (1972) 301
5. Nakamura, K., R. Endo and M. Takada. J. Polym. Sci., Polym. Phys. Ed. 14 (1976) 1287
6. Riess, G. and D. Rogez. ACS Polymer Preprint. 23 (1982) 19
7. Huynh-Ba-Gia, R. Jerome and Ph. Teyssie. J. Polym. Sci., Polym. Phys. Ed. 18 (1980) 2391
8. Averbukh, M.Z., N.I. Nikonorova, Ya.S. Rozinoer, I.I. Lushchikov, V.P. Shatalov, M.L. Gurari, N.F. Bakeev and P.V. Kozlov. Kolloid. Zh. 38 (1976) 419
9. Prasad, K.N., T.T. Luong, A.T. Florence, J. Paris, C. Vaution, M. Seiller and F. Puisieux. J. Colloid Interface Sci. 69 (1979) 225
10. Al-Saden, A.A., T.L. Whateley and A.T. Florence. J. Colloid Interface Sci. 90 (1982) 303
11. Oranli, L., P. Bahadur and G. Riess (in press)
12. Lally, T.P. and C. Price. Polymer 15 (1974) 325
13. Tuzar, Z., A. Sikora, V. Petrus and P. Kratochvil. Makromol. Chem. 178 (1973) 2743
14. Beamish, A., R.A. Goldberg and D.J. Hourston. Polymer 18 (1977) 49
15. Canham, P.A., T.P. Lally, C. Price and R.B. Stubbersfield. J.C.S. Faraday I 74 (1980) 1857
16. Philipps, M.C., M.N. Jones, C.P. Patrick, N.B. Jones and M. Rodgers. J. Colloid Interface Sci. 72 (1979) 98
17. Ikada, Y., H. Iwata, S. Nagaoka, F. Horii and M. Hatada. J. Macromol. Sci. Phys. B 17 (1980) 191
18. Stacy, C.J. and G. Kraus. Polym. Eng. Sci. 17 (1977) 627
19. Mandema, W., C.A. Emeiss and H. Zeldenrust. Makromol. Chem. 180 (1979) 2163
20. Krause, S. J. Phys. Chem. 68 (1964) 1448
21. Price, C. and D. Wood. Europ. Polym. J. 9 (1973) 827
22. Tanaka, T., T. Kotaka and H. Inagaki. Bull. Inst. Chem. Res., Kyoto Univ. 55 (1977) 206
23. Plestil, J. and J. Baldrian. Makromol. Chem. 174 (1973) 183 ; 176 (1975) 1009
24. Shibayama, M., T. Hashimoto and K. Hiromichi. Macromolecules 16 (1983) 16
25. Prochazka, A., M.K. Baloch and Z. Tuzar. Makromol. Chem. 180 (1979) 2521
26. Periard, J. and G. Riess. Europ. Polym. J. 9 (1973) 687
27. Tuzar, Z., A. Sikora, V. Petrus and P. Kratochvil. Makromol. Chem. 178 (1977) 2743
28. Price, C., P.A. Canham, M.C. Duggleby, T.V. Naylor, N.S. Rajab and R.B. Stubbersfield. Polymer 20 (1979) 615

29. Krause, S. and P.A. Reismiller. J. Polym. Sci., Polym. Phys. Ed. 13 (1975) 663
30. Schmolka, I.R. J. Amer. Oil Chemists' Soc. 54 (1977) 110
31. Averkukh, M.Z., N.I. Nikonorova, Ya.S. Rozinoer, I.I. Lushchikov, V.P. Shatalov, M.L. Gurari, N.F. Bakeev and P.V. Kozlov. Kolloid. Zh. 38 (1976) 419
32. Legrand, D.G. J. Polym. Sci., Polym. Sym. 63 (1978) 147
33. Booth, C., T.D. Naylor, C. Price, N.S. Rajab and R.B. Stubbersfield. J.C.S. Faraday I 74 (1978) 2352
34. Price, C., A.L. Hudd, C. Booth and B. Wright. Polymer 23 (1982) 650
35. Candau, F., J.M. Guenet, J. Boutillier and C. Picot. Polymer 20 (1979) 1227
36. Higgins, J.S., J.V. Dawkins and G. Taylor. Polymer 21 (1980) 627
37. Tuzar, Z., V. Petrus and P. Kratochvil. Makromol. Chem. 175 (1974) 3181
38. Selb, J. and Y. Gallot. Makromol. Chem. 181 (1980) 2605 ; 182 (1981) 1491, 1513
39. Horii, F., Y. Ikada and I. Sakurada. J. Polym. Sci., Polym. Chem. Ed. 12 (1974) 323
40. Price, C., A.L. Hudd, R.B. Stubbersfield and B. Wright. Polymer 21 (1980) 9
41. Mandema, W., H. Zeldenrust and C.A. Emeiss. Makromol. Chem. 180 (1979) 1521
42. Tuzar, Z. and P. Kratochvil. Makromol. Chem. 170 (1973) 177
43. Tuzar, Z., P. Bahadur and P. Kratochvil. Makromol. Chem. 182 (1981) 1751
44. Rameau, A., J.P. Lingelser and Y. Gallot. Makromol. Chem., Rapid. Commun. 3 (1982) 413
45. Tuzar, Z., B. Bednar, C. Konak, M. Kubin, S. Svobodova and K. Prochazka. Makromol. Chem. 183 (1982) 399
46. Wilson, D.J. and G. Riess (in press)
47. Schlienger, M. Thesis, Univ. Haute Alsace, Mulhouse, France (1976)
48. Barrett, K.E.J., Dispersion Polymerization in Organic Media (John Wiley and Sons, London, 1975)
49. Everett, D.H. and J.F. Stageman. Colloid Polym. Sci. 255 (1977) 293 ; Discuss. Faraday Soc. 65 (1978) 230
50. Tadros, Th.F. and B. Vincent. J. Phys. Chem. 84 (1980) 1575
51. Dawkins, J.V. and G. Taylor. Polymer Colloids, Vol. 2 (Ed. R.M. Fitch, Plenum Press, New York, 1980) p. 44
52. Dawkins, J.V. and G. Taylor. Europ. Polym. J. 15 (1979) 453
53. Dawkins, J.V. and G. Taylor. Polymer 20 (1979) 599
54. Dawkins, J.V. and G. Taylor. J.C.S. Faraday I 76 (1980) 1263
55. Schwab, F.C. US 3770712 (1973)
56. Powers, K.W. and R.H. Schatz (to Exxon) US 4252710 (1981)

214

57. Riess, G., S. Marti, J.L. Refregier and M. Schlienger. Polym. Sci. Technol. 10 (1977) 327
58. Street, N.L., K.H. Yochum and B. Mitacek, SAE, Technical Paper, 760267 (1976)
59. Echer, R. and J. Albrecht. Ger. Pat. 2156122 (1977)
60. Anderson, W.S. US 4036910 (1977)
61. St Clair, D.J. and R.K. Crossland. US 3965019 (1976)
62. Benda, R., H. Knoell, P. Neudoerfl and H. Pennewiss (to Roehm GmbH). Ger. Offen. 3001045 (1981)
63. Elliott, R.L. (to Exxon Research and Engineering Co.). Can. 1087343 (1980)
64. Ossenbach-Sauter, M. Thesis, Univ. Haute Alsace, Mulhouse, France (1981)
65. Lundsted, L.G. and I.R. Schmolka. Block and Graft Copolymerization Vol. 2 (Ed. R.J. Ceresa, John Wiley, New York, 1976)
66. Szymanowski, J., J. Myszkowski et al. Tenside Detergents 19 (1982) 7, 11, 14 ; 20 (1983) 23
67. Riess, G., J. Nervo and D. Rogez. Polym. Eng. Sci. 8 (1977) 634
68. Nervo, J., S. Marti and G. Riess. C.R. Acad. Sci. Paris Ser.C 279 (1974) 891
69. Marti, S., J. Nervo and G. Riess. Progr. Colloid Polym. Sci. 58 (1975) 114
70. Huynh-Ba-Gia, R. Jerome and Ph. Teyssie. J. Appl. Polym. Sci. 26 (1981) 343
71. Marie, P., Y. Le Herrenschmidt and Y. Gallot. Makromol. Chem. 177 (1976) 2773
72. Marie, P. and Y. Gallot. C.R. Acad. Sci. Paris Ser. C 284 (1977) 327
73. Clayfield, E.J. and D.G. Wharton. Theory and Practice of Emulsion Technology (Ed. A.L. Smith, Academic Press, New York, 1976)
74. Riess, G., J. Nervo and D. Rogez. Polym. Eng. Sci. 8 (1977) 634
75. Blackley, D.C. Emulsion Polymerization (Applied Science Publishers Ltd., 1975)
76. Rogez, D. Private Communication (1979)
77. Rogez, D. Thesis to be submitted, University of Haute Alsace, Mulhouse, France
78. Rogez, D., S. Marti, J. Nervo and G. Riess. Makromol. Chem. 176 (1975) 1393
79. Rossmy, G., H. Kollmeier, J. Lidy, H. Schator and M. Wiemann. J. Cell. Plastics 13 (1977) 26 ; Goldschmidt 58 (1983) 28
80. Plumb, J.B. and J.H. Atherton. Block Copolymers (Ed. D.C. Allport and W.H. James, Appl. Sci. Pub., London, 1973)
81. Owen, M.J., T.C. Kendrick, B.M. Kingston and N.C. Lloyd. J. Colloid Interface Sci. 24 (1967) 141
82. Owen, M.J. Chemtech. 288 (May 1981)

83. Riess, G., J. Periard and A. Banderet. Colloidal and Morphological Behaviour of Block and Graft Copolymers (Ed. G.E. Molau, Plenum Press, New York, 1971) p. 173

84. Periard, J., A. Banderet and G. Riess. Polymer Letters 8 (1970) 109

85. Periard, J., G. Riess and M.J. Neyer-Gomez. Europ. Polym. J. 9 (1973) 687

86. Marie, P. and Y. Gallot. C.R. Acad. Sci. Paris, Ser. C 284 (1977) 327

87. Marie, P. and Y. Gallot. Makromol. Chem. 180 (1979) 1611

88. Taubman, A.B. and L.T. Peregudova. Kolloid. Zh. 41 (1979) 604

89. Boutiller, J. and F. Candau. Colloid Polym. Sci. 257 (1979) 46

90. Gallot Y. Colloques Nationaux du CNRS, n° 938, Physicochimie des composés amphiphiles (CNRS, Paris, 1979) p. 149

91. Leong, Y.S., G. Riess and F. Candau. J. Chim. Phys. 78 (1981) 279

92. Leong, Y.S. Thesis, Univ. Louis Pasteur, Strasbourg, France (1983)

93. Gaillard, P., M. Ossenbach-Sauter and G. Riess. M.M.I. Press Symp. Vol. 3 (Ed. K. Solc, 1982) p. 289

94. Ossenbach-Sauter, M. and G. Riess. C.R. Acad. Sci. Paris Ser. C 283 (1976) 269

95. Ossenbach-Sauter, M. Thesis, Univ. Haute Alsace, Mulhouse, France (1981)

96. Flory, P.J. Principles of Polymer Chemistry (Cornell Univ. Press, Ithaca, New York, 1953)

97. Vrij, A. J. Polym. Sci. A 2 (1968) 1919

98. Vrij, A. and G.J. Roebersen. J. Polym. Sci., Polym. Phys. Ed. 15 (1977) 109

99. Marti, S. and G. Riess. Makromol. Chem. 179 (1978) 2569

100. Echte, A. Angew. Makromol. Chem. 58-59 (1977) 175

101. Echte, A., H. Gausepohl and H. Lutje. Angew. Makromol. Chem. 90 (1980) 95

102. Eastmond, G.C. and D.G. Phillips. Polymer Alloys (Polym. Sci. Technol. Ser., Vol. 10, Ed. D. Klempner and K.C. Frisch, Plenum Press, New York, 1977) p. 147

103. Lundsted, L.G. and I.R. Schmolka. Block and Graft Copolymerization Vol. 2 (Ed. R.J. Ceresa, John Wiley, New York, 1976)

RELATIONSHIPS BETWEEN MORPHOLOGY, STRUCTURE, COMPOSITION AND
PROPERTIES IN ISOTACTIC POLYPROPYLENE BASED BLENDS

Ezio Martuscelli

Istituto di Ricerche su Tecnologia dei Polimeri e Reologia,
C.N.R., Via Toiano 2, 80072 ARCO FELICE (Napoli), Italy.

INTRODUCTION

The properties of rubber modified non-crystallizable thermoplastic polymers are usually described and theoretically predictable
in terms of the mode and state of dispersion of the soft component
and of the adhesion between the matrix and the dispersed particles.
When the thermoplastic polymer is crystallizable many other factors
must be taken into account, as the addition of a second rubbery
component may induce drastic changes on some important properties
of the matrix. Due to reciprocal interactions between the major
and minor components in such blends the morphology, the structure
and size of spherulites and lamellar crystals, the spherulite
growth rate, the nucleation process, the melting and thermal
behaviour and the crystallinity often tend to be composition
dependent. Thus in order to understand the macroscopic properties
of such crystallizable blends one must have a knowledge of the
influence of the preparation and crystallization conditions, and
the nature and molecular mass of the soft components on the matrix
modifications, and of the mode and state of dispersion of the minor
component.

In the present paper blends having isotactic polypropylene
(iPP) as matrix and ethylene-propylene random copolymers (EPR),
polyisobutylene (PiB) and low density polyethylene (LDPE) as soft
components are considered. The characteristics of the polymers are
shown in Table 1.

218

TABLE 1

Characteristics of polymers used in the blend preparations.

Polymer and code	Source and trade name	Molecular mass	Viscosity Mooney ML(1+4) at 100°C	Ethylene/ Propylene ratio(W/W)
Isotactic polypropylene (iPP)	RAPRA	$\overline{M}_w=3.07\times10^5$ $\overline{M}_n=1.56\times10^4$		
Ethylene-propylene copolymers (EPR)	Dutral CO54 (Montedison)	$\overline{M}_w=1.8\times10^5$	43	60/40
	BUNA AP201 (HUELS)		44	50/50
	EPCAR 306 (Vistanex)	$\overline{M}_w=1.4\times10^5$	34.5	> 67/30
Ethylene-propylene-diene terpolymer (EPDM)	Dutral TER 054-E (Montedison)			
Polyisobutylene (PiB)	VistanexLM HM (Esso) PiB (LM)	$\overline{M}_v=6.6\times10^4$		
	Vistanex L120 PiB (MM)	$\overline{M}_v=1.6\times10^6$		
	EGA Chemie PiB (HM)	$\overline{M}_v=3.5\times10^6$		
Low Density Polyethylene (LDPE)	Fertene ZF5-1800 (Montedison)	$\overline{M}_w=4-5\times10^5$		
	Fertene CF5-2100 (Montedison)	$\overline{M}_w=1.2-1.5\times10^5$		

MODE AND STATE OF DISPERSION OF MINOR COMPONENT

Thin Films Isothermally Crystallized

The mode and state of dispersion of EPR and EPDM and its
evolution in time was studied as a function of composition,
molecular mass and crystallization temperature (Tc), in thin films
(about 10 μm thick) by using optical microscopy, during the
isothermal growth of iPP spherulites (1,2,3).

The following basic phenomena were observed:

—   At Tc, the rubber component in the melt is normally segregated
in domains of spherical shape. At low content of rubber and at the
early stages of spherulite growth these domains are first rejected
by the crystallizing front and then occluded in intraspherulitic
regions at longer crystallization times.
—   At higher concentrations of EPR coalescence of particles upon
rejection is first observed. The resulting domains are then
engulfed in intraspherulitic regions. Following this process the
rubber domains are oriented along a circular line (see Fig. 1).
—   The EPDM dispersed domains are so easily rejected that they
are even rejected into new boundaries created behind the occluded
particles. Hence the particles of EPDM, trapped in iPP spherulites,
lie radially on the same line. Deformation of rejected and occluded
particles is also observed (see Fig. 2).

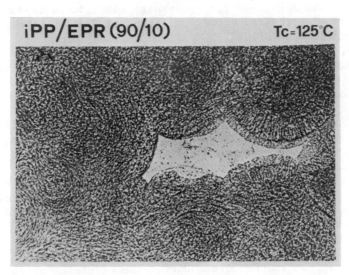

Fig. 1 – Optical micrograph of isothermally crystallized thin
        films of iPP/EPR (90/10) blend (Tc=125°C) Ref. 1.

220

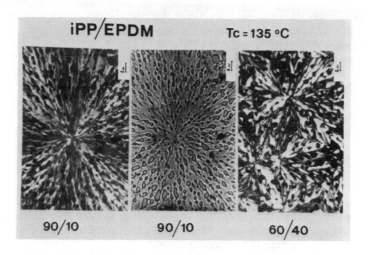

Fig. 2 - Optical micrographs of isothermally crystallized thin films of iPP/EPDM blends at different composition (Tc=135°C) Ref. 1.

In the case of iPP/PiB blends the mode and state of dispersion of PiB strongly depends on the molecular mass and the crystallization temperature. At low concentrations the polyisobutylene with lower molecular mass, for all Tc investigated, is rejected between spherulites. As the concentration increases drops of PiB are trapped inside spherulites.

In blends containing polyisobutylene with medium molecular mass distinct domains of PiB are observed in the melt. During crystallization the particles of PiB undergo rejection by all spherulite boundaries: those of spherulites as well as those generated inside spherulites behind occluded particles. The rejected particles are subjected to coalescence and occlusion. The described changes in the shape and distribution of the drops in the melt are clearly seen in Fig. 3a and 3b where the crystallized spherulite of iPP in the blend and the phase structure of the melt, after melting the spherulites, are seen. The melts of blends containing 10% or less of polyisobutylene with higher molecular mass do not show a droplike structure; during crystallization all the material is rejected into the interspherulitic regions (see Fig. 4). When the concentration of PiB increases droplike structures appear in the melt. During the crystallization of blends with higher amounts of PiB after rejection and coalescence the occlusion of drops occurs (see Fig. 4b and 4c). As the temperature of crystallization increases more material is rejected into inter-spherulite regions and less remains inside the spherulites. For all blends (at least under the conditions of preparation and crystallization investigated) the dispersed particles of the minor components have an average size that is noticeably smaller than that of iPP spherulites.

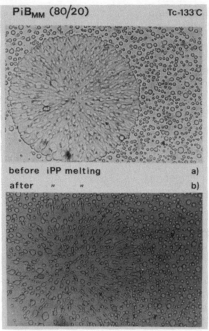

Fig. 3 – Optical micrographs of: a) iPP spherulite, surrounded by melt blend at the early stage of crystallization. b) The same region of film after melting of spherulite. Ref. 3.

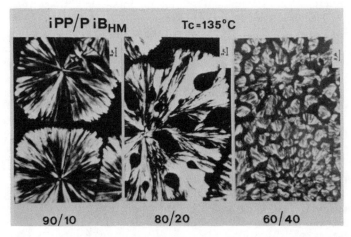

Fig. 4 – Optical micrographs of isothermally crystallized thin films of iPP/PiB(HM) blends with different composition (Tc=135°C) Ref. 2.

The presence of domains in the form of small or large droplets in the path of a crystallizing front can disturb the spherulite growth of iPP. Some amount of energy must be dissipated to perform the rejection, engulfing and deformation of the second component drops by growing spherulites. Such energies constitute new energy barriers controlling the growth of spherulites in blends. The equation that describes the dependence of the spherulite growth rate G on temperature, which for an homopolymer is given by the following expression:

$$G_1 = G_o \, EXP \, (-\Delta F^*/KT) EXP \, (-\Delta\phi^*/KT) \tag{1}$$

must be supplemented by new energy terms and written as:

$$G = G_1 \, EXP \, (- \frac{E_1 + E_2 + E_3 + E_4}{KT}) \tag{2}$$

where E_1 is the energy dissipated by the growing spherulite front for rejection of drops; E_2 is the kinetic energy needed to overcome the inertia of drops; E_3 is the energy needed to form new interface spherulite drops if the drops are engulfed and E_4 is the energy dissipated by the growing front for the deformation of engulfed second component drops. Expressions for the dissipation energy terms and for the corresponding growth rate of spherulites in a crystalli- zable – non-crystallizable polymer-polymer system were derived by Bartczak, Galeski and Martuscelli (3). The energy terms E_1, E_2, E_3 and E_4 were estimated in the case of iPP/rubber blends for which the radial growth rate as a function of the composition and crystallization temperature was already determined (1,2). The results are shown in Table 2.

TABLE 2

Energies dissipated by the crystallizing iPP spherulites to perform rejection, occlusion and deformation of dispersed particles.

Process	Energy (J/mol of iPP repeating units)
Rejection	$10^{+1} - 10^{+4}$
Kinetic energy of rejection	$10^{-15} - 10^{-4}$
Occlusion	$10^{-2} - 10^{-1}$
Deformation (only the surface change considered)	$10^{-1} - 10^{0}$

As can be seen, rejection may have considerable influence on the spherulite growth rate of iPP (3).

The dependence of the growth rate of iPP spherulites on the concentration of the second component was calculated taking into account a process of rejection of particles, for two different radii of dispersed domains (0.7 and 0.33 μm) and for a mean iPP spherulite size of 50 μm. The results showed that G should be significantly depressed at higher undercooling and that a finer dispersion of the second component should produce, at a given Tc, a relatively larger lowering in G (see Fig. 5).

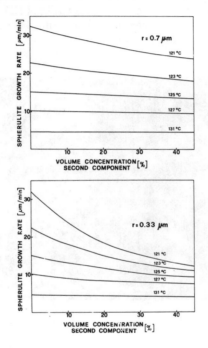

Fig. 5 – Prediction of growth rate depression for iPP/EPR blends in the case of rejection of the dispersed particles. Ref.3

The observed composition dependence of G is qualitatively in agreement with the theoretical curves for iPP/Dutral, iPP/BUNA and iPP/PiB (MM) blend although the rejection of rubber domains is usually followed by occlusion or by coalescence followed by occlusion. Plots of G versus the percentage of the second component show at higher undercooling a minimum in the case of iPP/EPCAR, iPP/PiB (LM), iPP/PiB(HM) and iPP/EPDM blends. This behaviour was attributed to the very fine dispersion of rubbery domains in blends at low concentration and/or to a process of partial miscibility of components in the melt (3).

Sheet Samples of iPP/PiB Blends

All samples of iPP/PiB blends, isothermally crystallized between Teflon sheets, show a double morphology (see Fig. 6). In the middle of the sheets (0.5 to 1.25 mm thick) iPP spherulites are formed whereas the edges show the presence of a region of trans-crystalline material (5). Such columnar or transcrystalline growth is due to the high surface nucleation density that severely limits crystallization in a sideways direction since impingement occurs. The fraction of the material that crystallizes according to a columnar morphology increases with an increase in the crystallization and quenching temperature for isothermally (IC) and non-isothermally (NIC) crystallized sheets respectively and it seems not to be influenced by the rubber content.

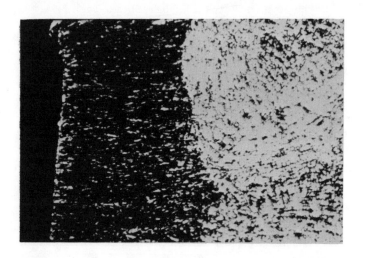

Fig. 6 – Optical micrograph of thin section of sheet of iPP/PiB (80/20) blend. (Tc=132°C; crossed polars). Ref. 5.

For IC sheets the density of particles of the dispersed phase is larger at the border of the spherulites. In such regions the particles undergo deformation and coalescence processes, induced by the crystallizing front, and are at the end of the crystallization oriented along concentric rings. In the case of NIC samples, due to the higher rate of crystallization, the domains of PiB, are distributed homogeneously all over the sample. The adhesion between the spherulitic and transcrystalline regions and among the spheru-lites is very poor in the case of IC samples (Figs. 7,8) and seems to increase for NIC blends. This effect, which is also confirmed by mechanical tests, is due to the increasing number of chains connecting crystallites and spherulites as soon as the crystallization rate is increased (5).

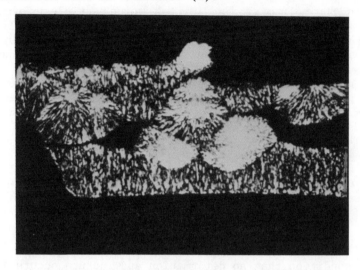

Fig. 7 – Optical micrograph of thin section of iPP/PiB (80/20) blend sheet crystallized at Tc=132°C (crossed polars). Ref. 5.

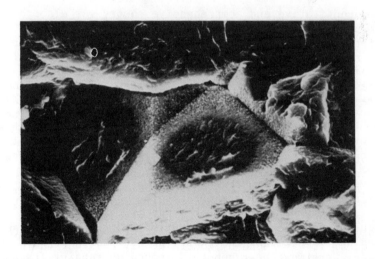

Fig. 8 – Scanning electron micrograph of fractured surface of iPP/ PiB (90/10) blend sheet crystallized at Tc=120°C. Ref. 5.

Factors Influencing the Particles Dimensions of Minor Components in the Melt

During the melt mixing an initial large particle or thread of the minor component, dispersed in a viscous melt matrix will be disrupted or disintegrated into a series of small droplets. Size reduction continues till the total force on the particle equals the surface tension (γ) which is the retracting force in the process. The overall process was mathematically treated by Taylor (6). The force τ_r acting on a sphere of radius r and viscosity μ_1 dispersed in a matrix with viscosity μ_o, subjected to a shear rate (dVx/dy) is given by the following expression:

$$\tau_r = c(dVx/dy) \; \mu_o \; f \; (\mu_o/\mu_1) \tag{3}$$

while the retracting force resulting from the interfacial surface tension is given by:

$$\tau_\gamma = 2 \; \gamma/r \tag{4}$$

Under the condition that no deformation and rupture occurs $\tau_r = \tau_\gamma$ and then

$$r = C^1 \; \gamma/\mu_o(dVx/dy)f(\mu_o/\mu_1) \tag{5}$$

A more quantitative and comprehensive theory was elaborated by Tomotika (7); however equation 5 is suitable for a qualitative description of the process. According to equation 5, for a given binary system, dispersed particles with a smaller size will be obtained when the shearing forces are higher and when the value of the phase viscosity ratio (μ_1/μ_o) is lower. The morphology in the melt may be preserved by rapid quenching or by crystallization at a high rate and can be investigated on thin sections of the solid blend. It was found that in the case of aPS/LDPE blends, obtained by melt mixing, at lower concentration of LDPE in aPS the particle sizes of LDPE are smaller than the particle sizes of aPS at lower concentrations in LDPE. This was explained, on the basis of the Taylor relation, taking into consideration the fact that in both regions the interfacial surface tension is the same and that at the temperature of mixing (150°C) the viscosity of aPS is twice that of LDPE (8). Recently, in qualitative agreement with equation 5, it was found by Kocsis et al. (9) that the number average particle size of EPDM, in iPP/EPDM blends increases with an increase of the value of the phase viscosity ratio. Thus it may be concluded that the viscosity of components, surface tension and shearing forces are the principal factors influencing the mode and state of dispersion of the minor component in the melt. The addition of surface active compounds during mixing may produce a lowering of the surface tension and in accordance with Taylor's formula a lowering of the dimensions of the particles. A transition from a stratified and lamellar structure to a more homogeneous and fine mode of dispersion of the minor component was obtained by adding an EPR rubber in the case of a binary iPP/HDPE blend extrudate (compare micrographs of Fig. 9). Such results indicate that EPR rubbers are able to act as emulsifying agents for iPP and HDPE (10).

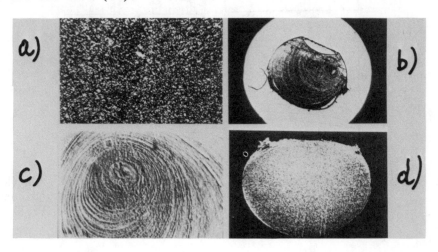

Fig. 9 – Optical micrographs of extrudate cross section of binary HDPE/iPP (25/75) and ternary HDPE/iPP/EPR blends:
    a) HDPE/iPP (25/75) blend (crossed polars) (350x);
    b) HDPE/iPP (25/75) blend (crossed polars) (80x);
    c) HDPE/iPP (25/75) blend (phase contrast) (800x);
    d) HDPE/iPP/Epcar (23.75/71.25/5) blend (crossed polars) (80x). Ref. 10.

STRUCTURAL, MORPHOLOGICAL AND MOLECULAR MODIFICATIONS OF iPP MATRIX INDUCED BY THE ADDITION OF A SECOND SOFT COMPONENT

Effect on the Primary Nucleation Process: Spherulite Dimensions

The addition of elastomers to iPP strongly influences the nucleation density and thus the number of spherulites per unit volume. For a given blend preparation procedure and thermal history, the type of effect and its extent depends on the chemical nature and molecular mass of the elastomer, crystallization temperature and composition (see Fig. 10).

228

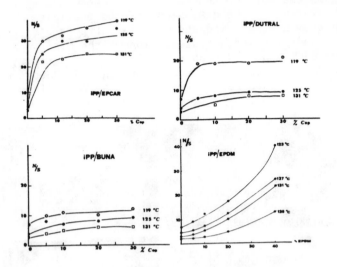

Fig. 10 – Variation of the number of spherulites per unit area (N/S) in thin films, isothermally crystallized at the indicated Tc, with the elastomer content for iPP/EPDM and iPP/EPR blends. Ref. 1,2.

In iPP/EPR mixtures it was observed that for the same Tc and composition, N/S increases with an increase of the ethylene content of the elastomer and with a decrease of the molecular mass (1,2). Nucleation effects were observed also in iPP/EPR, iPP/HDPE and iPP/aPS samples, non-isothermally crystallized, obtained by extrusion and by compression moulding (10,11). These observations indicate that, at least under the experimental conditions investigated, EPR, HDPE and atactic polystyrene (aPS) act as nucleating agents for iPP spherulites.

Recently a more systematic study of the influence of the presence of an EPR (Dutral) on the nucleation density of iPP was undertaken (12). Blends of iPP and Dutral were prepared by melt extrusion. The nucleation density was determined by optical microscopy on thin sections isothermally crystallized. For a given extrusion temperature (190°C) it was found that the final nucleation density of iPP (number of spherulites per unit area N/S) decreases with an increase of EPR content when the blend samples are kept, before cooling to Tc, at a temperature of precrystallization Tpc of 190°C.

The opposite behaviour is observed when Tpc is raised to 230°C and/or the blend samples are passed twice through the extruder (see Fig. 11a,b,c,d). Such behaviour may be probably accounted for by

assuming that, for lower values of Tpc, spherulites of iPP are
mainly nucleated by preexisting nuclei (self-seeded nucleation
process). Under these conditions the addition of EPR causes,
according to a process not yet clarified, suppression of a certain
number of preexisting iPP nuclei. On increasing the Tpc the number
of self-seeded nuclei decreases. iPP spherulites are nucleated
mainly by heterogeneous nuclei. The addition of EPR produces,
under these circumstances, an increase of heterogeneous nuclei with
the result of increasing the final nucleation density. The nucleation effect observed in a twice extruded iPP/EPR sample is probably
related to the fact that during this process a larger number of
heterogeneous nuclei may migrate from Dutral to the iPP melt.

No nucleation effects were observed in the case of thick
sheets of iPP/PiB, obtained by compression moulding, and isothermally
crystallized (5); while it was found that in thin films of iPP/PiB
blends, made by a different procedure, the PiB worked as an
effective nucleating agent (2). In particular the values of the
number of spherulites per unit area were found to be 30 times
higher than for pure iPP. Plots of N/S versus % of PiB content
show maxima whose values are more pronounced for blends containing
PiB with lower molecular mass. The disagreement may be attributed
to the different conditions of sample preparation and thickness, to
the fact that the starting polymers used in the preparation of
thick sheets were not purified and also to the fact that the
temperature of precrystallization was not the same (5).

A decrease in the number of spherulites, compared to plain iPP
was observed in isothermally crystallized, thin films, of iPP/LDPE
blends (13). The number of primary nuclei per unit volume $\overline{N}$ was
obtained using the relation:

$$K = 4/3 \ \pi (\rho c/\rho a) \frac{G^3 \ \overline{N}}{(1-\lambda)} \tag{6}$$

This assumes a spherical growth with instantaneous nucleation. In
equation 6, G is the spherulite growth rate, K is the overall kinetic
rate constant (both measured at the same Tc), ρ_c and ρ_a are the
densities of the crystalline and amorphous phase respectively and
$(1-\lambda)$ is the weight fraction of polymer that is crystalline at
$t=\infty$. It was found that for a given Tc and thermal history, $\overline{N}$
decreases drastically with an increase in the LDPE content of the
blend.

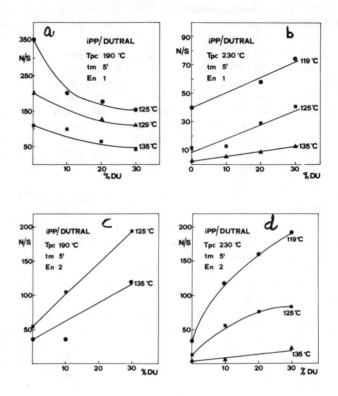

Fig. 11 – Number of spherulites per unit area N/S in iPP/Dutral
blends as function of EPR content (% W/W):
a) Samples once extruded; precrystallization temperature
Tpc=190°C; time at Tpc=5 sec.; b) Samples once extruded;
Tpc=230°C; time at Tpc=5 sec.; c) Sample twice extruded;
Tpc=190°C; time at Tpc=5 sec.; d) Sample twice extruded;
Tpc=230°C; time at Tpc=5 sec. Ref. 12.

It can be concluded that the addition of a soft component may
influence the process of primary nucleation of iPP. The kind and
extent of this effect is a function of a series of factors such as
the molecular structure and molecular mass of the minor component,
composition, precrystallization and crystallization temperature,
preparation conditions and the thermal history of blend samples.

In a process of non-isothermal crystallization the nucleation
density influences the values of the effective temperature of
crystallization T^e and of the effective supercooling $\Delta T = T_m^\circ - T^e$ (T^e
is defined as the temperature of the maximum overall rate of
crystallization, corresponding to the temperature of the peak of

the DSC exotherm of crystallization). It was found by Fraser et al. (14) that HDPE samples with a higher nucleation density crystallized at lower ΔT^e than samples with low nucleation density; consequently by affecting the supercooling at which crystallization occurs the nucleation density determines the thickness of the lamellar units and the melting point.

According to the study of Friedrich (15) the nucleation density, controlling the size of spherulites, plays an important role in determining the type of crack propagation in a polymeric material. In the case of samples characterized by small spherulites and by a fine and homogeneous dispersion of crystalline aggregates the crack propagation is controlled by crazing. The crazes run along spherulites boundaries as well as along radial or tangential trans-spherulitic paths which lie normal to the applied stress. When the morphology of the polymer sample shows larger spherulites then the crack essentially follows the boundary zones of the spherulites. The cracking behaviour can be described by a single interspherulitic craze which is followed by the crack.

Thus it may be concluded that the nucleation density is a determinant factor in crystallizable blends as it may influence many important matrix properties.

Modifications of iPP Matrix at Molecular and Structural Levels

For a given crystallization temperature, Tc, thin films of iPP/EPR mixtures have values of the observed melting temperature T'_m higher than that of plain iPP. The equilibrium melting temperature T_m, obtained by Hoffman plots, is higher in the case of blends and increases with the rubber content (1). The values of $\Delta T_m = T_m(\text{blend}) - T_m(\text{iPP})$ depend on the nature of the EPR copolymer. The larger effect in increasing T_m is obtained with EPCAR (1). The melting behaviour of iPP/EPR blends was explained by assuming that, during blending, the EPR copolymers are able to extract selectively from the iPP bulk molecules with an average higher concentration of chemical defects leaving a matrix of iPP molecules having, on average, a higher degree of stereoregularity. The polypropylene matrix is, as a matter of fact, characterized by a higher melting point and crystallinity when it crystallizes in the presence of EPR rubbers (1).

The lamella thickness (Lc), deduced by SAXS experiments (5), of iPP crystals, grown from iPP/PiB blend sheets, monotonically diminishes, for a given crystallization condition, with increased PiB content (see Fig. 12). Such a behaviour cannot be explained by thermodynamic considerations. In fact, if the PiB acted in the melt as a diluent for iPP (i.e. if the polymers were compatible) one should expect, at a given crystallization temperature, an increase of Lc with increasing rubber concentration because of the corresponding undercooling lowering. In the absence of appreciable nucleation effects it is likely that morphological and kinetic factors are responsible for the Lc depression. The inclusion of spherical elastomer domains in intraspherulitic regions during iPP spherulite growth may cause in fact some kind of hindrance to the lamellar crystal development.

Optical micrographs of iPP/LDPE films, taken at a crystallization temperature high enough to prevent any LDPE crystallization show that, during the growth of iPP spherulites, the melt LDPE is partially incorporated in intraspherulitic regions by forming distinct domains (13). The occlusion of LDPE particles introduces large changes in the internal structure of iPP spherulites (16). The LDPE domains hinder the spherulite growing front, causing distinct concavities (Fig. 13). Examination of surface replicas of the iPP/LDPE blends crystallized isothermally at 129 and 135°C shows that the lamellae, after passing the LDPE occlusions, turn and surround them. Since the lamellae from opposite sides of the occlusion do not coincide, they form a boundary beyond each LDPE drop (Fig. 14). In bulk samples the boundary beyond an LDPE occlusion (with respect to the iPP spherulite center and the direction of spherulite growth) resembles a line (not a plane as do boundaries between spherulites in bulk materials) since growing lamellae surround the LDPE drop on all sides around the particle circumference. Such boundaries formed beyond each LDPE occlusion introduce additional paths along which cracks can travel in impact tests. However, those boundaries are dead-ended, with soft occlusions. Scanning electron micrographs of the fracture surface of an iPP spherulite in an 80/20 iPP/LDPE blend broken in an impact test at room temperature shows that the drops of dispersed LDPE are revealed by the fracture. The area occupied by LDPE drops as seen on the micrograph ought to be equal to the volume concentration of LDPE if the fracture path is random. The measurements done for fracture fragments indicate that 34% of the total area of the micrograph is occupied by uncovered LDPE drops and voids instead of 20%, which is the approximate volume concentration of LDPE in the blend. These observations suggest that the fracture path in iPP is directed towards the embedded LDPE drops having the largest perimeters. Detailed examination of carbon replicas of fracture surface in a transmission electron microscope reveals intra-spherulitic boundaries lying on the fracture path (see Fig. 15). It is evident that the fracture tip is directed to boundaries and is channelled there losing its energy on meeting an occlusion of LDPE at the end. Such a process which explains the impact strength improvement of iPP/LDPE blends, compared to plain iPP, may represent a general mechanism of fracture for rubber modified semi-crystalline polymers (16).

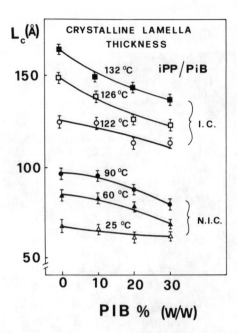

Fig. 12 – Crystalline lamella thickness, Lc, of iPP/PiB blends as
a function of elastomer content and temperature as
indicated.   Ref. 5.

234

Fig.13 – Polarizing optical micrograph of crystallizing iPP spherulite in a blend with 50% of LDPE (Tc=129°C; 200x) Ref.16.

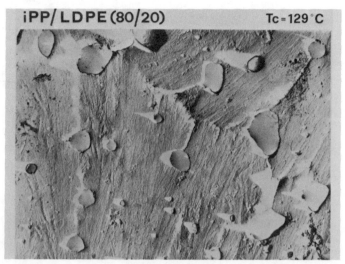

Fig.14 – Transmission electron micrographs of surface replicas of
iPP blend containing 20% LDPE (Tc=129°C; 7524x) Ref.16.

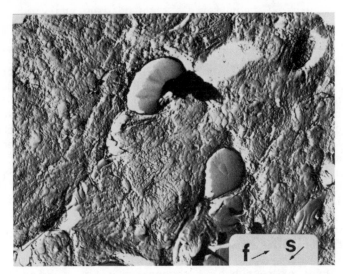

Fig.15 - Transmission electron micrograph of fracture surface of
iPP blend containing 20% LDPE (T$_c$=129°C, impact test conducted at room temperature; 4900x). Arrow S shows the
direction of spherulite growth, arrow f shows the direction
of fracture propagation. Ref. 16.

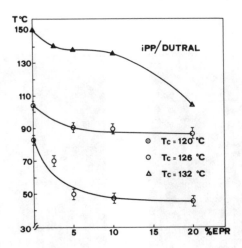

Fig.16 - Lower temperature at which iPP/Dutral specimens, on tensile deformation, are able to flow becoming ductile, as function of EPR content (W/W) (samples isothermally crystallized) Ref. 18.

TENSILE MECHANICAL PROPERTIES - COLD DRAWING BEHAVIOUR

Isotactic polypropylene/ethylene-propylene Random Copolymer Blends

From the trends of the stress strain curves of extrudate filaments of iPP/DUTRAL and iPP/EPCAR one can infer a decrease of the Young's modulus (E) and the yield stress (σ_y) with increased EPR content (17). The extent of such effects seems to depend on the nature of the copolymer added. The ratio between the modulus E and the overall crystallinity Xc is found to be dependent on the copolymer concentration, decreasing with an increase of EPR percentage. Such a result indicates that the lowering in the values of E cannot be attributed only to a decrease in the overall crystallinity of the blends. A drastic lowering in the values of the tensile strength (σr) is observed in the case of iPP/EPCAR blends while for iPP/DUTRAL mixtures σr is found to be almost constant when plotted against EPR content.

The type of response to mechanical tensile deformation of iPP/EPR blends was found to be strongly dependent on factors such as crystallization conditions, composition, temperature and rate of deformation (18). For a given crystallization temperature and rate of deformation it was found that the lower temperature at which the material is able to flow, becoming ductile on deformation, decreases with an increase in the Dutral content (Fig. 16). The WAXS diffractograms of samples of iPP and iPP/DUTRAL blends, which have been quenched and deformed to rupture at R.T., are compared in Fig. 17.

a)   b)

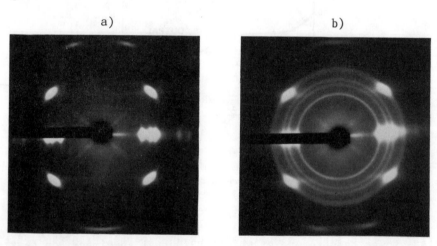

Fig. 17 - WAXS diffractograms of specimens of plain iPP a) and
         iPP/DUTRAL (80/20) blend b) quenched and deformed to
         rupture at R.T.   Ref. 18.

It can be observed that in the case of the deformed blend specimen the diffractogram shows both spots and circles indicating that in the sample, together with a well oriented iPP fibre structure, an isotropic iPP phase is still present. These observations indicate that the presence of Dutral makes the cold drawing process easier even though the morphological transition from spherulite to fibre, in the matrix, seems to be not completely achieved before rupture at R.T., in the case of blend samples obtained by quenching from the melt. Such behaviour could be related to the fact that, during melt mixing, parts of iPP molecules with different degrees of stereoregularity and/or molecular mass segregate forming regions with different molecular structure and different flowability. The WAXS diffractographs of samples of iPP and an iPP/PiB (80/20) blend isothermally crystallized at 120°C and deformed to rupture at 85°C show only the presence of spots. This indicates that the transition to fibre morphology is completely achieved even for the blend under such conditions.

Isotactic polypropylene/polyisobutylene Blends

The modulus of NIC samples, for the same crystallization condition, decreases almost linearly with increasing PiB content. Moreover for a given composition, it also diminishes with decreasing quenching temperature (T_q). These findings can be only partly attributed to a decrease in the overall crystallinity as plots of E/X_c versus PiB content show a decreasing trend. The strength behaviour is more complex because increasing T_q from 25 to 90°C and increasing the PiB content causes the specimens to become more and more brittle. In fact at 25°C all the specimens are very ductile undergoing rupture after neck formation and propagation (cold drawing). At 60°C pure iPP undergoes fibre rupture whereas the blend specimens containing 30% of PiB all break before yield. In between these two extreme behaviours the blend specimens containing 10 and 20% of PiB partly rupture after complete fibre formation and partly before. An analogous behaviour is shown at 90°C but in this case at 20% of PiB content the fibres are already no longer formed (5). Such findings indicated a progressive embrittlement of the blends as T_q increases and as the amount of PiB increases (see Fig. 18). The modulus of IC samples decreases with an increase in the percentage of PiB and it seems to be within experimental error, almost independent of the crystallization temperature. Both the results can be attributed to the overall crystallinity content of the blends which, of course, diminishes with increasing rubber content but stays almost constant with temperature in the considered range (122–132°C). The strength of the IC specimens, which all undergo brittle failure before yield, decreases with increasing PiB content. Furthermore, particularly at low PiB content, the strength decreases with increasing crystallization temperature. At higher PiB content such as effect is less and all the curves tend to merge together (see Fig. 19). The very brittle behaviour of the IC samples and the more ductile behaviour with decreasing quenching temperature for the NIC specimens, keeping the PiB content constant, can be attributed to a series of factors such as: the different number of tie-molecules, capable of carrying the load, trapped among the iPP crystallites at the different undercooling; the lower thickness and perfection of lamellar crystals grown at higher undercooling; the higher resistance to slipping and fragmentation processes of more perfect and thicker crystals grown at lower undercooling and finally to different sizes of spherulites and structure of boundary regions. On increasing the PiB content at constant Tc (Tq) the number per unit volume of rubber particles occluded in intraspherulitic regions increases in the system rendering the cold flow of the iPP matrix more and more difficult. This induces a higher and higher instability with rupture which, starting locally (where the stress concentration is higher), will produce more brittle fracture the greater the amount of the PiB. Furthermore an increase in the PiB content progressively reduces the cross-section of the matrix, which, in the absence of adhesion between rubber particles and matrix (as from morphological observations seems to be the case), carries all the load (5).

An interesting correlation can be found between the tensile mechanical properties of iPP/PiB blend sheets and the long spacing (L) and lamellar thickness (Lc) of the iPP matrix. At a given composition E increases with increasing L and Lc in the case of NIC specimens while it keeps almost constant for IC samples (see Fig. 20). As far as the strength is concerned it is observed that, for a given crystallization condition, the ratio $\sigma r/Xc$ (where Xc is the overall crystallinity of the blends) is an increasing function of Lc (see Fig. 21). As no nucleation effects have been found, at least under the experimental conditions used, these measurements should truly isolate lamellar thickness effects if morphology effects, due to different mode and state of dispersion of the minor component, are negligible. As shown in Fig. 21 the ratio $\sigma r/Xc$, for a given composition, decreases with increasing Lc. It must be noted however that the data points on the curves refer to specimens with probably different spherulite size and mode and state of dispersion of the minor component.

In the case of a semi-crystalline polymer the mechanical properties are related to the values which two independent variables, the size of spherulites and the lamellar thickness Lc, assume (19). According to the theory of Halpin and Kardos (20) a spherulitic polymer is regarded as a composite material with the lamellae acting as reinforcing fibres. The mechanical properties will be determined by the value of the aspect ratio ξ; defined as the ratio between the diameter of the spherulite and the lamella thickness. The theory predicts that the modulus should first

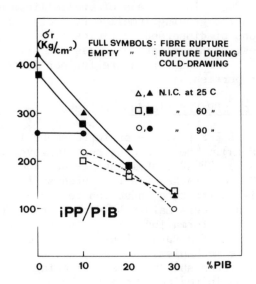

Fig. 18 – Strength (σr) of iPP/PiB blend sheets as a function of
PiB percentage for specimens non–isothermally crystallized
(NIC) at temperatures indicated. Ref. 5.

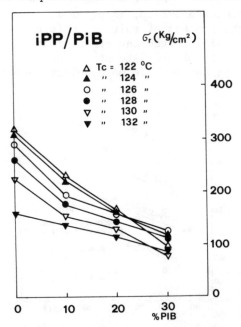

Fig. 19 – Strength (σr) of iPP/PiB blend sheets as a function of
PiB percentage for specimens isothermally, crystallized
(IC) at different temperatures as indicated. Ref. 5.

increase with ξ and then E should become constant for values of ξ greater than ca. 1000. In the case of crystallizable blends the defects induced in the growing lamellae by occlusion in intra-spherulitic regions of dispersed particles, the relative dimensions of the spherulites, the lamellar thickness and sizes of particles, and the adhesion at the boundaries are further factors determining the mechanical properties.

Impact Behaviour of iPP/EPR Blends

The addition of EPR produces an improvement in the impact resistance of iPP (4). The extent of such improvement at R.T. is larger in the case of EPCAR containing blends (see Fig. 22). Low magnification SEM micrographs of a fractured surface of pure iPP show that there is no fracture induction region and that the fracture propagation occurs mainly by direct crack formation without multiple craze formation. Such observations agree with a brittle mechanism of fracture. Binary iPP/copolymer blends show a different fracture mechanism. In fact, all the fractured specimens of such blends show a stress-whitening effect due to multicraze formation in regions close to the notch. The material volume involved in such a phenomenon increases with an increase of the copolymer content and is larger in the case of iPP/EPCAR blends. The larger R values observed in the case of iPP/EPCAR blends, together with the larger volume of material involved in the stress-whitening phenomenon, is probably due to the fact that, in such blends, the process of multicraze formation and termination is more effective owing to the larger interparticle distances and/or to the higher capacity of EPCAR to act as nucleant for iPP spheru-lites and to extract and occlude more defective molecules of iPP.

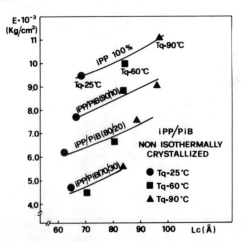

Fig. 20 – Modulus (E) versus lamellar thickness (Lc) for N.I.C. specimens of iPP/PiB sheet blends.

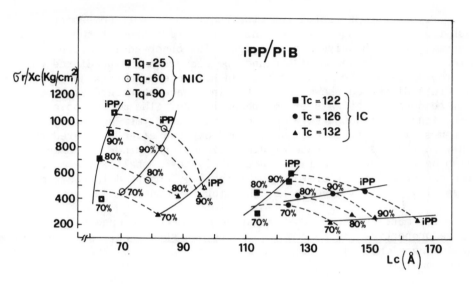

Fig. 21 – Ratio σr/Xc versus lamellar thickness Lc for IC and NIC
specimens of iPP/PiB sheet blends.

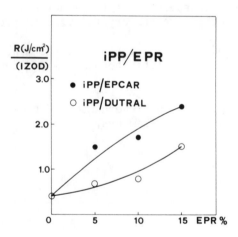

Fig. 22 – Izod impact strength (R) of iPP/EPR blends as a function
of the rubber content.

CONCLUDING REMARKS

We have demonstrated that the addition of soft components substantially changes the intrinsic characteristics of the iPP matrix. Thus the diversity of properties observed in various classes of iPP based blends cannot be described and predicted only on the basis of the crude idea of an homogeneous matrix and a discrete dispersed phase and on the interface structure as usually may be done in the case of amorphous blends. The macroscopic behaviour of blends with a semi-crystalline polymer as the matrix depends, in a very complex way, upon combined effects related not only to the mode and state of dispersion of the minor component but also on the supermolecular structure of the crystallizable matrix. The entire diversity of forms of morphology observed in the semi-crystalline matrix have to be taken into account.

The matrix must be then regarded as an heterogeneous body whose properties are determined by its own structural characteristics and changes induced by blend formulation, processing and crystallization conditions.

In conclusion it may be said that in order to characterize a blend with a crystallizable polymer as the matrix, the influence of composition, methods of formulation and crystallization conditions on a series of fundamental factors must be investigated and clarified. Such factors are:
a) Mode and state of dispersion of the minor component and its evolution in time during the crystallization process of the matrix.
b) The structure and size distribution of the spherulites, the structure of interspherulitic boundaries and their molecular nature.
c) The inner structure of the spherulites, i.e. lamellar and interlamellar thickness and the number of tie molecules between these structural elements.

REFERENCES

1. Martuscelli, E., C. Silvestre and G. Abate. Polymer 23 (1982) 229.
2. Martuscelli, E., C. Silvestre and L. Bianchi. Polymer 24 (1983) 1458.
3. Bartczak, Z., A. Galeski and E. Martuscelli. Polym.Eng.Sci., in print.
4. D'Orazio, L., R. Greco, C. Mancarella, E. Martuscelli, G. Ragosta and C. Silvestre. Polym.Eng.Sci. 22 (1982) 536.

5. Bianchi, L., S. Cimmino, A. Forte, R. Greco, E. Martuscelli, F. Riva and C. Silvestre. J.Mat.Sci., in print.
6. Taylor, G.I. Proc.R.Soc.London 146 (1934) 501.
7. Tomotika, S. Proc.R.Soc.London 150 (1935) 322. ibidem 153 (1936) 308.
8. Heikens, D. and W. Barentsen. Polymer 18 (1977) 69.
9. Karger-Kocsis, I., A. Kallo and V.N. Kuleznev. Polymer 25 (1984) 279.
10. D'Orazio, L., R. Greco, E. Martuscelli and G. Ragosta. Polym. Eng.Sci. 23 (1983) 389.
11. Martuscelli, E., C. Silvestre, R. Greco and G. Ragosta. Polymer Blends vol.1 (Plenum Press, 1980), edited by E. Martuscelli, R. Palumbo and M. Kryszewski.
12. Martuscelli, E. and H. Janik. work in progress.
13. Martuscelli, E., M. Pracella, G. Della Volpe and P. Greco. Makromol. Chem. 185 (1984) 1041.
14. Fraser, G.V., A. Keller and J.A. Oddell. J.Appl.Polym.Sci. 22 (1978) 2979.
15. Friedrich, K. Progr.Colloid and Polymer Sci. 64 (1978) 103.
16. Galeski, A., M. Pracella and E. Martuscelli. J.Polym.Sci.Polym. Phys.Ed., in print.
17. Greco, R., G. Mucciariello, G. Ragosta and E. Martuscelli. J.Mat.Sci. 15 (1980) 845.
18. Coppola, F., R. Greco, C. Mancarella and E. Martuscelli. work in progress.
19. Patel, J. and P.J. Phillips. Polym.Letters Ed. 11 (1973) 77.
20. Halpin, J.C. and J.L. Kardos. J.Appl.Phys. 43 (1972) 223.

This work was partly supported by: Progetto Finalizzato Chimica Fine del C.N.R.

RUBBER-RUBBER BLENDS

P J Corish

Dunlop Technology Division, Birmingham

Blending of two or more rubbers is carried out for three main reasons:

- Improvement in Technical Properties
- Better Processing
- Lower Compound Cost.

Many products in the rubber industry are based on blends in all or part of their construction. Tyres are typical examples of products in large-scale volume production.

Table 1 lists important Properties and Processing Characteristics for 4 components of tyres, viz. treads, casings, sidewalls and liners. The requirements may be seen to be complex and contradictory. Figure 1 shows the properties of 4 different types of rubber related to tyre properties expressed in terms of their glass transition temperature (Tg). It may be seen that cis polybutadiene has the lowest wet grip and air impermeability but has the best low temperature and heat build up performance. Chlorobutyl (ClIIR), however, has very good wet grip and air impermeability properties but poor (high) heat build and poor low temperature performance.

TABLE 1

IMPORTANT PROPERTIES AND PROCESSING CHARACTERISTICS
OF TYRE COMPOUNDS

	PROPERTIES	PROCESSING
Treads	Wear resistance Wet grip Heat build-up (Rolling resistance)	Extrudability (Die Swell, smoothness) Butt jointing
Casings	Heat build-up Non-reversion (Heat & Overcure stability) Cord/Wire Adhesion	Calendering ability Cord 'strike through'

Tack/green strength

Sidewalls	Flex) Ozone) Cracking	Extrudability
Liners	Air permeability "Cheese cutting" Fatigue resistance	Calendering) Extruding) Ability

TABLE 2

RUBBER PRICES (£/TONNE)

NR	835
SBR 1500	905
1712	755
Butyl	1380
Chlorobutyl	1482
EPDM	1100 - 1624
Neoprene	1310 - 1910
NBR low AcN	1350
medium "	1550
high "	1750

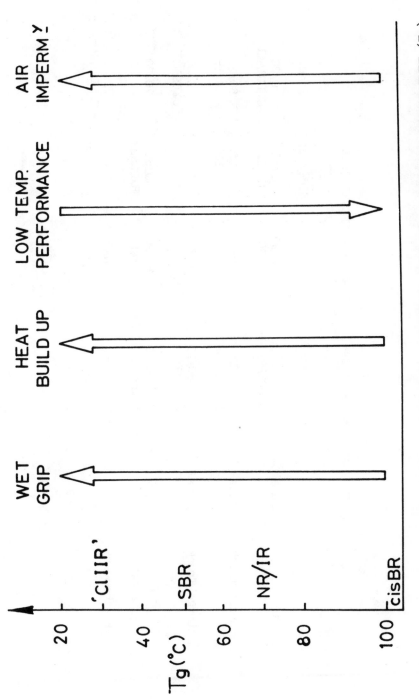

Fig.1. SOME IMPORTANT TYRE PROPERTIES OF RUBBERS vs GLASS TRANSITION TEMPERATURE (T_g)

TABLE 3

TYPICAL RUBBERS IN CAR, TRUCK AND EARTHMOVER TYRES

		Tread	Casing	Sidewall	Liner
Car	Radial	75/25 OESBR/SBR	70/30 NR/SBR	50/50 NR/BR	70/30 ClIIR/NR
Light Truck	Radial	33/33/34 NR/OESBR/OEBR	NR	50/50 NR/BR	80/20 NR/IR
Large Truck	Radial	80/20 NR/IR	NR	50/50 NR/BR	70/30 ClIIR/NR
	Cross-Ply	70/30 NR/IR or NR/BR	80/20 NR/IR	40/60 NR/OESBR	70/30 NR/SBR
Earthmover		NR	NR	NR	70/30 ClIIR/NR

Another cogent factor is the price of the rubber. Table 2 gives a reasonably up-to-date picture of comparative costs. The low price of oil-extended SBR 1712 is responsible for its widespread use and illustrates the point that compound cost is the real criterion and that volume cost should really be considered. Rubbers of low SG are therefore preferred. Bearing in mind these three reasons, rubber usage in tyre components is shown in Table 3 (1). For general processing and property considerations, most of the blends involve a proportion of natural rubber.

For car tyre treads, SBR is preferred because of its good wet grip property but it has a penalty in heat build up performance. In large tyres, where heat build up must be minimised, NR or blends of NR with IR are used. For inner liners, where air impermeability is a pre-requisite, chlorobutyl is normally preferred.

In the non-tyre area, two important property considerations for automotive rubbers are heat and oil resistance. These are illustrated in Figure 2 (2). By appropriate blending of rubbers, improvements in heat/oil resistance may be achieved, without too much cost penalty, with compounds of reasonable processability.

Blending of rubbers can be practically carried out in several different ways. These include latex and solution blending and mixing of solid rubbers - on mills, in internal mixers or during continuous mixing, sometimes in particulate or powder form. These methods have been described in a previous review (3) and will have been generally covered in Paper 7 but aspects relating to filler and curative distribution will be discussed later.

A novel method of blending is the 'Sol-Latex' Process (4) in which by use of an oppositely charged soap, a rubber latex and a solution of another rubber produce a solution blend.

Earlier papers in this course and in a previous publication (5), will have pointed out that blending of two different high molecular weight rubbers can never give rise to compatibility on the molecular scale. However, using some tests which are a guide to the performance of blends under dynamic service conditions, some rubber blends exhibit compatibility (6). Such tests include measurements of glass transition temperature, Tg, by dilatometry (static) or by dynamic tests at various frequencies, eg rolling ball spectrometer and dynamic damping. Using Tg as a test of compatibility, an incompatible blend has been defined as one which always exhibits two or more values of Tg, these being the same as those of the constituent rubbers in the blend. Some examples are given in Table 4. A compatible blend is defined as one which has a single intermediate value of Tg. To achieve this, the blend must consist of rubbers whose solubility parameters (δ) differ by less than 0.5. Some typical examples of such blends are shown in Table 5. However, such rubber blends do not always exhibit a single value of Tg.

250

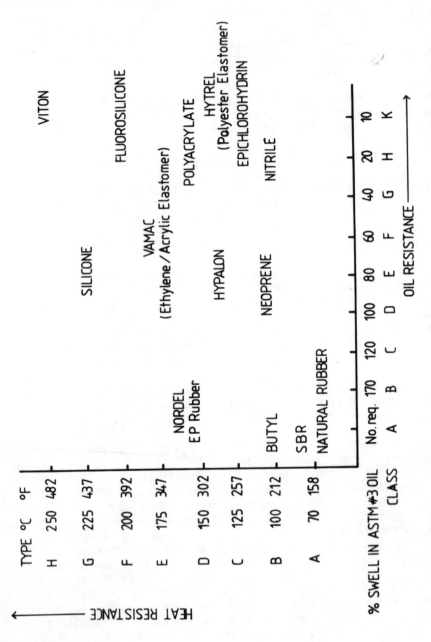

Fig. 2. HEAT / OIL RESISTANCE OF AUTOMOTIVE ELASTOMERS (SAE J200 Spec. System)

TABLE 4

INCOMPATIBLE BLENDS

	Tg°C	Solubility Parameter δ	$\Delta\delta$
NR	−68.5	8.25	
Medium High Nitrile NBR	−30	9.55	
50/50 NR/NBR	(−67.5 (−28		1.3
NR	−68	8.25	
Neoprene GNA	−43	9.26	
50/50 NR/Neoprene	(−65 (−44		1.01

TABLE 5

COMPATIBLE BLENDS

	Tg°C	Solubility Parameter δ	$\Delta\delta$
NR	−68.5	8.25	
Solprene 1204	−57	8.25	
50/50 NR/Solprene	−63.5		0.00
Diene 55NF	−95	8.17	
SBR 1712	−50	8.13	
58/42 SBR 1712/Diene 55	−78		0.04

252

Behaviour of blended rubbers in products cannot simply be considered in terms of rubber/rubber pairs. Three aspects of the
compounding of rubber/rubber blends will be discussed in detail:

- Filler distribution and level

- Curing aspects

- Effects of extenders, especially oils

- Filler Distribution

     It has been shown by electron microscopy studies (7) that
in a preblend of 50% NR with 50% cis BR, ISAF black preferentially locates itself in the cis BR phase. This hetero-
distribution affects vulcanisate properties giving rise to
optimum performance in wear, tensile and tear strengths and in
hysteresis/heat build up properties (Figures 3 and 4).

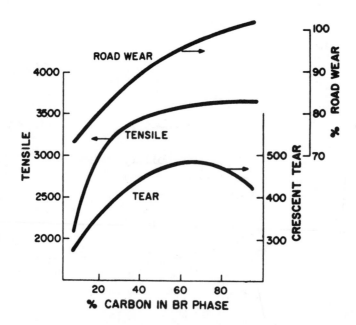

Fig.3. EFFECT OF CARBON BLACK DISTRIBUTION ON VULCANISATE
        PROPERTIES (40 phr ISAF black; 50/50 NR/cis BR)

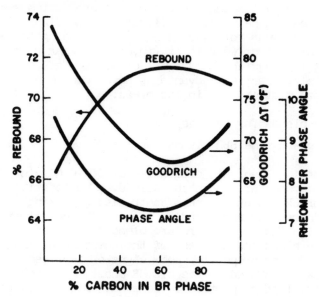

Fig.4. EFFECT OF CARBON BLACK DISTRIBUTION ON HYSTERESIS (40 phr
ISAF black; 50/50 NR/cis BR)

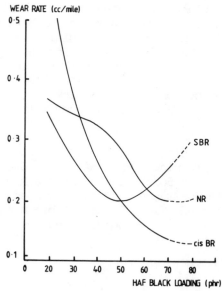

Fig.5. EFFECT OF HAF CARBON BLACK LOADING ON WEAR RATE FOR VARIOUS
RUBBERS

Rubbers have been shown (8) to have optimum filler (esp-ecially carbon black) loadings for important properties such as wear and tear resistance. This is illustrated in Figure 5 which displays graphs of wear rate versus HAF black loading for various rubbers. The maximum wear index (inverse of wear rate) occurs at lower black loadings for rubbers of higher Tg. Using this guiding principle, practical advantages have been found (9) for control of black distribution in two areas:

(a) NR/cis BR blends for casings

(b) cis BR/high-styrene SBR blends for treads

(a)   General vulcanisate properties of 50/50 NR/cis BR casing compounds are shown in Table 6. The location of the 40 phr GPF black has been controlled by mixing separate black motherstocks and blending these. The normal mix is one in which the two rubbers are blended in a Banbury before adding the black and the rest of the compounding ingredients. Tensile and tear properties show signs of improvement as more black is located in the cis BR phase. However, the general level of properties is somewhat inferior to the NR control.

Improvements in the general level of properties were achieved by using 75/25 NR/cis BR blends. The results (Table 7) again demonstrate increased tensile and tear strengths as more black is located in the cis BR phase. Dunlop Punch Fatigue results, shown in Table 8, confirm the dependence of properties on the initial black location in the blends. When all the black is located in the cis BR phase, a punch life equivalent to the NR control is obtained, appreciably better than that of the conventional mix. Thus, it may be concluded that 25% cis BR can be incorporated into the NR casing compound without impair-ment in heat build up behaviour.

(b)   50/50 blends of cis BR and a high styrene SBR (36% bound styrene), containing 50 phr HAF black overall, were prepared by blending a 50/50 cis BR/HAF black masterbatch and SBR gum. Both unextended and oil-extended versions of the rubbers were found (Table 9) to exhibit higher abrasion resistance than identical compounds mixed conventionally.

TABLE 6

50/50 NR/CIS BR CASING COMPOUNDS (Containing 40 phr GPF Black)

General Physical Properties (Optimum Cure)

	100% NR	All black in BR	50% black in BR	All black in NR	Normal mix
Tensile strength Kg/cm²	250	179	171	161	156
300% Modulus "	124	110	104	95	102
Elongation @ break %	511	435	438	439	415
RT Tear Resist. Kg/tp	19.1	15.1	15.6	11.5	11.3
Resilience @ 50°C %	87.5	86	84.5	84	86
Hardness °BS	59	57	59	56	60

75/25 NR/CIS BR CASING COMPOUNDS (Containing 40 phr GPF Black)

TABLE 7

General Physical Properties (Optimum Cure)

	100% NR	All black in BR	75% black in BR	50% black in BR	75% black in NR	All black in NR	Conventional mix
Tensile strength Kg/cm²	236	211	195	192	177	182	187
300% Modulus "	117	114	109	109	103	99	105
Elongation @ break %	509	452	451	452	444	463	456
RT Tear Resist. Kg/tp	27.0	25.2	20.1	18.4	18.8	23.4	17.7
Resilience @ 50°C %	87	89.5	88	86.5	86	86.5	86.5
Hardness °BS	58	55	57	58.5	58.5	57.5	58.5

TABLE 8

Dunlop Punch Fatigue Results (0.4 J/cc Work Input)

	100% NR	All black in BR	75% black in BR	50% black in NR	75% black in NR	All black in NR	Conventional mix
Punch life (log cycles to failure)	5.89	5.88	5.72	5.51	5.35	5.39	5.38
Av running Temp (°C)	124	114	126	130	145	140	132

TABLE 9

ABRASION INDICES

50/50 BR/36% Styrene SBR blends

Effect of incorporating all black in BR component initially

Cornering Coefficient	oil-free*		oil-extended**	
	m/b mix	std mix	m/b mix	std mix
0.2	131	105	136	113
0.3	110	96	124	96

*   SBR 1500/50 HAF = 100
**   SBR 1712/50 HAF = 100

TABLE 10

ABRASION INDICES

50/50 OEBR/32% Styrene OESBR blends

Effect of remilling 50/50 OEBR/Black m/b

Cornering Coefficient	No remill	2-min remill	5-min remill	10-min remill	Conventional compound
0.2	85	118	118	109	100
0.3	63	108	104	97	100

However, it was subsequently found that conditions of preparation of the high black cis BR masterbatch affect the abrasion results. Table 10 shows that, for 50/50 blends of oil-extended cis BR with a 32% styrene oil-extended SBR, remilling times of the masterbatch need to be controlled in order to optimise the abrasion resistance.

Also, the blending step itself can influence the abrasion results. Data shown in Table 11, for 50/50 blends of oil-extended cis BR with a 40% styrene oil-extended SBR, indicate that some blending is obviously necessary but that too much blending can reduce the abrasion advantage.

TABLE 11

ABRASION INDICES OF 50/50 OEBR/40% STYRENE OESBR BLENDS

Effect of time of blending of 50/50 OEBR/Black masterbatch and OESBR Gum

Cornering Coefficient	4' blend	8' blend	15' blend	std mix
0.2	110	125	112	100
0.3	106	120	111	100

TABLE 12

ABRASION INDICES OF 50/50 OEBR/40% STYRENE OESBR BLENDS

Effect of initial black location

Cornering Coefficient	OEBR/HAF OESBR/HAF	50/45 50/5	50/45 50/10	50/35 50/20	50/30 50/25	Conventional mix
0.2		115	120	107	109	100
0.3		121	123	120	110	100

When the proportion of black in the oil-extended cis BR and 40% styrene oil-extended SBR was varied systematically, it may be seen from Table 12 that highest abrasion resistance was developed when the maximum amount of black was located in the cis BR phase. These hetero-dispersed blends contained an additional 5 phr black to compensate for their slightly lower vulcanisate hardness.

In a more general study, similar trends were obtained in oil-extended cis BR and 40% styrene oil-extended SBR with high structure HAF and SAF blacks.

It would seem possible, therefore, by correct filler distribution in rubber blends, especially those involving cis BR, to enhance certain vulcanisate properties, eg abrasion resistance and heat build up characteristics. However, processing of high black-loaded cis BR master-batches is difficult on a large scale and availability of solution masterbatches would facilitate further progress.

It has been demonstrated that the surface polarity of the carbon black influences its distribution in rubber blends. Basic furnace blacks locate in the cis BR phase in cis BR/NR blends but acidic channel blacks locate in the more polar NR phase. Inorganic fillers also behave differently, eg silica locates in the NR phase of NR/cis BR blends (7).

Filler distribution also depends on the initial viscosity levels of the rubber phases and the blending methods involved. It is also possible by chemical and thermal promotion to locate a filler permanently in one rubber phase to obtain a preferred distribution. This area is therefore one of complexity but not without opportunity

● Curing aspects

In addition to the solubility parameter criterion, achievement of a single Tg in compatible rubber blends requires curing to develop interphase crosslinking. Loss peaks obtained (6) on a rolling ball spectrometer for a 50/50 blend of cis BR with NR are shown in Figure 6. Separate loss peaks for uncured and lightly cured blends merge together on progressive vulcanisation to provide a mechanically compatible blend. Blends of cis BR with emulsion SBR display (7) two dynamic mechanical loss peaks in the uncured state characteristic of the individual rubbers. These peaks merge quickly to form an intermediate loss peak on vulcanisation of the blends.

260

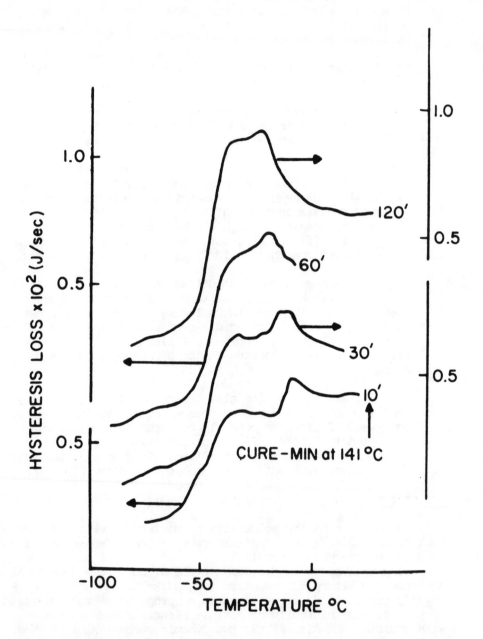

Fig.6. HYSTERESIS LOSS OF A BLEND OF CIS BR AND NR AS A FUNCTION
OF TEMPERATURE AND CURE TIME

TABLE 13

CRACKING AND CUT GROWTH PROPERTIES OF NR/EPDM BLENDS

	1	2	3	4
EPDM Crumb 5' Cure @ 135°C	39	–	–	–
EPDM Crumb 5' Cure @ 150°C	–	39	–	–
Std Uncured NR	61	61	61	–
Std Uncured EPDM	–	–	39	–
NR Std Control	–	–	–	100
Ozone Cracking*	Small to medium cracks overall	Small cracks overall	No cracking	V small cracks overall
Cut Growth (mm/hr)	3.36	4.53	0 to 0.1	0.1

* 18 hrs @ 50°C extension, 50 pphm O_3 @ 38°C

These differences in behaviour between the two pairs of blended rubbers are associated with cure compatibility: formation of interphase crosslinks is in competition with intra-phase crosslinking. In the NR blend, the faster curing NR appears to attain a fairly tight network before interphase crosslinking takes place.

Diffusion of soluble curative ingredients such as sulphur and accelerators occurs between dissimilar rubbers in contact. As solubility of these curatives is greater in high unsaturation rubbers than in low unsaturation rubbers, migration will occur into the former type (10). As the cure rate is faster in the high unsaturation rubbers, the imbalance will be accentuated. In view of the associated under- and over-cure of the phases, the vulcanisate properties of blends of high and low unsaturation rubbers may not attain the desired levels. Attempts have been made, particularly with blends of EPDM with high unsaturation rubbers, to improve compatibility and co-crosslinking. These include the use of long-chain hydrocarbon dithiocarbamate accelerators masterbatched into the EPDM (11) and the grafting of accelerators onto the EPDM (12).

However, mismatch of cure can be beneficial in some circumstances: advantages in ozone and cracking resistance of EPDM blends with high unsaturation rubbers is claimed (13) to be associated with the 'crack stopping' function of the relatively uncured EPDM phase.

To test out these claims and to examine another possible way of controlling the diffusion of curatives out of the EPDM phase and into the NR phase in EPDM/NR blends, some studies were carried out with lightly pre-cured EPDM compounds (14). Table 13 lists brief details of the compounds, including the pre-cures used for the EPDM masterbatches which were subsequently finely crumbed on a mill, using a tight nip, before incorporation into the NR masterbatch. Significant vulcanisate properties also listed include ozone cracking and cut growth performance. It is clear that although the NR/EPDM blend in which uncured compound masterbatches are blended together gives advantage in ozone cracking and possibly cut growth, blends containing the pre-cured EPDM compounds do not.

Control of the location of curatives is clearly important in optimising properties of blends.

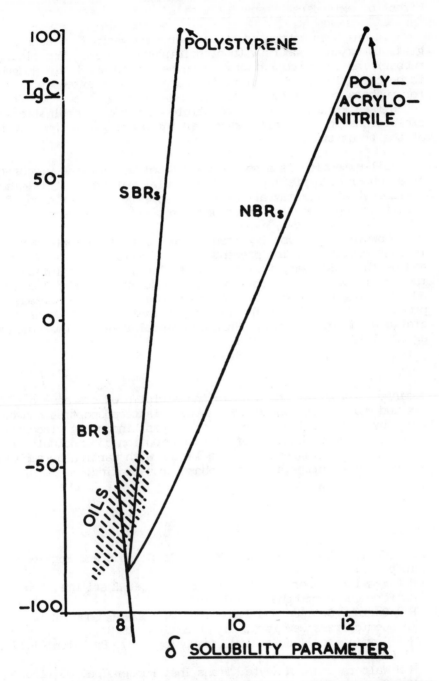

Fig.7. THE RELATION OF SOLUBILITY PARAMETER TO GLASS TRANSITION
TEMPERAUTRE FOR EXTENDER OILS AND SEVERAL RUBBER TYPES

● <u>Effect of extenders - especially oils</u>

It has been common practice over many years for technologists to extend rubbers with plasticisers or oils. Empirically, rubber/plasticiser combinations have been used which are judged to be compatible. The solubility parameters of extender oils for rubbers range from 7.5 to 8.5 and their Tgs from -90°C to -45°C (see Figure 7) (6). Compatible rubber/oil combinations exhibit one Tg, both oil and rubber influencing the temperature of the transition.

Oil extension is also one way of influencing the viscosity of a rubber of differing 'base' molecular weights. It is also possible that oils may affect the solubility and diffusion of curatives and hence the interphase crosslinking process.

However, the most important reason for use of extender oils is cost reduction. The present price of oils is ca 250 £/tonne and as their SGs vary from ca 0.88 to 1.05, they can influence the volume cost of the compound in a major way. The oil can be added during the compounding of the rubber or can be already present in the rubber. The latter alternative is often associated with production of a higher molecular weight base polymer, eg SBR 1712.

To summarise, rubber blends used in industrial products have a complex and multifunctional nature. Vulcanisate properties can be optimised by proper selection of rubber types, including viscosity considerations and by control of the distribution of insoluble and soluble compounding ingredients, the latter with particular emphasis on achieving good intra- and inter-phase crosslinking.

References

1. P J Corish, Encyclopedia of Materials Science and Engineering, in press.
2. R E Fenwick, Paper 17, Plastics and Rubber Institute, Rubber Conference, Birmingham, March 1984.
3. P J Corish, Science and Technology of Rubber, Chapter 12 (Academic Press New York, ed F R Eirich, 1978).
4. R P Campion and J F Yardley (to Dunlop Ltd), Br Patent 1,173,171 (1968).
5. M H Walters and D N Keyte, Trans Inst Rubber Ind, 38 (1962) 40.
6. P J Corish, Rubber Chem Technol, 40, (1967), 324.
7. J E Callan, W M Hess & C E Scott, Rubber Chem Technol, 44, (1971), 814.

8. D Bulgin and D L Walker (to Dunlop Ltd), Br Patent, 1,061,017 (1964).
9. P J Corish and M J Palmer, "Advances in Polymer Blends and Reinforcement", Inst Rubber Ind Conf, Loughborough (1969), 1, and unpublished work.
10. J B Gardiner, Rubber Chem Technol, 42, (1969), 1058.
11. R P Mastromatteo, J M Mitchell and T J Brett, Rubber Chem Technol, 44, (1971), 1065.
12. K C Baranwal and P N Son, Rubber Chem Technol, 47, (1974), 88.
13. Z T Ossefort and E W Bergstrom, Rubber Age (NY), 101(9), (1969), 47.
14. P J Corish, unpublished work.

PURE AND APPLIED RESEARCH ON INTERPENETRATING POLYMER NETWORKS AND
RELATED MATERIALS

L.H. Sperling

Polymer Science and Engineering Program
Department of Chemical Engineering
Materials Research Center
Coxe Laboratory 32
Bethlehem, PA 18015 USA

1. INTRODUCTION

Depending on interest and requirements, there are many ways to
prepare polymer mixtures. The simplest way involves mechanical
blending of two polymers, perhaps in an extruder or roll mill. In
the laboratory, the two polymers are sometimes dissolved in a solvent
(usually in separate containers) mixed together, and precipitated or
solvent removed. These blending techniques generally lead to
polymer I-polymer II systems which are not bonded to each other.

Important cases where the two polymers are bonded together
are the block and graft copolymers. In the block copolymer, the
chains are bonded together end-on-end. In the graft copolymer
system, polymer I serves as the backbone, and polymer II is the
side chain. Theoretically, these materials remain soluble and
thermoplastic.

There are two different ways of preparing thermoset
combinations of polymers. In one case, two polymers are grafted
together to form one network. Polymer II is bonded to polymer I at
both ends, or at various points along the chain. To distinguish
this type of grafting from the case where the product remains thermoplastic, this type of bonding is designated conterminous linking.

The second case, where two different networks are synthesized,
is called an interpenetrating polymer network, IPN. Up to a few
years ago, an interpenetrating polymer network, IPN, was defined
as a combination of two polymers in network form, at least one of

which was synthesized and/or crosslinked in the immediate presence
of the other (1). While this definition still fits most materials
called IPN's, there are now some hybrid compositions such as the
thermoplastic IPN's which call for a broader view. This paper will
review the IPN literature from 1979 to the present. The advent of
several reviews covering the literature up to 1979 dictates the
beginning date (1-4).

In all, about 125 papers and 75 patents had appeared by 1979.
This review, which will have to be incomplete because of space
limitations, will examine new papers and patents. Several IPN-based
products were identified by 1979. New products will be mentioned
in this review.

2. SEQUENTIAL IPN MORPHOLOGY

The main problems in sequential IPN morphology have centered
around the shapes and sizes of the phases and aspects of dual phase
continuity. During the 1970's, detailed studies of sequential IPN
morphology were carried out by Huelck, et al (5,6). Glass
transition studies suggested dual phase continuity for many of the
compositions studied, while transmission electron microscopy
suggested the presence of spherical domains. Most of the so-called
"spheres" were really somewhat ellipsoidal, however, as illustrated
in Figure 1 (5).

In spite of the slightly contradictory experimental picture,
Donatelli, et al (9), Michel, et al (10) and Yeo, et al (11) derived
equations showing how the phase domain diameters of the

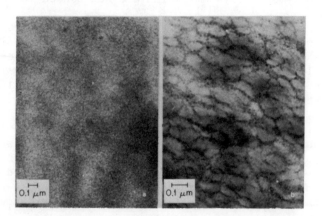

Figure 1. Morphology of sequential IPN's. (a) Poly(ethyl acrylate)/
poly(methyl methacrylate), showing significant miscibility.
(b) Poly(ethyl acrylate)/polystyrene, immiscible, showing a typical
cellular structure (5).

"sphere" of polymer II, D_2, varied with the crosslink density (concentration of network chains) of networks I and II (polymerized in that order), ν_1 and ν_2; the interfacial tension, ; each phase, Φ_1 and Φ_2; and temperature T. The model equation for a full sequential IPN may be written (11):

$$D_2 = \frac{4\gamma}{RT(A\nu_1 + B\nu_2)} \tag{1}$$

$$A = 1/2(\frac{1}{\Phi_2})(3\Phi_1^{1/3} - 3\Phi_1^{4/3} - \Phi_1\ln\Phi_1) \tag{2}$$

$$B = 1/2(\ln\Phi_2 - 3\Phi_2^{2/3} + 3) \tag{3}$$

Thus the domain diameters of the second polymerized polymer may be predicted solely on the knowledge of the networks and their inter-action. For midrange compositions, A>>B; that means that the first formed network dominates the final overall morphology.

The theory is compared to experiment for the IPN system poly(n-butyl acrylate)/polystyrene in Table I (11-14). The agreement between theory and experiment for this system as well as others is excellent, in fact, better than could be expected, noting the approximations. It must be pointed out that Yeo et al. (11) realized that spheres were a poor approximation for many sequential IPN's. However, equations based on cylinders or other shapes could not be easily derived, further developments are still required.

Widmaier and Sperling (15-17) approached the question of dual phase continuity through use of a decrosslinking and extraction route, using scanning electron microscopy techniques developed by Kresge (18). In brief, samples of poly(n-butyl acrylate) were crosslinked with acrylic acid anhydride, AAA. Then, sequential

Table 1. Domain Diameters of Poly(n-butyl acrylate)/Polystyrene
          IPN's (11)

P(nBA)/PS Volume Ratio	Diameters in Angstroms	
	D_2, Experiment	D_2, Theory
25/75	800	845
40/60	650	644
50/50	550	572

Molecular Characteristics:
ν_1 = 3.75x10^{-5} mole/cm^3
ν_2 = 21.8x10^{-5} mole/cm^3
γ = 3.65 dynes/cm

IPN's were made with polystyrene. On soaking in warm ammonium hydroxide, the AAA hydrolyzes, decrosslinking the polymer. The poly(n-butyl acrylate) was then extracted, and the sample examined by scanning electron microscopy. The porous structure shown in Figure 2 was found (15). The spherical type structures of about 0.1 micron in diameter correspond to the "spheres" mentioned above. However, clearly they are bound together to make a continuous structure. One such possibility is modeled in Figure 3.

Very recently, the morphology of polybutadiene/polystyrene sequential IPN's and semi-I IPN's was examined via small-angle neutron scattering (19). In this case, the polystyrene was deuterated to provide contrast. The results are summarized in Table II. Using the Debye plot, the correlation distance $\underline{a}$ was obtained. From that quantity, chord lengths, equivalent sphere diameters, and surface areas may be estimated. Except for the equivalent sphere diameters, no particular model of phase domain shape is required. In fact, the Debye model works best for completely random phase structures.

The results are compared to transmission electron microscopy studies on osmium tetroxide stained samples in Table II. The chord lengths and equivalent sphere diameters are in rough agreement with electron microscopy, especially the lower end of the size range reported. Surface areas in the range of 150–200 m^2/gm indicate true colloidal dimensions for the phase domains.

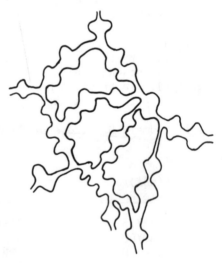

Figure 2. Decrosslinked and extracted PnBA(AAA)/PS(DVB) 50/50, viewed by scanning electron microscopy. The polystyrene phase remains, while the poly(n-butyl acrylate) was removed (15).

Figure 3. Model of sequential IPN morphology

Table II. PB/PS(D8) IPN's and Semi-I IPN Dimensions by SANS and
TEM Techniques (19).

Sample Code	Volume Fractions 1	2	Correlation Lengths $a(\text{Å})$	Intercept Lengths (Å) L_1	L_2	S_{sp} $(\frac{m^2}{gm})$	Diameters, (Å) $D^{(a)}$	$D^{(b)}$	Range
A	0.193	0.807	106	131	549	58.8	D_1 197		300–2000
							D_2 824		700–1500
B	0.270	0.730	53	73	194	149	D_1 109		150–1200
							D_2 294		350–600
C	0.230	0.770	58	75	252	122	D_1 113		150–1200
							D_2 378		500–1100
D	0.240	0.760	37.5	49	154	196	D_1 74		300–600
							D_2 234		400–800

(a) From SANS, $D_1 = \frac{3}{2}\frac{a}{(1-_2)}$; ; $D_2 = \frac{3}{2}\frac{a}{(1-_1)}$

(b) From TEM measurements

Subscript 1 = polybutadiene

Subscript 2 = polystyrene

Broadly defined, a sequential IPN is a material in which
network one is fully formed before monomer II is introduced.
Thus, the chains are extended by the swelling action of monomer II,
and the phase domain size is greatly reduced. In the case of
simultaneous interpenetrating networks, SIN's, both monomers (or
prepolymers) are introduced together. Ideally, both monomers are
reacted simultaneously but independently. In fact, this is rarely
done exactly. It seems much more convenient, and better materials
result, if the reactions are carried out more or less sequentially
after the monomers are introduced simultaneously. In an SIN,
ideally, the chains of neither polymer should be perturbed or
extended. By varying the reaction scheme, many materials may be
made that do not fit cleanly into one category or the other.

3. INTERFACIAL CHARACTERIZATION

Lipatov and coworkers have continued their characterization of sequential IPN's (20) and SIN's (21) via small-angle X-ray scattering, SAXS, and diffraction. Several quantities were determined which characterize the two-phased nature of these materials: specific surface area, S/V; distance of heterogeneity, l_c; reduced inhomogeneity length, l_p; average radius of the largest heterogeneous regions, R; and interlayer thickness, B.

The sequential IPN (20) was based on a polyurethane and polystyrene-DVB (3%). The SIN was based on a polyurethane and polyurethane acrylate. The range of values reported by Lipatov et al. (20,21) are summarized in Table III.

Of special interest are the values of B reported for the PU/PS IPN. As more PS was added, B became smaller, indicating that mixing between the phases was relatively greater at low PS levels than in mid-range compositions. The specific surface areas reported for the PU/PUA SIN are of the same order of magnitude as reported by Sperling, et al. (19), using small-angle neutron scattering, see Table II.

4. TRULY MISCIBLE IPN's

The entropy of mixing two polymeric species is very small because of the very long chains involved. Coupled with the usually positive heat of mixing, the resulting free energy of mixing is

Table III. Domain Characterization of IPN's and SIN's via SAXS Techniques (20,21).

Composition	% Second Polymer	l_p, Å	B, Å	l_c, Å	R, Å	S/V, $\frac{m^2}{gm}$
PU/PS	3.6	20	37	320	---	---
IPN	8.6	24	30	120	---	---
	31.4	72	23	180	---	---
	35.4	80	21	190	---	---
PU/PUA	0	--	--	550	---	---
SIN	5	--	--	440	360	10
	10	--	--	475	480	40
	40	--	--	454	450	290
	50	--	--	533	730	120
	75	--	--	123	430	320
	90	--	--	134	540	100
	100	--	--	280	---	---

positive, causing these materials to phase separate. This tendency to phase separate is true for polymer blends, grafts, blocks, and IPN's alike. The one known way to achieve truly miscible polymer blends is to have negative or zero heats of mixing, i.e., the two polymers attract one another.

All of the IPN systems described above are thought to have positive heats of mixing because they phase separate. The blend of poly(2,6-dimethyl-1,4-phenylene oxide), PPO, with polystyrene is known to have a negative heat of mixing, and indeed these two polymers are miscible in all proportions (22). Frisch and coworkers (23,24) decided to make SIN's out of this polymer pair, so to achieve better interpenetration of the two networks.

Frisch and coworkers (23,24) crosslinked the PPO by brominating the polymer, followed by reaction with a tertiary amine. The polystyrene was crosslinked with DVB. The PPO with ethylene diamine was dissolved in styrene monomer containing DVB and AIBN, and heated to 70°C for 24 hours. Semi-SIN's and linear blends were also synthesized. As perhaps expected, all of these materials exhibited a single T_g which varied systematically with composition, and no phase separation was observed via electron microscopy, leading to the conclusion that truly interpenetrating polymer networks were produced.

The study of homo-IPN's bears on the topic of maximizing molecular interpenetration. A homo-IPN is one in which both networks are identical in composition. Several teams have studied homo-IPN's of polystyrene/polystyrene, and this work was recently reviewed (25). While molecular interpenetration was extensive, there was still a slight degree of segregation. The segregation, of the order of 50-100Å, may be caused by random fluctuations in network I. By and large, there was little evidence for added crosslinks in the system, probably because both networks dilute each other. However, there was evidence for domination of the physical properties by network I, which behaves as if it is more continuous in space.

Along the lines of polymer/polymer miscibility in IPN's, Sperling and Widmaier (26) recently pointed out that while two polymers become less miscible as their molecular weights are raised to infinity, as crosslinks are added the polymers become more miscible. The thermodynamic explanation awaits an effective definition of the molar volume in a polymer network. For linear polymers, this is given by the volume occupied by one mole of polymer. How shall this be defined for a network? Sperling and coworkers tentatively identified this quantity with M_c, the molecular weight between crosslinks.

5. SEMI-COMPATIBLE IPN's

Adachi and coworkers have made a detailed study of the synthesis and properties of three sequential IPN systems: (1) poly(ethyl acrylate)/poly(methyl methacrylate) PEA/PMMA, (27), (2) polyoxyethylene/poly(acrylic acid) (28), and (3) poly(acrylonitrile-co-butadiene)/poly(methyl methacrylate) (29). In each case, semi-IPN's were compared with full IPN's.

In the case of the PEA/PMMA IPN's, two synthetic routes were taken: (1) The crosslinked PEA (polymer I) was swelled with MMA monomer plus EDMA and AIBN for various periods of time, and the MMA polymerized. This is the SW-series. This synthetic method was very similar to that employed by Huelck, et al.(5,6). (2) The PEA was fully swollen with monomer, about a factor of three. In each case, the polymerization was permitted to proceed for a short period of time, and the remaining monomer removed by evaporation. This is called the CV-series.

Amazingly, the two series were not alike. The SW-series, like the studies of Huelck, et al. (5,6) were transparent materials which exhibited one broad glass transition. The CV-series, however, exhibit opaque optical behavior, with two peaks in the tan δ - temperature plots.

Adachi and Kotaka (27) concluded that PMMA microgels of various sizes were presumably formed within the PEA network. The PMMA microgels eventually interconnect, and develop into a fully interpenetrating network having a microheterogeneous nodular structure. The reader is referred to Figure 3 for a model of such a structure. Adachi and Kotaka (27) were able to isolate this structure at an early stage of development by interrupting the reaction.

Most interestingly, this study applies to theories of network formation. While important theories suggest that a network polymerization from a vinyl and divinyl reaction is random, at least three theories have suggested an early microgel formation (30-32). In each of these theories, a mixture of monomer and linear polymer is formed, with microgel floating around in the mixture. As the polymerization proceeds, the gels eventually aggregate, and connect together. In the case of polymerizing monomer II in the presence of network I, the viscosity of the polymerizing mixture is very high. Perhaps this causes the microgel aspects to be emphasized.

Lim, et al. (33) prepared three-component IPN's made from polyurethane/poly(n-butyl methacrylate)/poly(methyl methacrylate), PU/PBMA/PMMA. In various combinations, either the PU/PMMA or the PU/PBMA were made in SIN form, and the third polymer network added by sequential polymerization modes. The glass transition behavior of the three component IPN's showed three separate but broad T_g's for the (PU/PMMA)/PBMA IPN's, one broad T_g for the PU transition and the other at an intermediate temperature between the PMMA transition and the PBMA transition. Both three component IPN's showed high damping characteristics, with tan δ being in the range of 0.2-0.4 in the temperature range of -30 to +90°C.

Lee and Kim (34) prepared PU/PMMA SIN's under high pressure, up to 20,000 kg/cm^2. The polyurethane phase domain sizes decreased from about 300 to about 30Å as the pressure was increased up to the 20,000 kg/cm^2 mark. The phase continuity also changed with pressure. The PU phase became continuous at higher pressures. The SIN's prepared at 20,000 kg/cm^2 exhibited a very broad glass transition ranging from about 0°C to about 100°C, and the materials were optically transparent.

The PU/PMMA SIN's prepared under more standard conditions by Morin, et al. (35) exhibited two loss peaks in tan δ vs. temperature, but the transitions were shifted inwards significantly, with broadening. It was concluded that these materials exhibited incomplete phase separation.

Morin, et al. (35) studied a range of mechanical behavior of these SIN's. Table IV shows the ultimate elongation of some of these materials, as a function of the crosslinker concentration in each network. It appears that lower crosslinking in the PMMA phase and or higher crosslinking in the PU phase tends to increase elongation. Only above the transition temperature does crosslinking appear to reinforce a polymeric material.

Table IV. % Ultimate Elongation of PU/PMMA SIN's with Crosslinker Content (35).

K	Percent TRIM		
	5	1	0
0.77	46	---	70
0.89	99	---	56
1.01	99	---	223
1.07	184	264	336
1.13	191	---	220

PUR content: 34 percent

K = NCO/OH ratio
TRIM = trimethylol-1,1,1-propane trimethacrylate

276

6. NOISE AND VIBRATION DAMPING MATERIALS

One of the leading teams in this area is at Lancaster University, headed by Hourston (36-39). In one study on acoustical damping, Hourston and McCluskey (36) swelled a crosslinked poly (vinyl isobutyl ether) polymer with methyl acrylate and divinyl benzene to make a sequential IPN. The IPN's exhibited high damping over a broad temperature range, and had relatively constant loss peaks (E") over the temperature range of -20 to +20°C.

In another paper, Hourston and Zia (37) made what can best be described as a semi-II SIN. They mixed polyurethane precursors with unsaturated polyesters, styrene, and methyl acrylate. A linear polyurethane was formed at 20°C. Then the temperature was raised to 70°C and the polyester-styrene-methyl acrylate network was formed. While these materials were not specifically examined for noise and vibration damping, they seemed semi-compatible, or in the newer terminology, semi-miscible, see Figure 4 (37).

In newer work, Hourston and Zia (38) made both semi-I and semi-II versions of SIN compositions of polyurethane and poly (methyl acrylate). Using the Davies equation, they found that both phases exhibited some degree of phase continuity. Hourston and coworkers (39) are also studying the behavior of latex IPN's made from acrylics. These materials have a core and a shell, and are synthesized sequentially from two different monomers and their respective crosslinkers in latex form.

7. ON DIFFERENT SYNTHETIC ROUTES FOR MAKING SIN's

In the above, sequential IPN's were distinguished from SIN's. However, there are three different routes to making an SIN after mixing the two monomers or prepolymers: Monomer I may be polymerized first, the two might be truly simultaneously polymerized, or monomer II may be polymerized first.

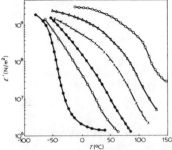

Figure 4. The dynamic storage modulus, E', of polyurethane (linear) and poly(ester-styrene-methyl methacrylate) (crosslinked) semi-II SIN's. The midrange compositions have very broad glass transitions (37).

Suzuki and coworkers (40) made a systematic study of the effects of each of these three ways using an acrylate and an epoxy system. UV light and heat were used to control the two reaction rates, respectively. The three products were, as might be expected, significantly different. The "normal" SIN, with the acrylic polymerized first, was heterogeneous, while the other two behaved much more homogeneously. In all three cases, significant deviation of the densities from the additivity rule was observed, and all three, again, were different.

8.   IPN-BASED ION EXCHANGE RESINS

Effective ion exchange resins have their cation and anion exchange portions in close proximity. However, while the two charges should be close together in space, they need to be phase separated or else the two charges may react with one another and cause coacervation.

IPN's can be made into ion exchange resins in several ways.
In one case, one of the networks has the anionic exchange capacity, and the other the cationic ion exchange capacity. Alternately, only one of the networks is charged, the other holding it in place, preventing excess swelling. One particularly effective mode has been developed by Kolarz (41,42), who used poly(styrene-co-DVB) for network I, and poly(methacrylic acid-co-DVB) for network II. The interesting feature of these IPN's is that the polystyrene network is formed by suspension polymerization chemistry in the presence of aliphatic solvents. These aliphatic solvents dissolve the monomer mix, but on polymerization, the polystyrene precipitates. A high concentration of DVB is required, because the polystyrene must form a network before precipitation ensues. After polymerization of network I is complete, the structure is a highly porous bead. Then, monomer mix II is swollen in, most of it going in the pores. On polymerization, an IPN is formed. In the present case, only the methacrylic acid portion serves as an ion exchange resin. Kolarz (42) determined the concentration of elastically active chains, and found that the volume fraction of interpenetration ranged from 0.06 to 0.17.

Alternatively, the ionic IPN can be made into a membrane possessing good transport properties toward strong mineral acids. Such a system was developed by Pozniak and Trochimczuk (43,44). They employed a membrane of polyethylene into which was swollen mixtures of styrene and DVB. After polymerization, the polystyrene portions were reacted via a chloromethylation procedure and subsequent amination with diethylamine to form a weak-base type of membrane.

9. SIN's FROM SPECIAL FUNCTIONAL TRIGLYCERIDE OILS

Most triglyceride oils, such as corn oil or soy bean oil, have only double bond functionality. A few oils, however, have other functional groups. For example, castor oil has one hydroxyl group per acid residue, allowing the synthesis of polyesters and polyurethanes, the latter being commercial. These oils are also fine candidates for making SIN's. A special reason for research in this area is that these oils are renewable resources.

SIN's made from castor oil and polystyrene (45) made a series of tough plastics or reinforced elastomers, depending on the relative quantities of each product.

Research in this area then shifted to Vernonia oil, which comes from a wild plant growing in Kenya, Africa. Each acid residue of vernonia oil has an epoxy group (46). On polymerization with sebacic acid, tough polyester SIN's were made.

Most recently, research in this area shifted to Lesquerella palmeri, a desert wildflower growing in Western Arizona (46). Lesquerella oil has hydroxyl groups, like castor oil, but not quite so many. However, the acid residue is a little longer, so the T_g from lesquerella-sebacic acid rubber networks is lower, about -55^oC.

10. SIN-BASED RIM SYSTEMS

The spectacular rise of reaction injection molding, RIM, in the automative and other industries was caused by the high speed of the process, and high energy efficiency. Current RIM materials are based on polyurethane technology, because hydroxyl groups and isocyanate groups react rapidly and completely, and the finished urethanes can be made with high moduli and strength. However, the industry would like RIM plastics with still higher moduli. While reinforced RIM, RRIM, with glass fibers or other materials has become important, the high viscosities and abrasiveness of the filled systems present some disadvantages.

A novel solution of the problem is the use of SIN-based RIM systems being developed by Frisch and coworkers (47-49). For example, an elastomeric polyurethane can be used in combination with a plastic epoxy resin to produce a rapidly reacting SIN (47,48). Depending on the composition, Pernice, et al. (47,48) made their IPN-based RIM materials by curing in the mold at 100^oC for five minutes, followed by post-curing at 121^oC for one hour.

RIM materials made in this way developed good phase separation, as indicated by their glass transition temperatures, see Table V

Table V. Glass Transition Temperatures of Urethane/Epoxy SIN–RIM
Materials (48).

Composition PU/Epon 828	Glass Transition Temperatures, °C		
	Tg_1	Tg_2	Tg_3
100/0	−99	+95	---
90/10	−91	+82	---
80/20	−89	+96	+114
70/30	−93	+85	+119
60/40	−92	+86	+118
0/100	---	---	+119

(48). Three glass transitions are observed via DSC, two for the
soft and hard segments of the polyurethane, and one for the epoxy.
The inward shifting of the glass transition temperatures provides
some indication of the extent of actual phase mixing. The Frisch
IPN–RIM system is visualized as an alternative to RRIM (49).

Manson and Sperling (50) suggested making RIM systems from
castor oil-based SIN's. As a natural triglyceride, castor oil is
unusual in having three hydroxyl groups per molecule, one for each
acid residue. Castor oil is already used to make commercial ure-
thane elastomers, especially foam rubber. Castor oil can be combined with TDI, styrene, DVB, and initiators, and reacted to make
tough plastics and elastomers by the SIN route. Other oils that
can be used include Vernonia (epoxy bearing) and Lesquerella
(hydroxyl bearing) (45,46,51). What is required is the development
of catalytic systems to react these materials with the requisite
rate.

11. GRADIENT IPN MICROENCAPSULATION

11.1. Controlled Drug Delivery Systems

While most of the applications described above relate to
mechanical behavior, IPN's find several other uses. Anionic/
cationic ion exchange resins have already been mentioned. A new
application involves controlled drug delivery.

Conventional oral dosage forms of water-soluble drugs consist
of coated tablets. After dissolution of the coating in the stomach
they disintegrate more or less rapidly. As a result, drug concentrations in the blood quickly reach a sharp peak, followed by a
decrease at a rate determined by their metabolic half-life in the
body. For many purposes, a controlled steady drug delivery is
desirable.

One solution to the problem of controlled drug delivery utilizes insoluble hydrogel beads. The beads are then loaded with the drug. Then, delivery depends on the hydrophilicity of the polymer, the diameter of the bead, and the diffusion rate. This last can be controlled by utilizing coatings or shell structures, forming multilayered beads with different diffusion rates. Since generally it is desired to retard the delivery rate, the outer layer(s) should have a low permeability towards the drug.

A simple method of preparing a multilayered bead involves gradient IPN technology (52-56). There are two ways of making a gradient IPN. For chain polymerization, network I is briefly swollen with monomer II mix, and polymerized before diffusion creates a uniform solution. For step polymerizations such as polyurethanes, one of the component monomers is swollen into the bead first, and the second component added later.

Mueller and Heiber (57-58) prepared a gradient IPN based on hydrogel beads utilizing bead substrates of 2-hydroxyethylmethacrylate (HEMA) and N-vinylpyrolidone (NVP). Polymer I was crosslinked with poly(n-butylene oxide) capped with isophorone-diisocyanate, which reacts with the hydroxyl groups on the HEMA. The beads are swelled with a diol or triol and a polyurethane gradient IPN is formed by reaction with 2,4,4(2,2,4)-trimethylehexane-1,6-diisocyanate, TMDI, by diffusion controlled reaction.

An important feature of gradient IPN's is the concentration of network II as a function of distance from the surface. Mueller and Heiber (57) noted that the boundary between IPN-modified surface regions and the unmodified core was sharpest when polymer network I was more polar, and therefore had a lower affinity for TMDI. The nonuniform polyurethane distribution was studied by SEM X-Ray microprobe methods, using brominated diols, see Figure 5 (57).

Oxprenolol HCl, highly water and ethanol soluble, was used as a model compound. Mueller and Heiber found that the hydrophilicity rate of swelling and the permeability of the IPN was highest for the most hydrophilic polymer matrice. Thin network II layers synthesized on the surface of network I act as membranes and retard the release of oxprenolol-HCl.

In another system, Raghunathan, et al. (59) used a gel-type divinyl benzene-sulfonic acid cation exchange resin, which held phenylpropanolamine in an acid-base complex. Ethyl cellulose was used as the diffusion barrier material. There are several advantages of adsorbing basic nitrogen-containing drugs onto sulfonic acid ion exchange resins for dosage forms: (a) prolonged drug release from the complex, (b) reduced toxicity by slowing drug absorption, (c) protection of the drug from hydrolysis or other degradation, (d) improved palatability, (e) ease of formulation (60).

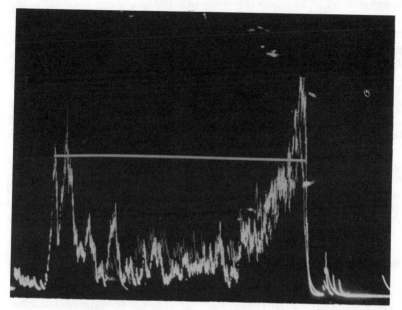

Figure 5. SEM X-ray linescan for bromine in a suspension IPN of poly(2-hydroxyethyl methacrylate-co-N-vinyl pyrrolidone)/dibromo-neopentylglycol-TMDI polyurethane, 75/25. Bar indicates diameter of the bead, 1.1 mm (57).

Ion-exchange resin beads have also been microencapsulated acacia and gelatin (61).

Microencapsulation also serves to improve the gritty and astringent taste of finely divided polystyrene-divinyl benzene copolymer anion-exchange resins for swallowing. One method of doing this is by coating the particles with an acrylic polymer crosslinked with allyl sucrose to make a surface-modified sequential IPN (62).

11.2 Miscellaneous Applications

In addition to interfacial polymerization, which produces a core-shell structure, multiwalled structures have been made. For example, a three-layer structure composed of epoxy resin, polyurea, and polyamide can be prepared by emulsifying an aqueous polyamine in toluene, and adding the epoxy, isocyanate, and acyl chloride components sequentially (63). The final microcapsule has an aqueous droplet in the center.

An IPN-related process involves a gelatin-gum arabic coacer-vates, which are finally crosslinked with formaldehyde. This product serves as a wall, with an oily core (63).

The applications of microcapsules, besides controlled drug delivery, include carbonless copy paper, containment of perfumes, food flavorings, and decongestants, marine feeding, artificial blood cells, and augmenting artificial kidney function (62). While it is not obvious that all of these applications utilize IPN's or semi-IPN's, the potential if not reality is clear.

The first group of the above depend on breakage of the shell for release of contents. Thus, carbonless copy paper released ink with applied pressure. Food flavorings are contained until chewed. Artificial blood cells, containing sheel hemolysate, require diffusion in both directions. For augmenting the artificial kidney function, microcapsules containing activated carbon are ingested. Intestinal adsorption of creatinine and uric acid are of interest.

12. THE PATENT LITERATURE

A number of IPN patents have been issued recently. Skinner, et al. (65) made SIN coatings, some grafted, of vinyl/urethane compositions which are, respectively, radiation and thermally polymerized. Frisch, et al. (66) obtained a patent on their IPN's. Sperling, et al. (67) patented tough plastic compositions based on polymerized castor oil.

Foscante, et al. (68,69) have a patent on a semi-SIN composition of epoxy/siloxane, with the epoxy crosslinked and the siloxane linear. The Foscante, et al. IPN material is intended for use as a coating material, supplied as a two-package system, and applied to a surface by conventional techniques such as spraying or brushing. Acid and solvent resistance are mentioned. Possible applications cited include coating steel structures of chemical processing plants, oil refineries, and the internal surfaces of tanks of petroleum tankers.

Sebastiano (70) invented a transparent polyurethane/poly-acrylate/SIN coating for safety glass. The monomer-polymer or prepolymer solutions can be applied on the inside glass of automobiles by spraying, thicknesses of 100 to 190 microns yielding good optical properties and tear resistance.

Prosthetic teeth and other dental appliances made from a sequential acrylic IPN composition were invented by Tateosian and Roemer (71). In one example, crosslinked poly(methyl methacrylate) in suspension-sized particles, linear poly(methyl methacrylate), and methyl methacrylate monomer were mixed with initiator and pigment, molded, and polymerized to an impact and wear resistant tooth structure.

A thermoplastic semi-IPN consisting of a rubbery ionomer and poly(dimethyl siloxane) was invented by Lundberg, Phillips and Duvdevani (72). Improved melt flow and a lower coefficient of friction are cited.

De La Mare and Brownscombe (73) describe an epoxy/styrene-crosslinker RIM SIN material. Cure times of one to ten minutes are reported. Catalysts are selected from Group I or Group II salts for the epoxy component.

13. POLYPROPYLENE OR POLYETHYLENE/EPDM THERMOPLASTIC IPN's

A new class of IPN was invented by Fischer (74) of Uniroyal in the period of 1973-74. His materials were primarily blends of iso-tactic polypropylene and EPDM. During the blending operation, while the mix was subjected to a shearing action, the EPDM component was partly crosslinked. The resultant materials, dynamically partly cured, retained a thermoplastic nature. On cooling, the polypropy-lene component crystallizes, forming a type of physical crosslink. These materials are called thermoplastic IPN's, see Table VI, No. 2.

It soon became clear that thermoplastic IPN's could be made in a variety of ways. Physical crosslinks can be incorporated through the use of semi-crystalline polymers, block copolymers, or ionomers. The main idea is that at service temperature the polymer should behave as if it were crosslinked, while at some higher temperature it should flow.

Also, dynamically shearing a polymer in blend form during crosslinking (as in the case of the EPDM) results in a two dimen-sional structure of cylindrical shape, really elongated droplets.

Table VI. Polypropylene of Polyethylene/EPDM Thermoplastic IPN Patents

No.	Major Feature	Company	Patent No.
	Dynamically Partly Cured		
1.	EPDM precured, then blended	Uniroyal	U.S.3,758,643
2.	EPDM blended, then cured	Uniroyal	U.S.3,806,558
3.	EPDM of high molecular weight	Uniroyal	U.S.3,835,201
4.	Dynamically Fully Cured	Monsanto	U.S.4,130,535
5.	High Ethylene Length Index (with PE)	Goodrich	U.S.4,046,840
6.	High Ethylene Sequence Index (with PP)	Goodrich	U.S.4,036,912
7.	70-85% by wt. ethylene EPDM with dicyclopentadiene	Uniroyal	U.S.4,031,169

284

The cyclindrical structure interpenetrates the other polymer, to form a material with two continuous phases. These materials are actually hybrids between polymer blends and IPN's, exhibiting some of the properties of both.

These ideas naturally led to further developments. Coran and Patel (75-78) learned how to prepare thermoplastic vulcanizates with the rubber completely cured, see Table VI, No. 4. If the EPDM were prepared with high ethylene sequence lengths so that it crystallized slightly, no crosslinking was required at all, see Nos. 5 (79) and 6 (80). Having a high ethylene content in the EPDM also allows partial crystallization, see No. 7 (81). Of course, the EPDM can be precured, then blended, No. 1, or of high molecular weight, No. 3. Thus, there are several different ways of preparing thermoplastic IPN's of this type. The subject has recently been reviewed (82).

REFERENCES

1. L. H. Sperling, "Interpenetrating Polymer Netowrks and Related Materials," Plenum, 1981.
2. Yu. S. Lipatov and L. M. Sergeeva, "Interpenetrating Polymeric Networks," Naukova Dumka, Kiev, 1979.
3. Yu. S. Lipatov and L. M. Sergeeva, Russian Chem. Revs. 45(1), 63 (1976); translated from Uspekhi Khimii, 45, 138 (1976).
4. D. Klempner, Angew. Chem., 90, 104 (1978).
5. V. Huelck, D. A. Thomas, and L. H. Sperling, Macromolecules, 5, 340 (1972).
6. V. Huelck, D. A. Thomas, and L. H. Sperling, Macromolecules, 5, 348 (1972).
7. A. A. Donatelli, L. H. Sperling, and D. A. Thomas, Macromolecules, 9, 671 (1976).
8. A. A. Donatelli, L. H. Sperling, and D. A. Thomas, Macromolecules, 9, 676 (1976).
9. A. A. Donatelli, L. H. Sperling, and D. A. Thomas, J. Appl. Polym. Sci., 21, 1189 (1977).
10. J. Michel, S. C. Hargest, and L. H. Sperling, J. Appl. Polym. Sci., 26, 743 (1981).
11. J. K. Yeo, L. H. Sperling, and D. A. Thomas, Polymer, 24, 307 (1983).
12. J. K. Yeo, L. H. Sperling, and D. A. Thomas, Polym. Eng. Sci., 21, 696 (1981).
13. J. K. Yeo, L. H. Sperling, and D. A. Thomas, J. Appl. Polym. Sci., 26, 3283 (1981).
14. J. K. Yeo, L. H. Sperling, and D. A. Thomas, Polym. Eng. Sci., 22, 190 (1982).

15. J. M. Widmaïer and L. H. Sperling, Macromolecules, 15, 625 (1982).
16. J. M. Widmaïer and L. H. Sperling, J. Appl. Polym. Sci., 27, 3513 (1982).
17. J. M. Widmaïer, J. K. Yeo, and L. H. Sperling, Coll. & Polym. Sci., 260, 678 (1982).
18. E. N. Kresge, in "Polymer Blends," D. R. Paul and S. Newman, Eds., Academic Press, NY, 1978, Vol. 2.
19. L. H. Sperling, A. Klein, A. M. Fernandez, and M. A. Linne, Polym. Prepr. 24(2), 386 (1983).
20. Yu. S. Lipatov, V. V. Shilov, V. A. Bogdanovitch, L. V. Karbanova, and L. M. Sergeeva, Vysokomol. Soyed. A22(6), 1359 (1980); Polym. Sci. USSR 22, 1492 (1980).
21. V. V. Shilov, Yu. S. Lipatov, L. V. Karbanova, and L. M. Sergeeva, J. Polym. Sci. Polym. Chem. Ed., 17, 3083 (1979).
22. W. J. MacKnight, F. E. Karasz, and J. R. Fried, in "Polymer Blends," Vol. I, D. R. Paul and S. Newman, Eds., Academic Press, NY, 1978.
23. H. L. Frisch, D. Klempner, H. K. Yoon, and K. C. Frisch, in "Polymer Alloys II," D. Klempner and K. C. Frisch, Eds., Plenum, NY, 1980.
24. H. L. Frisch, D. Klempner, H. K. Yoon, and K. C. Frisch, Macromolecules, 13(4), 1016 (1980).
25. D. L. Siegfried, D. A. Thomas, and L. H. Sperling, in "Polymer Alloys II," D. Klempner and K. C. Frisch, Eds., Plenum, NY, 1980.
26. L. H. Sperling and J. M. Widmaïer, Polymer Eng. Sci., 23, 693, 1983.
27. H. Adachi and T. Kotaka, Polym. J., 14(5), 3791 (1982).
28. H. Adachi, S. Nishi, and T. Kotaka, Polym. J., 15, 985 (1983).
29. H. Adachi and T. Kotaka, Polym. J. 15(4), 285 (1983).
30. E. G. Bobalek, E. R. Moore, S. S. Levy, and C. C. Lee, J. Appl. Polym. Sci., 8, 625 (1964).
31. S. S. Labana, S. Newman, and A. J. Chompff, in "Polymer Networks," A. J. Chompff and S. Newman, Eds., Plenum, 1971.
32. K. Dusek, in "Development in Polymerization-3," R. N. Haward, Ed., Applied Science Pub., London, 1982.
33. D. S. Lim, D. S. Lee, and S. C. Kim, IUPAC Proceedings, University of Massachusetts, July 12-16, 1982, p. 223.
34. D. S. Lee and S. C. Kim, Macromolecules, 17, 268 (1984).
35. A. Morin, H. Djomo, and G. C. Meyer, Polym. Eng. Sci., 23, 394 (1983).
36. D. J. Hourston and J. A. McCluskey, Polymer, 20, 1573 (1979).
37. D. J. Hourston and Y. Zia, Polymer, 20, 1497 (1979).
38. D. J. Hourston and Y. Zia, J. Appl. Polym. Sci., 28, 2139, 3745, 3849 (1983); 29, 629 (1984).
39. D. J. Hourston and coworkers, Roy. Soc. Chem. Meeting, Lancaster, England, April, 1983.
40. Y. Suzuki, T. Fujimoto, S. Tsunoda, and K. Shibayama, J. Macromol. Sci.-Phys., B17(4), 787 (1980).

41. B. M. Kolarz, Angew. Makromol. Chem., 90, 167 (1980).
42. B. M. Kolarz, Angew. Makromol. Chem., 90, 183 (1980).
43. G. Pozniak and W. Trochimczuk, Angew. Makromol. Chem., 92, 155 (1980).
44. G. Pozniak and W. Trochimczuk, Angew. Makromol. Chem., 104, 1, (1982).
45. (a) L. H. Sperling and J. A. Manson, JOACS, 60(11), 1887 (1983).
    (b) L. H. Sperling, J. A. Manson, Shahid Qureshi, and A. M. Fernandez, I&EC Prod. Res. & Dev. 20, 163 (1981).
46. M. A. Linne, L. H. Sperling, A. M. Fernandez, Shahid Qureshi, and J. A. Manson, Polym. Mater. Sci. Eng. Prepr., 49, 513 (1983).
47. R. Pernice, K. C. Frisch, and R. Navare, in "Polymer Alloys III," D. Klempner and K. C. Frisch, Eds., Plenum, 1983.
48. R. Pernice, K. C. Frisch, and R. Navare, J. Cellular Plast., p. 121, Mar/Apr (1982).
49. Anon. Plastics Technology, p. 76, January 1983.
50. J. A. Manson, L. H. Sperling, Shahid Qureshi, and A. M. Fernandez, "Technology Transfer," Preprints of 13th National SAMPE Technical Conference, 13, 81 (1981).
51. N. Devia, J. A. Manson, L. H. Sperling, and A. Conde, Macromolecules, 12, 360 (1979).
52. G. Akovali, K. Biliyar, and M. Shen, J. Appl. Polym. Sci., 20, 2419 (1976).
53. M. Shen and M. B. Bever, J. Mater. Sci., 7, 741 (1972).
54. G. C. Berry and M. Dvor, Am. Chem. Soc. Prepr. Div. Org. Coat. Plast. Chem., 28, 465 (1978).
55. C. F. Jasso, S. D. Hong, and M. Shen, Am. Chem. Soc. Advan. Chem. Ser., Multiphase Polymers, 444 (1979).
56. G. C. Martin, E. Enssani, and M. Shen, J. Appl. Polym.Sci., 26, 1465 (1981).
57. K. F. Mueller and S. J. Heiber, J. Appl. Polym. Sci., 27, 4043 (1982).
58. K. F. Mueller and S. J. Heiber, (a) European Patent Application 0046136 (1982); (b) U.S. 4,423,099 (1983).
59. Y. Raghunathan, L. Amsel, O. Hinsvark, and W. Bryant, J. Pharmaceutical Sci., 70(4), 379 (1981).
60. J. W. Keating, U.S. 2,990,332 (1961).
61. S. Motycka and J. G. Nairn, J. Pharmaceutical Sci., 68(2), 211 (1979).
62. T. J. Macek, C. E. Sloop, and D. R. Stauffer, U.S. 3,499,960 (1970).
63. A. Watanabe and T. Hayashi, in "Microencapsulation", J. R. Nixon, Ed., Marcel Dekker, NY, 1976.
64. J. R. Nixon, Ed., "Microencapsulation", Marcel Dekker, NY, 1976.
65. E. Skinner, M. Emeott, and A. Jevne, U.S. 4,247,578 (1981).
66. H. L. Frisch, K. C. Frisch, and D. Klempner, U.S. 4,302,553 (1981).

67. L. H. Sperling, J. A. Manson, and N. Devia-Manjarres, U.S. 4,302,553 (1981).
68. R. E. Foscante, A. P. Gysegem, P. J. Martinich, and G. H. Law, U.S. 4,250,074 (1981).
69. R. E. Foscante, A. P. Gysem, P. J. Martinich, and G. H. Law, Int. Pat. No. PCT/US79/00833.
70. F. Sebastiano, Int. Pat. No. 0035130 (1981).
71. L. H. Tateosian and F. D. Roemer, Int. Pat. No. WO82/02556 (1982).
72. R. D. Lundberg, R. R. Phillips, and I. Duvdevani, U.S. 4,330,447 (1982).
73. H. E. De La Mare and T. F. Brownscombe, U.S. 4,389,515 (1983).
74. W. K. Fischer, U.S. 3,806,558 (1974).
75. A. Y. Coran, B. Das, and R. P. Patel, U.S. 4,130,535 (1978).
76. A. Y. Coran and R. Patel, Rubber Chem. Tech., 53, 141 (1980).
77. A. Y. Coran and R. Patel, Rubber Chem. Tech. 54, 91, 892 (1981).
78. A. Y. Coran and R. Patel, Rubber Chem. Tech., 55, 1063 (1982); 56, 210 (1983).
79. C. J. Carman, M. Batuik, and R. M. Herman, U.S. 4,046,840 (1977).
80. P. T. Stricharczuk, U.S. 4,036,912 (1977).
81. H. L. Morris, U.S. 4,031,169 (1977).
82. G. E. O'Connor and M. A. Fath, Rubber World, 185(3), 25 (1981); 185(4), 26 (1982).

ACKNOWLEDGMENTS

The author wishes to thank the National Science Foundation for financial support through Grant No. DMR-8106892, Polymers Program.

FRACTURE TOUGHNESS EVALUATION OF BLENDS AND MIXTURES AND THE USE OF
THE J METHOD

J G Williams and S Hashemi

Mechanical Engineering Department, Imperial College, LONDON SW7.

INTRODUCTION

The process of blending of two or more polymers is frequently
carried out in order to enhance the toughness of the final product
when compared with the original materials. It is therefore of great
importance to specify very clearly what is meant by toughness (and
the closely related property of strength) when one sets out to
achieve some goal. In addition the characterisation of toughness
must be precise if the development of the new system is to follow
a logical course. That this is not so in most chemically based
laboratories where such work is usually performed is common
knowledge. The reasons are various ranging from the predominance
of chemistry in the training of those involved which traditionally
includes very little mechanics to the apparent lack of a suitable
scheme even if one made the effort to seek it. The upshot of all
this is the dominance of toughness characterisation by the Izod
number. This is a reasonable test in basic concept since it
subjects a material to a set of circumstances which are likely to
promote brittle failure and thus low energy absorption. These
conditions are the combination of a notch and a high rate of
loading and the performance is expressed as energy absorbed scaled
on the specimen size. Comparisons by experienced workers can be
very helpful using this method and indeed it may be refined to
give rather detailed descriptions of material behaviour using
various notch sharpness values and temperatures.

The basic difficulty with the test is that it is essentially
empirical and although the chosen conditions do yield useful
numbers they are not fundamental material properties and are
related to such properties in an undefined way. Illogical

and unsatisfactory results are thus common occurrences and there
is no way of establishing what has gone wrong. Indeed such effects
are often not noticed and much effort is wasted in following false
trails. Is there a remedy for this situation? The theme of this
paper is that there is and that it is provided by the subject of
Fracture Mechanics. Here the subject is considered in great detail
and a framework of analysis has been established within which
toughness can be defined precisely and measured accurately as a
material property. In this review we will consider how this is
done and pay particular attention to those factors of relevance
to blends.

It is worth summarising what is to follow by stating that the
basic analysis is that of elastic solids containing sharp cracks
which grow in a brittle manner. This is the well established
Griffith theory [1], which has undergone considerable modification
in order to encompass less brittle, and thus tougher fracture
in materials. Polymer blends are often very tough indeed which,
of course, is why they are produced and because of their toughness
they pose difficult problems to the theory. These have been
addressed, however, and there is considerable experience now
available on characterising very tough polymer systems using the
more recent developments of the subject. Here we will consider the
linear analysis and then go on to its extension to include the non
-linear effects which are necessary for the description of very
tough materials. It will be shown that such methods have great
utility in describing two-phase polymer systems.

2.0   SOME BASIC ASSUMPTIONS

Fracture mechanics assumes that all real materials contain
flaws which are usually modelled as cracks. The subject then seeks
to describe when and in what manner these cracks grow. The theory
does not include any mechanisms for flaw generation from a perfect
system but it can include the description of cracks which do not
grow for long periods and then proceed to do so. The distinction
between a very small flaw which becomes bigger and is then seen
and one which was created from no flaw at all is often a topic of
controversy but experience suggests that the notion of a
pre-existing flaw to be the more useful.

The description of the initiation and growth processes is
carried out in terms of the energy per unit area to form new
surfaces. Care must be taken to distinguish genuine fracture which
only occurs when new surface area is created and large deformation.
The latter is a continuum process where the material remains
continuous at all times while the first, fracture, is not and
involves the violation of continuity in the crack tip region.
This very important notion will now be developed in considering the
fracture of elastic bodies.

3.0 ELASTIC FRACTURE

In elastic fracture we consider the growth of a crack in an elastic body and we assume that whatever local deformations are going on at the crack tip they are on a sufficiently small scale that they do not effect the overall elastic load-deflection behaviour. Fig. 1. shows a cracked plate of uniform thickness B in which we consider the increase in crack length a. The shaded area at the crack tip represents the "plastic zone" to be discussed later [section 5.0] and is the region of high, non-linear, deformation where the energy is absorbed and it is this which must be small compared with the specimen dimensions, W, B and a.

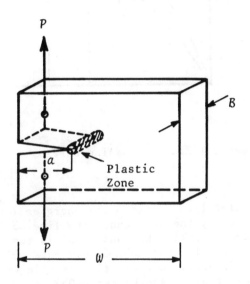

Fig. 1. A Loaded, Cracked Body.

Let us suppose that the specimen is loaded with the original crack length a as shown in Fig. 2. The loading line would be that of line (i) and in general, it need not be linear. Suppose now that at some point the crack grows an amount δa then the specimen would become less stiff and, in general, the load would decrease and the displacement increase, say to a point S'. If the specimen were now unloaded it would return to O since it is elastic (line (ii)) and this would be the loading line for the body containing a crack of length $a + \delta a$.

292

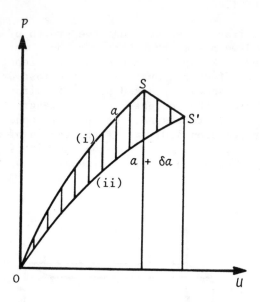

Fig. 2. The General Elastic Loading Curves.

It is now helpful to consider the energies involved in this process. That given by the area under the line OS is the original stored elastic energy while that under OS' is the final stored energy. The part of this under SS' is new external work done on the body and it goes directly into the final strain energy. The difference between the two stored terms OSS' shown shaded in Fig. 2. is thus released in the sense that the system undergoes a decrease in energy when the crack grows. Thus we are dealing with a non-conservative system in which energy is not conserved (for static conditions). The rate of this energy release per unit of crack area, termed G, may be calculated from;

$$G = -\frac{1d}{Bda} \cdot [\int_0^u P\, du]\ u\ const. \tag{1}$$

This general form can be applied to non-linear elastic systems and is used, for example, in describing fracture in elastomers which have large non-linear deformations up to fracture. By far the most common case, however, is that of linear elastic behaviour where the load deflection diagram can be characterised by the compliance C which is a function of only crack length;

$$C(a) = u/P \qquad\qquad (2)$$

and the stored energy becomes;

$$U = \int P\, du = \frac{1}{2} P u = \frac{1}{2} u^2/C$$

Substitution in equation (1) and differentiating gives,

$$G = \frac{1}{2B} \cdot \left(\frac{u}{C}\right)^2 \cdot \frac{dC}{da} = \frac{P^2}{2B} \frac{dC}{da} = \frac{U}{B} \frac{1}{C} \frac{dC}{da} \qquad (3)$$

From these equations it is thus possible to calculate the energy release rate at fracture by measuring either u, P or U and then applying the appropriate function of the compliance. This is the way the specimen is calibrated and may be done either as part of the test, or more conveniently as a separate exercise in which $C(a)$ is found and hence dC/da.

G is now used in fracture analysis by supposing that the energy per unit area to produce new surfaces, the fracture toughness, is G_c and that fracture will occur for

$$G > G_c$$

and fracture initiation may be defined as when $G = G_c$. Thus the required parameter for evaluation can be defined, in principle, for any geometry via the calibration of $C(a)$. If either u, P or U is measured at fracture G_c can be found.

An example for which C can be calculated and which is frequently used is the double cantilever beam (DCB) specimen shown in Fig. 3. Hence the deflection of each arm can be found from simple beam theory and this is half of the total deflection;

i.e.
$$u/2 = \frac{4P\, a^3}{E\, BD^3} \quad ; \quad C = \frac{8\, a^3}{E\, BD^3}$$

and
$$dC/da = \frac{24\, a^2}{E\, BD^3}$$

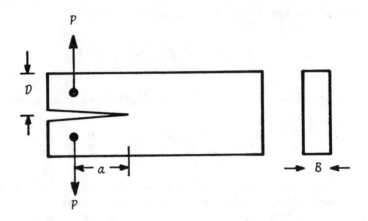

Fig. 3. The Double Cantilever Beam Specimen.

Thus for a test at which fracture occurs at a load P_c, for example we have;

$$G_c = \frac{12\ P_c^2\ a^2}{E\ B^2\ D^3} \tag{4}$$

A point of some importance in later sections is the question of stability. If $G > G_c$ then there is an excess of energy over that required to form the crack so the crack will continue to accelerate until it achieves some dynamically controlled stable state close to the Rayleigh wave speed. This instability criterion can be written as;

$$\frac{dG}{da} > \frac{dG_c}{da} \tag{5}$$

Different specimens and loading conditions have different dG/da values so that stability effects vary greatly. For example in the DCB case at constant load

$$\frac{dG}{da} \Bigg| P \text{ const.} = \frac{24 \ P^2 \ a}{E \ B^2 \ D^3}$$

while at constant displacement,

$$\frac{dG}{da} \Bigg| u \text{ const.} = -\frac{3}{4} \frac{E \ D^3 \ u^2}{a^5}$$

i.e. for a constant G_c value ($dG_c/da = 0$) the first would be unstable while the second would be stable.

If we have stable growth for whatever reason, then $G = G_c$ throughout and we need not consider the energy release rate part of the argument but simply say that the energy between the two loading lines in Fig. 2. goes directly to G_c. For many tough materials G_c is not a constant but increases as the crack grows giving what is termed a rising resistance curve. dG_c/da is thus positive and stable behaviour much more likely.

4.0 STRESS INTENSITY FACTORS

It is useful at this stage to consider the solution for G for an infinite plate loaded with a constant stress σ and containing a crack of length $2a$ (or a if it is an edge crack). This is the case analysed by Griffith and is important since it is a useful first approximation for any stressed part containing a small flaw. The result is;

$$G = \frac{\pi \ \sigma^2 \ a}{E} \tag{6}$$

and Griffith equated G with 2γ where γ is the true surface work of inorganic glasses. For less brittle cases the energy to form new surfaces is much greater than 2γ because of the plastic work which is performed and so subsequent development of the idea to encompass tougher materials involved replacing 2γ by G_c where G_c includes this extra plastic work.

Equation (6) really embodies the whole essence of fracture mechanics in that it can be written as;

$$K_c^2 = E \ G_c = \pi \ \sigma^2 \ a$$

An experiment may now be performed in which a is varied and σ at fracture measured. From these values EG_c can be found and if E is known then G_c can be calculated. However, for the purposes of calculation, there is no need to separate E and G_c and they may be lumped together as K_c^2 as shown. For design purposes if K_c, called the stress intensity factor, is known then for any flaw size a limiting stress can be found and vice-versa.

K_c can thus be regarded as a convenience but it does have physical significance. It arises because of the singular stresses set up at the tip of an elastic crack which can be written as,

$$\sigma = \frac{K}{\sqrt{2\pi\hbar}} \cdot f(\theta) \quad , \quad f(0) = 1 \qquad (7)$$

where $\hbar$ and θ are polar coordinates at the crack tip as shown in Fig. 4. This stress state is dominant close to the crack tip and clearly as $\hbar \to 0$, $\sigma \to \infty$ and such a scheme has no value as a fracture criterion. However equation (7) shows that the local stress field is the same for all loadings and that only its intensity changes which is determined by K. Since the product $\sigma \sqrt{2\pi\hbar} = K$ is constant in the crack tip region the criterion $K = K_c$ was suggested as a fracture criterion by Irwin. [2]. In fact it can be seen that $K^2 = EG$ as mentioned earlier so that K_c and G_c are equivalent criteria. This is not surprising since G is a consequence of the violation of continuity caused by the singularity and K defines the strength of that singularity.

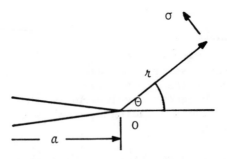

Fig. 4. Coordinates at a Crack Tip.

It turns out, in addition, that it is often simpler to calibrate specimens via K than G. The singular term in the stresses of a computation can be isolated so that K can be found. This form of calibration is usually expressed as,

$$K^2 = Y^2 \sigma^2 a \qquad (8)$$

where Y^2 is the calibration factor (or some equivalent form) and is given as polynomials in (a/W) for most specimen geometries. We may find K_c (and if E is known, G_c) simply by a set of specimens of various a values, measuring σ at fracture and then plotting $Y^2 \sigma^2$ versus a^{-1} which should result in a straight line from which K_c can be found.

In many cases this works very well as illustrated in Fig. 5. which shows data for a rubber modified polystyrene but Fig. 6., which gives data for ABS shows that it does not work here. The reason is the higher toughness of ABS and we must now consider why this non-linearity occurs.

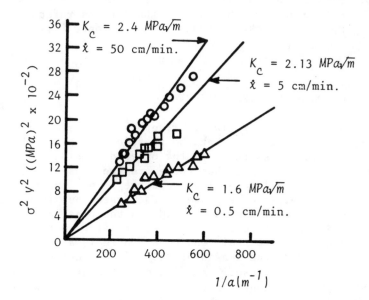

Fig. 5. Data for the Determination of K_c - rubber modified polystyrene at 20°C (data from reference [3]).

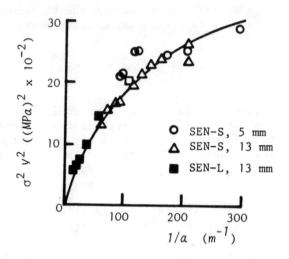

Fig. 6. Data for ABS showing pronounced curvature - SEN is
single edge notch test and the lengths refer to
specimen thickness (data from reference [3]).

5.0 PLASTIC ZONES AND SIZE CRITERIA

The use of the linear elastic fracture mechanics parameters
[G_c and K_c] (LEFM) requires that the plastic zone size, shown
in Fig. 1., is much less than all the other dimensions of the
specimen. If this is not so then there are two consequences.
The first is that the load-deflection curve is no longer fully
determined by the elastic response of the system but is changed by
the softening effect of the lower loads on the plastic zone.
As would be expected the effect is rather similar to the crack being
somewhat longer than its real length and there is a correction that
can be made by adding the length of the zone to the crack length
and then performing the usual calculations. The size of the zone
can be estimated from equation (7) since it will start when
$\sigma = \sigma_y$, the yield stress, so that at fracture;

$$r_p = \frac{1}{2\pi} \left(\frac{K_c}{\sigma_y}\right)^2 \qquad (9a)$$

and the corrected crack length is given by $(a + r_p)$. Fig. 7. shows

the data from Fig. 6. replotted with such a correction and clearly
a reasonable result can be obtained. The method is only of value,
however, for modest degrees of non-linearity and when r_p becomes
large compared with a then it is not adequate.

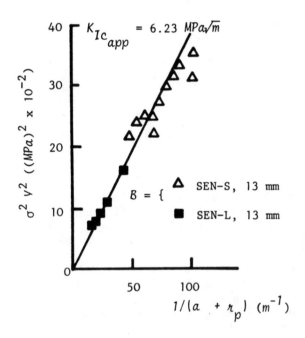

Fig. 7. ABS data corrected for plastic zone effect.
(S and L refer to different specimen sizes,
data from reference [3]).

The second effect which a large r_p induces is to change the
local stress state at the crack tip. The fracture toughness changes
with stress state and is lowest when the material being deformed is
most heavily constrained. This is because to absorb energy it must
move but when constrained it is limited in the motions it can
achieve. When the plastic zone is small the surrounding elastic
material restricts any lateral contraction parallel with the crack
front and the zone is then in a state of <u>plane strain</u> which is
usually regarded as an effective minimum of material toughness and
the K_c or G_c values are quoted as plane strain values. As r_p increases
with respect to thickness then this constraint decreases and the
measured toughness increases until, in the limit, it reaches the

fully unconstrained plane stress condition which is an upper bound. In all fracture mechanics testing we try to achieve the plane strain condition since it gives a true measure of a material's toughness when brittle cracks form, for whatever reason. It may, of course, be useful to know a value for some lesser degree of constraint for a particular purpose but, in general, the only safe course is to use the plane strain value.

To achieve both of these effects ASTM have suggested criteria for metals of the form, [4]

$$B > 2.5 \left(\frac{K_c}{\sigma_y}\right)^2$$

(9b)

and

$$W > 5 \left(\frac{K_c}{\sigma_y}\right)^2$$

The first guarantees plane strain while the second ensures an adequate width and hence linearity. Although these are empirical to some extent (they are simply some number of plastic zone sizes, e.g. ~ 16 for B) they have been found to be very good estimates for polymers. This is illustrated in Fig. 8. which shows K_c as a function of B for a polypropylene copolymer and clearly shows K_c increasing at B less than the critical value. Fig. 9. shows K_c as a function of W in which K_c decreases for W less than the critical value. This is because the necessary load for the LEFM solution cannot be achieved and a low value results.

With these restrictions in mind it is clear that valid tests can only be performed over a limited set of conditions. For many blends and two phase systems $K_c \sim 6$ MPa$\sqrt{m}$ and $\sigma_y \sim 50$ MPa so that the critical thickness is 36 mm, an unreasonable value to achieve in most polymer molding operations. Testing at low temperatures can frequently help since K_c changes very little while σ_y increases so that if σ_y increases to 100 MPa, for example, then the size is 9 mm which is quite reasonable. What happens if one needs a toughness measurement at a set of conditions where the size requirement is too large to make and it is not acceptable to cool it down or speed it up? There is an answer in a scheme which allows smaller specimens to be used, which is the J method which we will now consider.

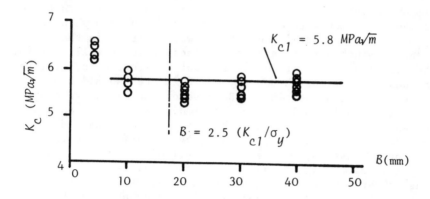

Fig. 8. Effect of specimen thickness on the fracture toughness of polypropylene in three-point bend at 60°C.
$\sigma_y = 70$ MN/m^2, $W = 20$ mm; (data from reference [5]).

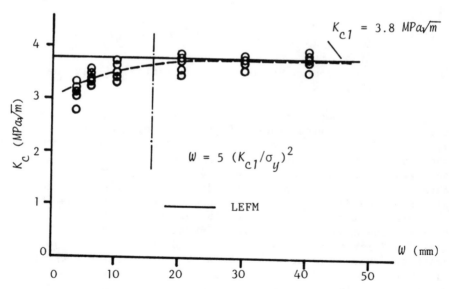

Fig. 9. Effect of specimen width on the fracture toughness of polyacetal in three-point bend at 20°C.
$\sigma_y = 68$ MN/m^2; (data from reference [5]).

6.0 THE J METHOD

It has been observed that in certain configurations of
notched specimens the constraint remains high even when the plastic
zone is large and indeed even when it includes the whole of the
uncracked ligament $(W - a)$. This is the case for deeply notched
specimens subjected to bending so that if, for example, a three
point bend specimen with $a/W = 0.5$ is used then even when it is
fully plastic plane strain conditions will pertain in the
ligament. In such a test the load deflection curve will certainly
not be linear nor will it be elastic but if a crack is initiated
under these conditions the energy per unit area should be the same
as the elastic plane strain G_c. Because of the extensive plasticity
crack initiation is followed by stable ductile tearing and, as
discussed previously, in this stable case we can find the energy
per unit area, which we shall term J_c by performing a calculation
equivalent to that for G_c discussed previously. The test can be
done by loading two specimens with slightly different crack
lengths, observing the load-deflection diagrams up to crack
initiation and hence deducing J_c from the area between the two
curves. This is not very convenient, although it has been used
successfully, and a better method is to use the total energy form
equivalent to that for G_c given in equation (3). For the deeply
notched bend case it can be shown that [6]

$$J_c = \frac{2U}{B(W - a)} \tag{10}$$

where U is total energy up to crack initiation so only one specimen
need be used to find J_c.

Here we encounter a further problem in that it is very
difficult to determine crack initiation in opaque materials. In
general it occurs first in the centre of the original notch, where
the constraint is highest, and is thus very difficult to see.
An extrapolation scheme has been developed by ASTM [7] to overcome
this in which a set of nominally identical specimens are loaded to
different, post initiation, displacements which are less than that
for total failure. These may then be broken open by first freezing
them and the amount of slow crack growth, Δa, measured. For each
test the J for each Δa can be calculated from equation (10) and a
graph of J versus Δa drawn which may be extrapolated back to
$\Delta a = 0$ to give the initiation value. The scheme is shown
diagrammatically in Fig. 10 where one further complication, the
blunting line, is shown in the final diagram. This extra correction
is necessary because in these highly ductile situations the
initially sharp crack will blunt significantly during loading prior
to crack initiation. The blunting arises from the crack opening

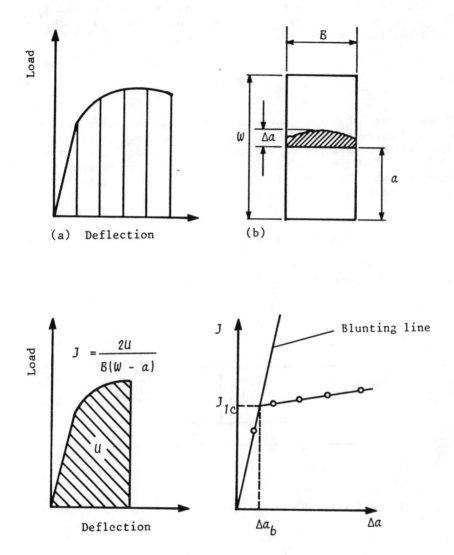

$$J = \frac{2U}{B(W - a)}$$

Fig. 10. Procedure for measuring J; (from reference [8]).

displacement (COD) which occurs at the crack tip and this can be calculated from J since J is an energy per unit area at the crack tip and the local <u>force</u> per unit area is the yield stress σ_y. The energy per unit area, or work done per unit area is this σ_y times the local displacement δ.

i.e.
$$J = \sigma_y \cdot \delta \tag{11}$$

If we assume that the crack tip blunts smoothly, i.e. without discontinuities, then the shape will be as shown in Fig. 11.and there will be an apparent moving forward of the crack tip of $\Delta a_b = \delta/2$. True initiation will occur from this blunted tip as shown but what is measured will include Δa_b so we must extrapolate back to the blunting line given by

$$J = 2 \sigma_y \cdot \Delta a_b \tag{12}$$

to obtain the true initiation condition.

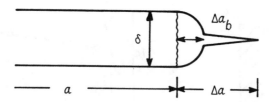

Fig. 11. Crack tip blunting.

This scheme has been applied to several toughened polymers and some data for a toughened Nylon, a toughened PVC and an ABS are given here. The $J - \Delta a$ curves for the Nylon are shown in Fig. 12. for two temperatures and a similar set for the PVC are shown in Fig. 13. In general the linearity is good and there are no problems in the extrapolation. The load-deflection diagram for Nylon is shown in Fig. 14 marked with the points at which the different specimens were unloaded. Also shown are the Δa values plotted

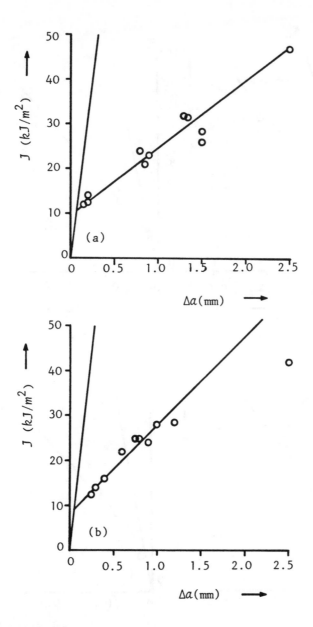

Fig. 12. J versus Δa curves for a toughened Nylon at two temperatures, (a) T = 20°C, (b) 0°C, B = 11 mm, W = 15 mm, a/W = 0.58.

306

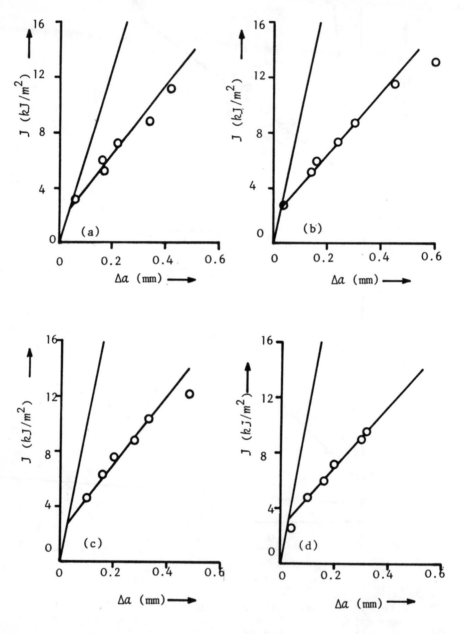

Fig. 13. J versus Δa curves for a toughened PVC at five
temperatures; (a) T = 20°C, (b) 0°C, (c) - 10°C
(d) - 20°C and (e) - 30°C. B = 20 mm, W = 5 mm,
a/W = 0.55.

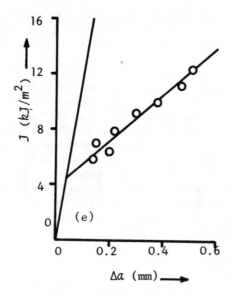

Fig. 13. (Continued).

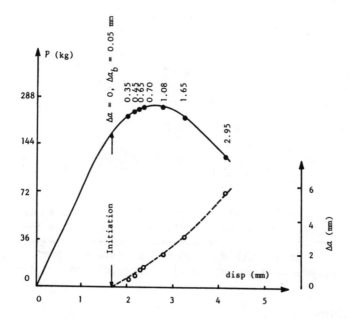

Fig. 14. Load deflection curve and Δa deflection data for the toughened Nylon at 20°C.

versus deflection and here again there is reasonable linearity. The initiation point is also shown and it is seen to be well before maximum load which is typical of the behaviour of this type of material.

The J_c values for the three polymers are shown as functions of temperature in Fig. 15. The closed points are derived by the method given here while the open points at the lower temperatures are G_c values determined using the usual LEFM methods since valid LEFM specimens could be made at those temperatures. There is reasonable continuity of results from the two methods which supports the notion that the stress states are indeed the same. Rather marked transitions are apparent in both ABS and PVC but less so in the considerably tougher Nylon. In Fig. 16. the same data is presented as K_c values and although the pattern is similar the difference is less marked. The ABS value is rather lower than that given in Fig. 7. which was for the maximum load and not the true initiation as given here.

The size criterion for J testing by ASTM below which the plane strain conditions will not hold is, [9].

$$B > 25 \left(\frac{J_c}{\sigma_y}\right) \tag{13}$$

and

$$B = \frac{W}{2}$$

This is approximately equivalent to;

$$25 \cdot \left(\frac{K_c^2}{E\sigma_y}\right) = 25 \, e_y \left(\frac{K_c}{\sigma_y}\right)^2$$

where e_y is the yield strain. In polymers $e_y \sim 0.03$ so the factor is 0.75 giving a decrease of a factor of about 3.5 on the LEFM condition. For the tough materials mentioned previously $J_c \sim 12 \, kJ/m^2$ so that with $\sigma_y = 50 \, MPa$ we have $B = 6mm$, a considerable reduction and a practical proposition for testing.

7.0   CONCLUSIONS

This very brief review of the use of Fracture Mechanics to characterise the toughness of polymer blends has shown that a very powerful method exists which will give a real material property independent of specimen test conditions. Because of the very

309

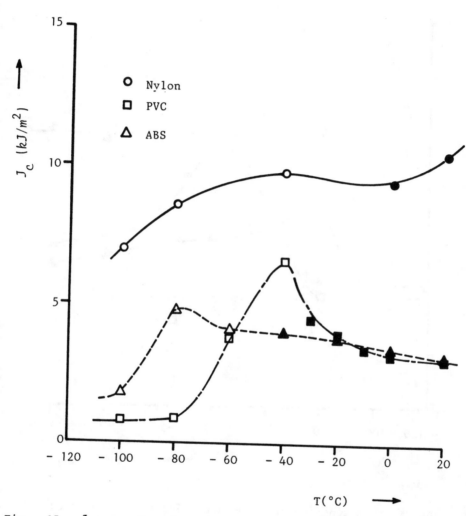

Fig. 15. J_c as a function of temperature for three toughened polymers – closed points from J_c tests – open points from LEFM ($J_c = G_c = K_c^2/E$)

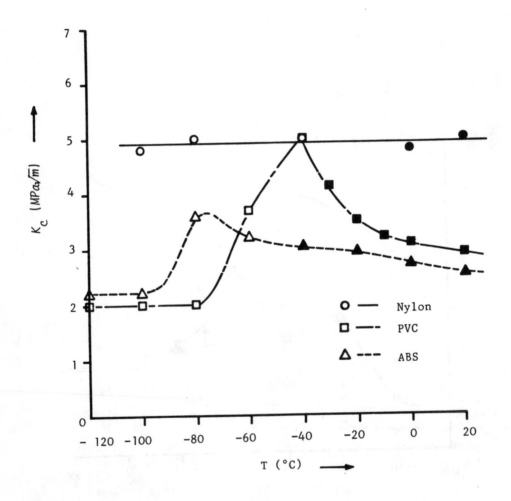

Fig. 16. K_c as a function of temperature for three toughened polymers, - closed points converted from J_c values ($K^2 = E J_c$) - open points from LEFM tests.

toughness which the blending and mixing achieves it becomes difficult to achieve appropriate size conditions in toughness tests but newly developed methods such as the J test do provide some promise in overcoming this problem. It is hoped that this resumé will encourage people to use these methods and abandon the empirical and archaic schemes of yesteryear.

8.0 REFERENCES

[1] Griffith, A.A. Phil. Trans. R. Soc. A221 (1920) 163-198.

[2] Irwin, G.R. J. App. Mech. 29 (1962) 651-654.

[3] Williams, J.G., "Fracture Mechanics of Polymers" (Ellis Horwood - Wiley, 1984).

[4] Brown, W.F. and J. Srawley ASTM STP 410 (1966).

[5] Hashemi, S. and J.G. Williams J. Mat. Sci., in press.

[6] Rice, J.R., P.C. Paris and J.G. Merkle ASTM STP 536 (1973) 231-245.

[7] Landes, J.D. and J.A. Begley ASTM STP 560 (1974) 170-186.

[8] Chan, M.K.V. and J.G. Williams Int. J. Fract. (1983) 145-159.

[9] ASTM Committee E-24, Task Group E 24.01.09, Recommended Procedure for J_{Ic} determination, Task Group meeting in Norfolk, Virginia, USA (March 1977).

CRAZING AND CRACKING IN GLASSY HOMOPOLYMERS

R.P. Kambour

Polymer Physics and Engineering Branch, Corporate Research and Development GENERAL ELECTRIC COMPANY, Schenectady, New York 12301

Abstract: Crazing and cracking in glassy homopolymers are reviewed with particular emphasis on the influence of shear flow on craze initiation and growth and on crack propagation.

Crazes, like true cracks grow normal to the direction of largest tensile stress, reflect light and eventually give rise to complete fracture if stressed sufficiently (1-4). Unlike cracks, crazes are load bearing (Figure 1) as a result of a web of

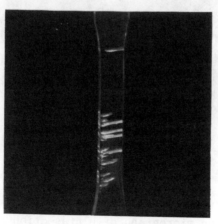

FIGURE 1   A bar of BPA polycarbonate containing several unusually large crazes that have run through the entire cross-section without specimen failure, which thus demonstrates that crazes are load-bearing. (Ref. 2)

micro-fibrils (Figure 2) plastically stretched between the walls of what would otherwise be a crack. Crazes are important for several reasons: a) cracks initiate in crazes, b) during the plane strain crack propagation more crazes are formed at the crack tip thus constituting a plastic zone that blunts the crack and c) they are a major source of ductility in rubber modified plastics.

1 CRAZE COMPOSITION AND STRUCTURE

The craze is a thin layer of resin in which plastic deformation (e.g. 60%) and elastic deformation (e.g. 200%) in the stress direction have occurred without lateral contraction on a gross scale. As a result the void fraction in the craze $1-\lambda^{-1}$ = 40 to 75% or greater.

The craze tip may be only 10nm thick while the mature craze body may be 10^2 to 10^4 times thicker (Figure 2). When formed far

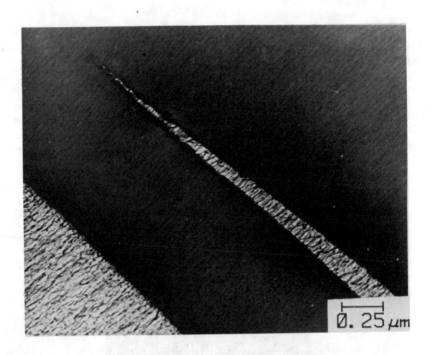

FIGURE 2   Electron micrograph of ultrasection through craze tip and part of another craze in poly(2,6-dimethyl-1, 4-phenylene oxide). Note fibril orientation, sharp boundary between craze and bulk polymer, and uniformity of structure everywhere except at youngest part of tip. (Ref. 2)

below T_g the thickening occurs primarily by drawing in more resin at the bulk polymer interface; thus, the void content is fairly uniform throughout the craze. Thickening at high temperatures involves a coarsening of the structure through fibril breakdown (Figure 3).

Formed well below T_g the fibrils are 6 to 30nm in diameter (Figure 2). When stretched taut they confer a nearly linear stress strain curve on the craze albeit the low polymer volume fraction makes for a low modulus. Stressing from the disoriented state produces a rather ductile stress-strain curve (Figure 4).

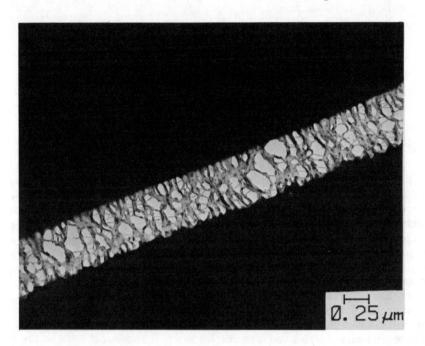

FIGURE 3   Polystyrene craze with coarse weak structure grown in plasticizing fluid near T_g. Note dark spots on fibril which appear to be the residues of fibrils that broke and retracted. (Ref. 2)

316

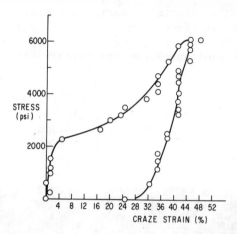

FIGURE 4   Cyclic stress strain curve for craze in BPA polycarbonate
            after aging at low stress for several months. (Ref. 2)

2 CRAZE STRESSES

The fibrils of the body of the craze may sustain quite large
loads. Nevertheless, a concentration of the major principle stress
of the order of 1.2 to 1.7 exists at the craze tip (Figure 5)
depending on the resin, the environment, etc. A concentration in
the major principle stress implies similar concentrations in minor
principle stress and also in the resolved shear stress. In thin
films and surface layers of relatively craze resistant resins the
shear stress can be great enough to stop crazes by the production
of shear bands growing forward from the craze tip (Figure 6a). In
the center of a thick specimen the direction of maximum shear
stress is different and occasionally shear zones may emanate in the
x-y plane (Figure 6b).

3 CRAZE INITIATION MECHANICS AND KINETICS

Craze initiation is a time-, temperature-, and stress-dependent
process that follows an Eyring stress-biased activation expression
(2). However, the Eyring parameters differ quantitatively from
those of shear flow often by small degrees. Thus, apparently small
changes in test parameters may produce shear flow instead of
crazing or vice versa. With some resins - apparently the less
craze resistant ones - lowering of the test rate or applied stress
tends to allow shear flow to occur instead. On the other hand,
with the glassy thermoplastics considered to be craze resistant,
lowering the test rate greatly or limiting the applied stress
allows crazes to form before or below the onset of shear flow.

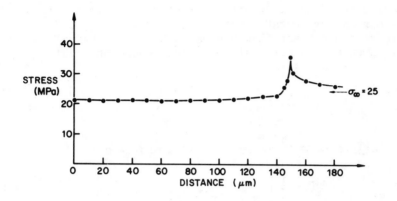

FIGURE 5   Stress distribution in the bulk polymer at the boundary
of a craze in polystyrene of M_w = 200,000 and M_w/M_n =
1.06.   Reproduced with permission of Kramer.
(Ref. 4)

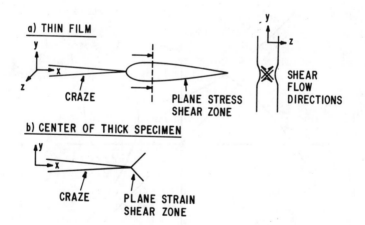

FIGURE 6   Shear zones in front of craze tips in a) thin film,
b) center of thick specimen.   (Refs. 2 and 4).

318

3.1 Stress Field Effects

Crazing and shear flow differ greatly in their sensitivity to the shear and hydrostatic components of stress (5,6,7). In an isotropic specimen shear flow begins when the octahedral shear stress τ_{oct}^{*} reaches a critical, pressure dependent level:

$$\tau_{oct}^{*} = \tau_{o} - \mu P \tag{1}$$

where

$$\tau_{oct}^{*} = \frac{1}{3} [\sigma_{x} - \sigma_{y})^2 + (\sigma_{y} - \sigma_{z})^2 + (\sigma_{z} - \sigma_{x})^2]^{1/2} \tag{2}$$

$$P = \frac{1}{3} (\sigma_{x} + \sigma_{y} + \sigma_{z}) \tag{3}$$

and σ_{x}, σ_{y} and σ_{z} are the principal stresses.

τ_{o} is the pure shear stress under which shear flow begins. The coefficient $\mu = 0.1$ approximately. Thus, shear flow is made slightly more difficult when P is negative (compressive). The biaxial stress field locus for Eq. 1 is shown in Figure 7.

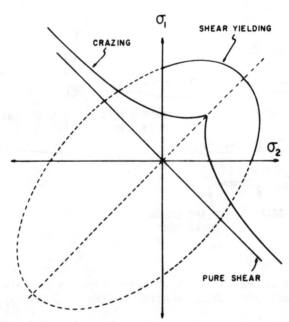

FIGURE 7   Biaxial stress field loci for a) shear flow and b) craze initiation in polymethylmethacrylate at 60°C. Reproduced with permission of Sternstein et.al. (Refs. 5 and 6).

By contrast craze initiation follows a very different locus
(Figure 7). Crazing never occurs when P is negative and in fact
crazing resistance asymptotically approaches infinity as P
approaches zero (i.e. stress becomes pure shear). Thus, the stress
state (i.e. ratio of P to τ_{oct}) has a profound effect on whether a
given resin at a particular test rate and temperature exhibits
crazing or shear flow. The crazing locus has been fitted by a
number of expressions the first of which was (5)

$$\sigma_y - \sigma_x = \frac{A}{(\sigma_y + \sigma_x)} + B \tag{4}$$

where A and B are time and temperature dependent.

The intrinsic significance of all of the existing craze locus
equations is doubtful. Most of the experiments have been performed
on machined and (sometimes) polished surfaces. The effects of the
scratches, polymer orientation from the pressure of the abrading
particles and possibly even from imbedded grit make for interpretation difficulties.

3.2 Kinetics

Often, as mentioned above, craze initiation rates are fitted
to an Eyring stress biased activation framework (2). Thus,

$$k = k_o \exp (-E_a + \beta\sigma/RT)$$

where k the rate is the reciprocal of the initiation time, σ is
the applied stress, E_a the apparent activation energy and β is a
constant.

Under constant strain testing a so-called critical strain $_c$
is found; that is, an applied strain below which crazes do not
seem to initiate even at infinite time (2). Most craze testing,
partly for reasons of convenience, is done at constant strain.

3.3 Polymer Orientation

Craze resistance is well-known to be raised by polymer
orientation when the stress is applied in the orientation direction.
Recent evidence (8) indicates that as little as 1% plastic strain
can be an effective suppressant of craze initiation. Crazes in
more ductile materials often stop growing because they hit pre-
existent shear bands or because shear bands are spawned at the
craze tips themselves (2).

3.4 Molecular Effects

While craze initiation resistance is independent of resin molecular weight, it is linearly dependent on the difference ΔT between the test temperature and T_g and also on cohesive forces. The dependence of craze resistance on the product (ΔT) (CED) where CED is the cohesive energy density is linear up to the point where plastic flow begins (9) (Figure 8). The precise reason for the dependence on CED is not clear. Intuitively the cavitational aspect of crazing seems to be involved. However, the probable role of foreign particles in all craze initiation except at crack tips suggests a debonding mechanism.

3.5 Crazes Spawned by Plain Strain Shear Flow: The Ductile/ Brittle Transition

In contrast to the retarding effects of shear flow cited above, crazes can be initiated at the tip of a zone of plane strain shear flow when the latter occurs along curved slip lines (10-13). This is the situation that occurs with a notch or crack in a thick specimen of a resin that exhibits substantial ductility in uniaxial testing (Figure 9). The constraints imposed by the notch cause the level of major principle stress σ_{yy} necessary to keep $\tau_{oct} = \tau^*_{oct}$ to increase as the zone grows. This rise in σ_{yy} and the attendant

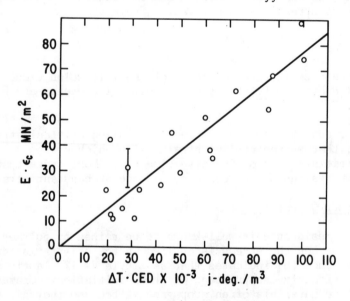

FIGURE 8   Modulus E times critical applied strain ε_c for craze initiation in air of a wide variety of resins vs. $(T_g - T_{test})$ times cohesive energy density CED. (Ref. 15)

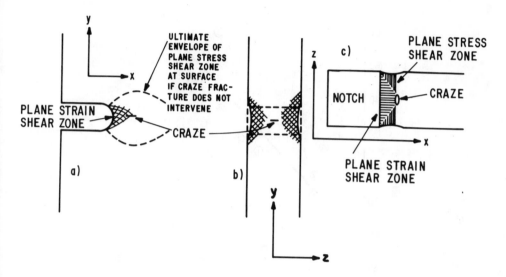

FIGURE 9   Shear zones and craze beyond notch in ductile plastic.
a) Plane strain shear zone and craze in mid-plane of
specimen. b) Section through specimen at craze showing
invading plane stress shear zone. c) Section through
specimen at notch mid-plane showing craze and both
shear zones.

rise in σ_{xx} the minor principle stress operating in the mid-plane
result in a rise in P (see Eq. 3). Thus, the ratio P/τ_{oct} rises as
the deformation proceeds and the shear zone grows further into the
specimen. Eventually, σ_{yy} and/or P become large enough to cause
craze initiation at the zone tip. These stresses are both calcul-
able from if the distance of craze initiation below the notch and
the resins uniaxial yield stress are measured and so-called slip-
line field plasticity theory is assumed applicable. This procedure
has been effected recently for a large number of glassy thermoplastics at temperatures where they are sufficiently ductile. The
results are found to be correlated by (ΔT) (CED) again (Figure 10)
suggesting that the mechanism of internal craze nucleation is the
same as that for surface craze nucleation (14).

If the loading of the thick notched specimen is increased
further the craze breaks eventually and the brittle crack runs
through the remainder of the specimen. If the specimen is thin
enough or the yield stress low enough (for whatever reason) plane
stress shear flow from the surfaces invades the craze nucleation
position, craze fracture is prevented and the remainder of the
failure process occurs by plane stress flow, i.e. with high

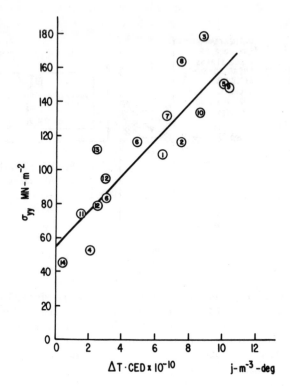

FIGURE 10   Major principal stress σ_{yy} at point of craze
initiation beneath notch vs. (ΔT) (CED) for fourteen
different thermoplastics. (Ref. 14)

toughness. Hence the brittle to ductile transition. This transition is sensitive to temperature (15), rate (16) and the thermal
aging (17) because of their different effects on craze and shear
flow resistance, and to molecular weight because of its effect on
craze strength (2) and thermal aging (17). The transition is also
dependent on notch radius and specimen thickness (16) through their
influence on the relative ease with which the plane stress zone
reaches the crazing position.

4   CRAZE GROWTH

Crazes grow in planar area much more rapidly than in thickness.
Their growth rates are constant under constant uniaxial stress in
some cases (particularly in polystyrene). More frequently the
rates decelerate with time under load (2). No quantitative studies
of growth rate kinetics for crazes at notches have been made;

however qualitative observations indicate that much faster rates
are attainable. The deceleration in growth rates may be related
to aging effects in the craze (18) or perhaps to simultaneous
matrix creep (2).

The best model of craze tip growth is that of Argon and
Salama (19) – the so-called Taylor meniscus instability model
(Figure 11) which is applicable to any situation where one "fluid"
(e.g. polymer) is being displaced by a less viscous one (e.g. air).
Beyond a certain velocity the fluid/fluid interface becomes
unstable and develops a waviness that leads to a breakup that
leaves behind "fingers" of material. A variant of this model has
been developed by Kramer to rationalize the craze thickness process
(4). [Meniscus instabilities can be seen by naked eye in the
elastomer at the junction between pressure-sensitive adhesive tape
and the substrate when the tape is peeled off slowly].

5 CRACK INITIATION IN CRAZES

As alluded to above the brittle fracture of a glassy polymer
starts from the breakdown of a craze (2). The craze can fail in at
least two different ways. Far from T_g localized cracks grow
separately in the craze, (primarily at the craze/bulk resin interface) and link up eventually. Closer to T_g (and/or at lower

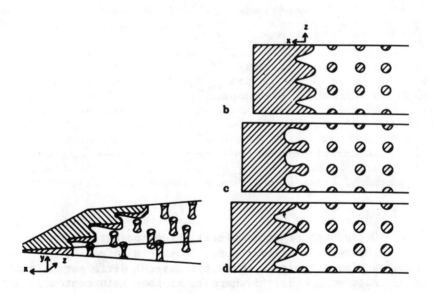

FIGURE 11   Argon-Salama meniscus instability model of craze
             growth. With permission of Kramer. (Ref. 4)

324

stresses probably) a craze may deteriorate by a process of fibril
breakdown throughout that leads to a general coarsening and
weakening of the craze structure. As with craze initiation, the
kinetics of craze breakdown can be rationalized apparently in an
Eyring framework. Breakdown stresses also depend on the true
stress carried by the fibril which in turn depends on the natural
draw ratio λ in the fibril. The strain hardening rate determines
λ and this rate in turn appears to rise with chain entanglement
density (4) (Figure 12). In short, plastics with greater entanglement densities should have stronger crazes.

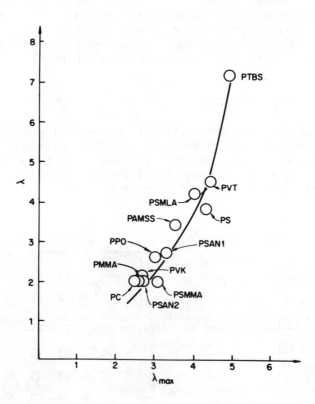

FIGURE 12   Dependence of craze fibril extension ratio λ on
theoretical maximum possible extension ratio, λ_{max} of
a single isolated chain for several different polymers.
$\lambda_{max} = l_e/k(M_e)^{1/2}$ where l_e is the chain contour
length, k is a constant peculiar to each resin and M_e
is the molecular weight between entanglements.
Reproduced with permission of E.J. Kramer. (Ref. 4)

6 CRACK PROPAGATION

Crack propagation in glassy polymers, as in metals, can occur in two modes basically: so-called plane stress and plane strain failure (20).

The plane stress form – which usually consumes more energy – involves a thinning of the region in front of the crack tip via shear flow – as in Figure 6a. Because the cross-sectional area of the shear flow zone is proportional to the square of the specimen thickness the energy consumed in crack propagation tends also to be proportional to the square of the thickness. More precisely the strain energy release rate rises linearly with thickness (21-23). Plane stress crack propagation to a first approximation occurs only in "decidedly ductile" materials. In the context of polymers, "decidedly ductile" means resins that, in the form of uniaxial tensile specimens, exhibit under the same rate and temperature conditions shear failure (yielding) followed by drawing to some tens of percent before fracture occurs.

The plane strain mode – brittle or semi-brittle crack propagation – occurs in plastics, that, in the form of uniaxial tensile specimens exhibit crazing and usually crack initiation at stresses below those at which shear failure would otherwise occur (2). The plane strain mode of propagation tends also to occur at higher crack velocities, at lower temperatures and in grooved specimens and/or thicker specimens of materials exhibiting limited shear ductility in uniaxial tension at room temperature and low speeds. In fact, it is very likely true that plane strain propagation can be induced in sufficiently thick specimens of all rigid thermoplastics.

Plane strain propagation in thermoplastics is resisted primarily by the formation of crazes – from one to 50 or more – at the crack tip (2,23,24). In fact, propagation seems to involve the continued initiation and termination of crazes (Figure 13). The initiation step occurs very close to the crack tip apparently. Most of the new crazes grow forward on curved trajectories that take the craze tips away from the crack plane and thus out of the region of maximum stress (23,24). Hence the need for continued initiation as the crack progresses.

The number of crazes at the crack tip determines crack toughness. This number is a function of crack speed as well as material properties like craze strength. In many or perhaps all of the uniaxially ductile resins a great many crazes exist at the low speed crack tip and thus plane strain crack resistance is quite high. However, a so-called crack instability affects all these materials apparently (23,24). Above a threshold speed the crack tip jumps through the halo of crazes and runs at high speed

FIGURE 13   Matched halves of poly(2,6-dimethyl-1,4-phenylene
             oxide) specimen split by low speed plane strain crack.
             Note multitude of crazes veering away from crack plane
             and terminating. Light border of each fracture surface
             is evidence of plane strain shear flow. Transmitted
             polarized light. (Ref. 24)

with attendant production of two, one or perhaps no crazes – and
with very low crack toughness [e.g. 0.2% of the low speed value in
polycarbonate (25) (Figure 14)].

The mechanism of the instability in thermoplastics is poorly
understood. It presumably involves an inability for new crazes to
initiate close to the crack tip within the craze zone at a rate
sufficient to replace those crazes that have terminated by veering
away from the crack plane. That this is true is supported by
various observations that rubber particle incorporation tends to
ameliorate or wipe out entirely the crack instability. However,
the mechanism(s) by which new crazes are initiated inside the
plastic zone at the glassy polymer crack tip is not known at
present.

In thick sheets of ductile thermoplastics low speed propagation
involves a mix of the plane stress shear mode and the high energy
plain strain crazing mode. The former mode occurs in the surface
layers while the latter occurs in the interior where geometry makes
shear flow more difficult (24,25). Since plane stress propagation
usually consumes more energy than even the high-energy multiple-
craze plain strain mode, mixed mode crack propagation with thick
plane stress surface layers is second only to pure plane stress
propagation in total energy consumption. The thickness of each
surface layer reflects the competition between shear resistance
and the stresses around the plain strain zone. The larger is this
zone, the larger is the stress at the elastoplastic boundary

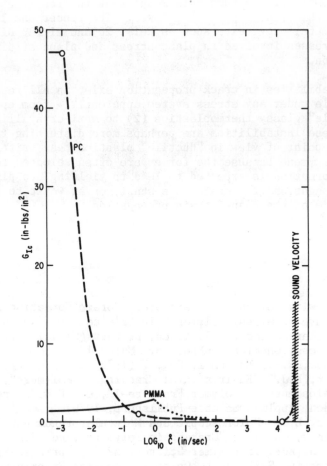

FIGURE 14   Crack speed dependence of the strain energy release
             rate under plane strain conditions G_{Ic} of polycarbonate
             compared to that in polymethylmethacrylate. (Ref. 25)

necessary for plane strain propagation, and the larger is the
thickness of the surface layers in which τ_{oct} has reached τ_{oct}^*.

When the plane strain portion of a mixed mode crack experiences
a crack instability and the crack runs forward at much greater speed
the plane stress surface layer thickness becomes much smaller (25).
This happens for two reasons probably: first, shear flow resistance
usually increases with rate much more than does the craze formation
stress. Second, because the plane strain zone size is smaller, the
stress at the plane strain elastoplastic boundary necessary to
cause the stress at the crack tip to exceed the craze strength is

now much smaller. In turn this means that the resolved shear
stress at this zone tip is also lower. The reader should consult
standard metallurgy texts such as Refs. 20 and 21 for discussions
of the stresses involved in plane stress and plane strain mode
competition.

Instabilities in crack propagation exist in all resins that
are ductile under any stress system apparently – from epoxies (26)
to "brittle" glassy thermoplastics (2) to semi–crystalline plastics
(27). These instabilities are perhaps more disturbing from a
practical point of view in "ductile" plastics than brittle ones
like polystyrene because the former are often selected for uses in
which deformation is expected to lead to yielding, to plane stress
crack propagation if a crack does start, or at worst to the high
energy mode of the plane strain propagation.

REFERENCES

1. Rabinowitz, S. and P. Beardmore. "Craze Formation and
    Fracture in Glassy Polymers" in Critical Reviews in Macro-
    molecular Science, Vol. 1, ed. E. Baer (Cleveland, Ohio,
    CRC Press, Chemical Rubber Co., March 1972).
2. Kambour, R.P. Macromol. Rev. 7 (1973) 1.
3. Kramer, E.J. "Environmental Cracking of Polymers", Ch. 3 in
    "Developments in Polymer Fracture-1", ed. E.H. Andrews
    (London, Applied Science Publishers, 1979).
4. Kramer, E.J. "Microscopic and Molecular Fundamentals of
    Crazing", Ch. 1 in "Crazing in Polymers", ed. H.H. Kausch.
    (Berlin, Adv. in Polymer Science, 52/53, Springer-Verlag, 1983).
5. Sternstein, S.S. and L. Ongshin. Polymer Preprints, Am. Chem.
    Soc., Div. Polymer Chem., 10 (1969) (2) 1117.
6. Sternstein, S.S. and F.A. Meyers. J. Macromol. Sci.-Phys.,
    B8 (1973) 539.
7. Oxborough, R.J. and P.B. Bowden. Phil. Mag. 28 (1973) 547.
8. Kambour, R.P. Polymer Communications 25 (1983) 130.
9. Kambour, R.P. Polymer Communications 24 (1984) 292.
10. Mills, N.J. J. Materials Sci. 11 (1976) 363.
11. Ishikawa, M., I. Narisawa and H. Ogawa. J. Polymer Sci.,
    Polymer Phys. Ed. 15 (1977) 1791.
12. Ishikawa, M., H. Ogawa and I. Narisawa. J. Macromol. Sci.
    Phys. B19 (1981) 421.
13. Ishikawa, M. and I. Narisawa. J. Materials Sci. 18 (1983)
    2826.
14. Kambour, R.P. and E.A. Farraye. Polymer Communications, 1984,
    in press.
15. Allen, G., D.C.W. Morley and T. Williams. J. Materials Sci.
    8 (1973) 1449.

16. Morecroft, A.S. and T.L. Smith. J. Polymer Sci., Part C Symposia No. 32 (1971) 269.
17. LeGrand, D.G. J. Applied Polymer Sci. 13 (1969) 2129.
18. Verheulpen-Heymans, N. Polymer 21 (1980) 97.
19. Argon, A.S. and M.M. Salama. Phil. Mag. 36 (1977) 1217.
20. Tetelman, A.S. and A.J. McEvily, Jr. Fracture of Structural Materials, Chs. 2 and 8 (New York. J. Wiley and Sons, Inc., 1967).
21. Srawley, J.E. and W.F. Brown, Jr. In "Fracture Toughness Testing and its Applications", ASTM Special Technical Publication No. 381, p. 14 (Philadelphia, PA., American Society for Testing and Materials, 1965).
22. Knott, J.F. Mat. Sci. Eng. 7 (1971) 1.
23. Kambour, R.P. and S.A. Smith. J. Materials Sci. 11 (1976) 823.
24. Kambour, R.P. and S.A. Smith. J. Polymer Sci., Polymer Phys. Ed. 20 (1982) 2069.
25. Kambour, R.P., A.S. Holik and S. Miller. J. Polymer Sci., Polymer Phys. Ed. 16 (1978) 91.
26. Kinloch, A.J. and J.G. Williams. J. Materials Sci. 15 (1980) 987.
27. Kambour, R.P. Unpublished observations.

THE MECHANICAL PROPERTIES OF HOMOGENEOUS GLASSY POLYMER BLENDS

R.P. Kambour

Polymer Physics and Engineering Branch, Corporate Research
and Development GENERAL ELECTRIC COMPANY, Schenectady,
New York 12301

Abstract: The mechanical properties of a number of homogeneous
glassy polymer blends are reviewed. To a first approximation these
properties are additive. In many cases negative deviations in
ductility are seen that can be attributed to suppression of glassy
state relaxations, which, in turn, are thought to arise from
negative volumes and heats of mixing. Two blend systems are known,
however, for which the measured properties appear to be additive.
Probable requirements for suppressing negative deviations in
ductility and the small probability of finding homogeneous glassy
systems manifesting positive deviations in ductility are discussed.

To a first approximation the mechanical properties of a
homogeneous glassy polymer blend tend to be volume averages of the
properties of the individual components in the blend. There are
often noticeable deviations from linear averages, however.
Furthermore, the transition from the ductile (i.e. shear flow)
mode of failure to the brittle (i.e. craze formation and breakdown)
mode of failure is rather easily effected in all glassy polymers
by any of several changes (1). Thus, these small deviations from
linear averaging in basic mechanical properties of blends often
result in marked changes in toughness. The results of studies of
the mechanical properties of small molecule/glassy polymer blends
have formed a useful basis for understanding the nonlinearities
found in polymer/polymer blends and these results will be briefly
summarized first.

EFFECT OF SMALL MOLECULE DILUENTS ON THE MODULUS, YIELD STRESS AND CRAZING RESISTANCE OF GLASSY POLYMERS

All glassy polymers exhibit secondary relaxations below T_g. Their effect is to reduce both the modulus (2) and yield stress (3) for shear flow at higher temperatures (Figures 1 and 2). A large number of small molecule diluents are known that lower T_g but nevertheless reduce the magnitudes of these glassy state relaxations (4) (Figure 3). Because modulus and shear flow resistance are strongly coupled these so-called anti-plasticizers <u>added</u> <u>in</u> <u>minor amounts</u> produce a rise in shear resistance before the reduction in T_g becomes great enough to begin to reduce the ambient modulus (4) (Figure 4).

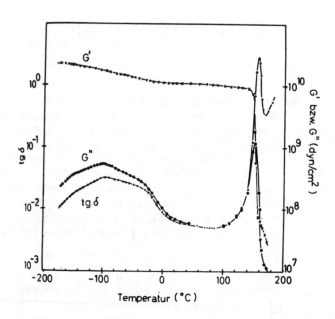

FIGURE 1   Dependence of storage modulus G', loss modulus G" and loss tangent tg δ on temperature for BPA polycarbonate. Reproduced with permission of Illers and Breuer (2).

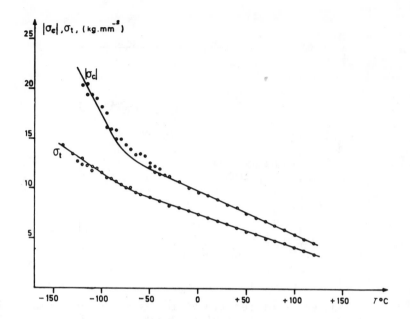

FIGURE 2   Dependence of yield stress in tension σ_t and in
compression σ_c on temperature for BPA polycarbonate.
Reproduced with permission of Bauwens (3).

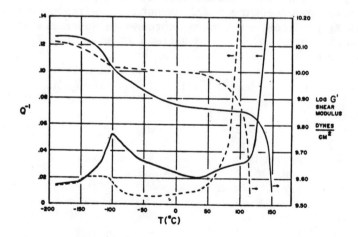

FIGURE 3   Temperature dependence of shear modulus G' and loss
factor Q^{-1} on temperature for BPA polycarbonate neat
(solid line) and anti-plasticized with 30% Arochlor
5460 (dashed line). Reproduced with permission of
Robeson and Faucher (4).

334

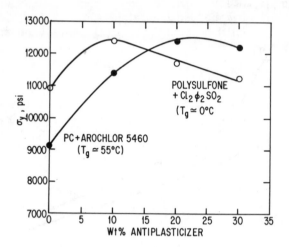

FIGURE 4   Dependence of tensile yield stress on anti-plasticizer
concentration for BPA polycarbonate and the polysulfone
from BPA and dichlorodiphenyl sulfone (4).

It appears moreover that the immobilizing of relaxations
occurring above ambient temperature as well [which occurs to
polystyrene (5) on adding small amounts of o-dichlorobenzene (DCB)
(Figure 5)] produces a relative increase in shear resistance at
lower temperatures (6) when comparison is made on a T_g-shifted test
temperature scale (Figure 6).

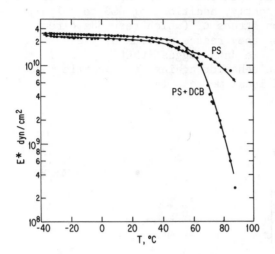

FIGURE 5   Temperature dependence of the complex modulus E* for
polystyrene neat (upper line) and containing 7.5 wt.%
o–dichlorobenzene (lower line)(5).

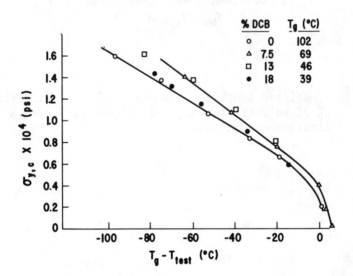

FIGURE 6 T_g–shifted temperature dependence of compressive yield
stress for polystyrene/o–dichlorobenzene mixtures (6).

336

However, in the PS/DCB case at least the T_g-shifted temperature scale is capable of uniting the brittle fracture resistance of the plasticized resin with that of the neat resin (6) (Figure 7). Expressed differently, additions of DCB to polystyrene reduce the craze initiation stress σ_c (which responds directly to T_g reduction) much sooner than they reduce modulus or yield stress (6) (Figure 8). Thus, the available data suggest that many small molecule diluents bias deformation behavior against the ductile mode through an immobilizing of local motions in the glass.

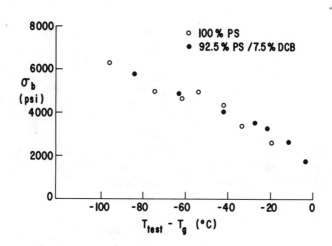

FIGURE 7 T_g-shifted temperature dependence of tensile strength σ_b of polystyrene neat and containing 7.5 wt.% dichlorobenzene (6).

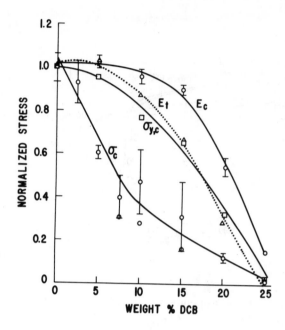

FIGURE 8   Dependence of normalized crazing stress σ_c, compressive yield stress $\sigma_{y,c}$, tensile modulus E_t and compressive modulus E_c on concentration for polystyrene/o-DCB mixtures, all at ambient temperature (6).

BIASES AGAINST SHEAR DEFORMATION IN POLYMER/POLYMER BLENDS

Immobilization of secondary relaxations and attendant increases in yield stress are well documented effects in the case of some miscible polymer/polymer blends as well. The most completely studied system is that of polystyrene (PS) and poly(2,6-dimethyl-1, 4-phenylene oxide), trivially named poly(xylenyl ether)(PXE). Dynamic mechanical temperature data (7) show that even small amounts of PS added to PXE raise its modulus (Figure 9) and reduce its losses (Figure 10) at low temperatures substantially. The room temperature modulus (Figure 11) thus shows positive deviations from additivity (8). These effects have been suggested to arise from the negative heat and volume of mixing in this system.

In any case the result is a substantial positive deviation in (shear) yield stress (8) (Figure 12). By contrast the crazing resistance of these blends is additive or even less than additive in composition (8,9). In addition, the extension ratio of fibrils in crazes in these blends appear to show negative deviations (10)

338

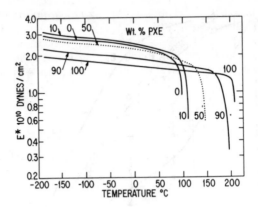

FIGURE 9   Temperature dependence of storage modulus of
            PS/PXE mixtures (7).

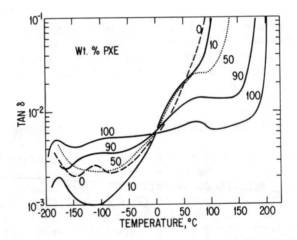

FIGURE 10   Temperature dependence of loss tangent of
            PS/PXE mixtures (7).

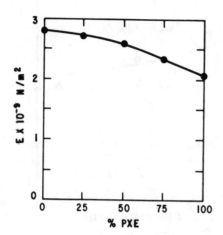

FIGURE 11   Dependence of modulus at ambient temperature on
             composition for PS/PXE blends (8).

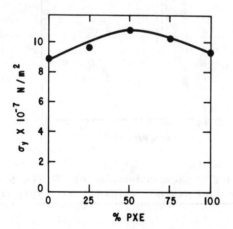

FIGURE 12   Dependence of compressive yield stress σ_y at ambient
             temperature on composition for PS/PXE blends (8).

suggesting a reduction in ductility even here (Figure 13). This
seemingly small bias against ductile deformation is probably
responsible for the rather noticeable reduction in impact toughness
of these blends (11) (Figure 14).

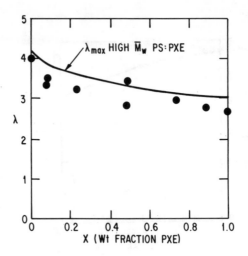

FIGURE 13 Craze fibril extension ratio in PS/PXE blends vs. composition. Reproduced with permission of Kramer (10).

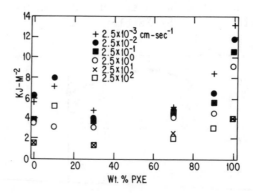

FIGURE 14 Notched Izod impact toughness of PS/PXE blends vs. composition at ambient temperature (11).

Further support for these correlations is afforded by a study (8) of low speed plane strain crack propagation in PS/PXE blends. The steady state plane strain crack propagation energy R_{max} of PXE is suppressed strongly by the addition of small amounts of polystyrene (Figure 15). The source of this suppression has been traced, at least in part, to the negative effects of PS addition on the stress intensity factor K_{Icc} for craze formation at the crack tip (8) (Figure 16).

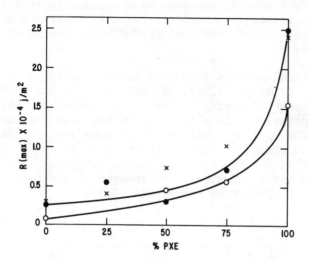

FIGURE 15   Steady-state low speed plane strain crack propagation
energy R_{max} vs. composition for PS/PXE blends (8).

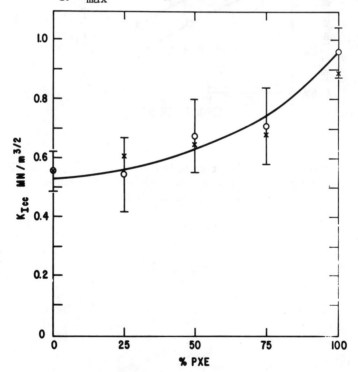

FIGURE 16   Composition dependence of the stress intensity factor
K_{Icc} for craze initiation at the crack tip in PS/PXE
blends (8).

Noticeable biases of the same kind against shear deformation, elongation and impact toughness can be seen in the mixtures of BPA polycarbonate and a copolyester studied by Joseph et.al. (12) (Figures 17-19).

In both blend systems discussed above negative deviations from volume additivity exist (7,12). In addition for the PS/PXE system an exothermic heat of mixing is now well established (13,14). Although amounting to only about 1/2% of the cohesive energy density of either polymer this heat – or more properly the molecular interaction responsible for this heat – is considered to be responsible for the reduction of mobility that leads to the higher modulus and yield stress and lowered elongation, craze resistance and toughness in the blends (7,8).

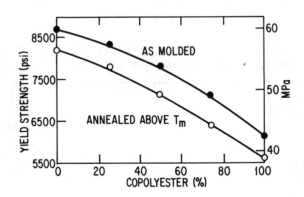

FIGURE 17   Yield stress vs. composition for mixtures of BPA polycarbonate and a copolyester. Reproduced with permission of Joseph et.al. (12).

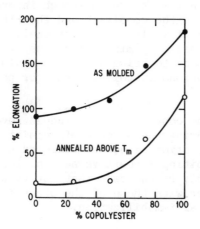

FIGURE 18   Ultimate elongation vs. composition for Figure 17
             blends. Reproduced with permission of Joseph et.al.
             (12).

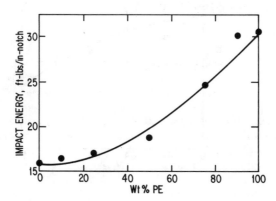

FIGURE 19   Notched Izod impact energy vs. polyester content
             for Figure 17 blends. Reproduced with permission
             of Joseph et.al. (12).

        If true, it might be thought that any blend system close to
the point of phase separation (i.e. to a state in which the heat
of mixing is zero or even slightly positive) might be one in which
interactions between the two polymers are sufficiently weak that
no loss in internal mobility would occur on mixing. One study (15)
exists that currently suggests that this criterion (zero heat of
mixing) is not a sufficiently restrictive one: Mixtures of PXE and
a styrene/p-chlorostyrene copolymer containing just enough styrene
to retain miscibility of the copolymer with PXE show positive

deviations in modulus and yield stress (Figures 20 and 21). These deviations suggest then that there is enough interaction between the 43% styrene units in the copolymer and the PXE to retard its internal mobility significantly. This rationalization is consistent with the observation that only minor fractions of an anti-plasticizer are necessary to bring about the maximum reduction in low temperature relaxations in polycarbonate or polysulfone (Figure 4).

If the PXE/styrene copolymer study indeed reveals a general principle, then countering the adverse effects of negative segment-segment heats of interaction cannot be effected by copolymerization with a comonomer exhibiting a positive heat of interaction. That is, it may unfortunately be the case that interactions between all segments of one polymer with those of the other must be athermal in order to avoid segmental immobilization.

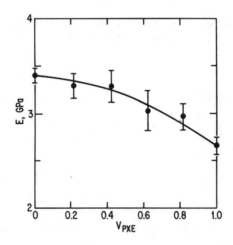

FIGURE 20   Composition dependence of the modulus of blends of PXE with a 43% styrene/57% p-chlorostyrene copolymer. Phase separation occurs for copolymer styrene contents less than 33%. Reproduced with permission of Fried et.al. (15).

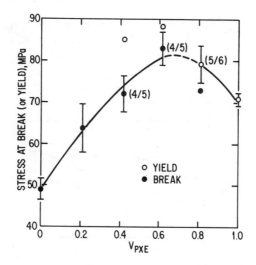

FIGURE 21   Composition dependence of strength of Figure 20 blends.
             Filled circle—brittle, open circle—ductile. Reproduced
             with permission of Fried et.al. (15).

    While such modification may not be realizable in practice
very often, it is nevertheless encouraging to note that there is at
least one blend system known in which adverse property deviations
appear not to exist. This system is composed of blends of BPA
polycarbonate with bisphenol chloral polycarbonate, injection
moldings of which show <u>negative</u> deviations in modulus and yield
stress (Figure 22), and additivity in ultimate elongations (16).
Another system that is at least ambiguous in this regard is
composed of polystyrene and tetramethyl BPA polycarbonate. This
system is known to have a lower critical solution temperature of
250°C implying very nearly athermal mixing (17). According to Yee
and Maxwell specific volumes in these blends deviate negatively
from additivity and yet the mechanical losses exhibited by the
polycarbonate above room temperature seem to be reduced in the
blends in proportion to concentration roughly (11). Unfortunately,
high stress data for this system appear not to have been published.

346

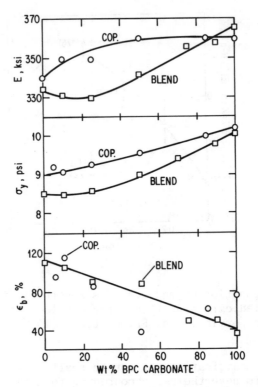

FIGURE 22   Composition dependence of mechanical properties of
             blends of BPA polycarbonate with bisphenol chloral
             (BPC) polycarbonate and of BPA/BPC copolycarbonates.

CONCLUSIONS

     It appears that in many homogeneous glassy blends the exother-
mic interactions to which miscibility is attributed give rise to
reductions in internal mobility and consequent adverse effects on
toughness. At present it seems doubtful that these adverse effects
can be countered by use of copolymers containing a mixture of
negatively and positively interacting segments. That is, achieving
additive mechanical properties seems to require that the segment-
segment heat of interaction be approximately zero for all segment
pairs. Only in that case will strength properties be additive.

     Finally, since the low temperature relaxations giving rise
to reductions in modulus and yield stress in a given polymer are
never seen to be enhanced by additives, positive deviations in
ductility and toughness in miscible glassy blends seem unlikely
to be found in the future.

REFERENCES

1. Kambour, R.P. Preceeding paper in this workshop.
2. Iller, K.H. and H. Breuer. Kolloid-Z. 176 (1961) 110.
3. Bauwens, J.C. J. Materials Sci. 7 (1972) 577.
4. Robeson, L.M. and J.A. Faucher. J. Polym. Sci. B, 7 (1969) 35.
5. Yee, A.F. Unpublished work.
6. Kambour, R.P. In "Mechanisms of Environment Sensitive Cracking of Materials", eds. P.R. Swann, F.P. Ford and A.R.C. Westwood (London, Metals Society, 1977).
7. Yee, A.F. Polym. Eng. and Sci. 17 (1977) 213.
8. Kambour, R.P. and S.A. Smith. J. Polym. Sci., Polym. Phys. Ed. 20 (1982) 2069.
9. Kambour, R.P. Polymer Communications 24 (1983) 292.
10. Kramer, E.J. In "Crazing in Polymers", Adv. in Polym. Sci., 52/53, ed. H.H. Kausch (Berlin, Springer-Verlag, 1983).
11. Yee, A.F. and M.A. Maxwell. J. Macromol. Sci.-Phys. B17 (1980) 543.
12. Joseph, E.A., M.D. Lorenz, J.W. Barlow and D.R. Paul. Polymer 23 (1982) 112.
13. Weeks, N.E., F.E. Karasz and W.J. MacKnight. J. Appl. Phys. 48 (1977) 4068.
14. Maconnachie, A., R.P. Kambour, D.M. White, S. Rostami and D.J. Walsh. Macromolecules, in press.
15. Fried, J.R., W.J. MacKnight and F.E. Karasz. J. Appl. Phys. 50 (1979) 6052.
16. Factor, A. and C.M. Orlando. J. Polym. Sci., Polym. Chem. Ed. 18 (1980) 579.
17. Casper, R. and L. Morbitzer. Angew. Makromol. Chem. 58/59 (1977) 1.

MECHANICAL PROPERTIES OF HIGH-IMPACT POLYMERS

Clive B. Bucknall

Cranfield Institute of Technology, Cranfield, Bedford,
MK43 OAL.

1 STRUCTURE AND MORPHOLOGY

High impact polymers are two-phase composites. In a typical
case, the major component is a rigid thermoplastic or thermoset
which forms a continuous matrix, and the minor component is an
elastomer which is dispersed uniformly throughout the material in
the form of small particles. The volume fraction of particles is
rarely if ever equal to the volume fraction of the elastomeric
phase. Electron microscopy reveals a variety of complex
morphologies within the rubber particles, which may contain very
large amounts of the rigid phase, as illustrated in Fig. 1.
Polystyrene sub-inclusions account for 75 to 80 percent of the
total volume of the composite rubber particles in standard
commercial grades of HIPS.

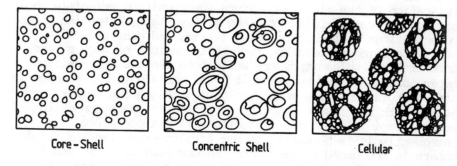

Core-Shell   Concentric Shell   Cellular

Fig. 1 Complex morphologies in HIPS (high-impact polystyrene).

Rubber particle volume fraction RPVF in HIPS depends upon stirring conditions during polymerisation (1,2), and upon such factors as the choice of initiator, which affects grafting (3), and the use of styrene-butadiene block copolymer rubbers (3, 4). Greater control can be achieved over structure by sequential polymerisation in emulsion: examples are ABS, which is made by adding styrene and acrylonitrile to a polybutadiene latex, and toughened PMMA, which is made in a three- or four-stage process beginning with a rigid core, which is successively overcoated with a layer of acrylate rubber and an outer layer of PMMA. Small sub-inclusions can sometimes be seen within the rubber particles in ABS emulsion polymers, although the resulting effects upon rubber particle volume fraction are much smaller than in HIPS. Toughened epoxy resins may have very low RPVF values as a result of incomplete precipitation of the elastomer from solution, or high values caused by molecularly dispersed and reacted epoxy resin or by sub-inclusions (5). Because of the wide range of RPVF that can be obtained from a single composition in all of these procedures for making toughened polymers, it is important to check volume fractions before drawing conclusions about other factors that might affect mechanical properties, e.g. the size or internal structure of rubber particles. The volume fraction of the load-bearing matrix is the most important factor determining both the moduli of the two-phase composite and the magnitude of the stress concentrations within it.

Examination of dynamic mechanical loss spectra shows that the magnitude of the loss peak due to the glass transition in the rubber phase increases with the volume fraction of composite rubber particles, although the relationship appears not to be linear (1). The temperature of the loss peak should be as low as possible, not only to ensure good low-temperature properties, but also to improve impact energy absorption at room temperature. Cracks can propagate at speeds of up to 500 m/s in polymers, which means that a distance of 5 microns is covered in 10 nanoseconds. Application of the time-temperature superposition principle shows that rubbers cannot relax from the glassy state, establish stress concentrations in the surrounding matrix, and generate energy-absorbing deformation from these regions of high stress concentration, unless they are well above the glass transition temperature indicated by tests at 1 Hz. An additional problem is that polybutadiene, the most widely used toughening rubber, and the one with the lowest Tg, is susceptible to oxidation which causes a rapid rise in its glass transition temperature. Consequently it is necessary to use ethylene-propylene or acrylate rubbers in toughened polymers for outdoor applications, and in materials that have to be processed at high temperatures. These rubbers represent a compromise: initially, the Tg is higher than that of polybutadiene, but the upward temperature shift due to oxidation is greatly retarded.

2   MECHANISMS OF TOUGHENING

Microscopy has played a key role in the study of toughening mechanisms. As a result of numerous studies, it is known that the major mechanism of energy absorption is deformation of the matrix polymer, although some energy is taken up in associated deformation of the rubber particles. The feature common to all rubber-toughened plastics is that the rubber lowers the tensile yield stress. In HIPS, the dominant mechanism of yielding is multiple craze formation which causes a measurable increase in volume (6, 7). By contrast, the mechanism of deformation in commercial rubber-toughened PMMA is shear yielding, which occurs at constant volume (8, 9). Both mechanisms have been observed to occur simultaneously in a number of rubber-modified plastics, notably blends of HIPS with poly(phenylene oxide) (PPO) (7, 10).

Although electron microscopy is an essential tool in the study of deformation mechanisms, it does have its limitations. In particular, the observation of shear yielding in two-phase materials is difficult, especially at high rubber contents. A quantitative analysis of the relative importance of crazing and shear mechanisms in any deformation where both are present is virtually impossible using microscopy. An alternative and very effective method for studying mechanisms of toughening is to measure volume changes in specimens under tensile load. The earliest method was the simultaneous measurement of elongation and lateral contraction in a parallel-sided specimen; the method is restricted to specimens that do not neck, and has mainly been used for creep experiments up to about 5% strain, although it can be extended to higher strains by using suitable extensometers (5-10). An alternative, which avoids the problem, is to surround the specimen with a suitable fluid and measure volume changes by direct dilatometry (11, 12). This method works well up to high elongations, but involves some uncertainty about the strain in the region undergoing the volume change, since this region reaches into the shoulders of the tensile specimen, and its length is ill-defined.

Volumetric data obtained by these methods can be used in several different ways. The simplest is to plot volume strain against elongation, as illustrated in Fig. 2. Materials deforming entirely by crazing have a slope of 1.0, whilst materials deforming exclusively by shear yielding have a slope of zero. Intermediate slopes indicate a combination of dilatational and shear mechanisms. It is important to recognise that not all volume increases are due to craze formation. Where crazing in the matrix is accompanied by fibrillation of the associated rubber particles, as in ABS and HIPS (13), there is little point in making a distinction between the two contributions to strain.

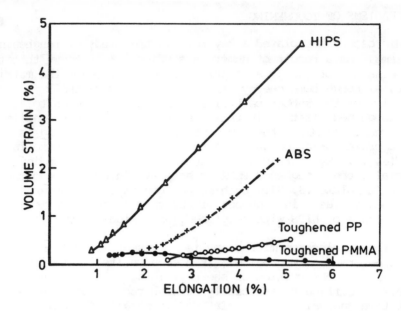

Fig. 2   Relationship between volume strain and elongation

On the other hand, in toughened PVC and toughened epoxy resins
shear yielding of the matrix is the dominant deformation
mechanism, but a small volume increase is also observed, which can
be shown by optical and electron microscopy and by light
scattering techniques to be due to formation of single cavities
within the rubber particles (14-16). A combination of volumetric
and microscopical techniques is desirable in any study of
deformation mechanisms.

The chemical character of the matrix is one of the most
important factors affecting the mechanism of toughening. This
point has been demonstrated in experiments on a series of blends
containing 50% HIPS with various mixtures of PS (polystyrene) and
PPO. As the matrix polymers are compatible with each other, the
nature of the matrix can be varied whilst the size and concentration of rubber particles are kept constant. Blends having a pure
PS matrix deform exclusively by multiple crazing, whereas blends
with a high PPO content exhibit simultaneous crazing and shear
yielding (10). Higher stresses are required to maintain a given
rate of strain in the PPO blends, indicating that the change in
mechanism is due principally to a greater resistance to craze
formation in these blends, which is probably associated with the
higher glass transition temperature of the blends (17).

Rates of crazing and shear yielding are affected differently
by test conditions, including temperature and strain rate, and by
materials structure, including orientation and particle size.
When the two rates are comparable, the mechanism of toughening can
be quite sensitive to changes in any of these variables. On the
other hand, when one mechanism is several orders of magnitude more
rapid than the other, a change in test conditions or in structure
may have no observable effect. The results of slow tensile tests
on ABS polymers at room temperature provide good examples of the
former pattern of behaviour. Plots of volume strain against
elongation for a typical ABS emulsion polymer are shown in Fig. 3.
At low stresses and strain rates shear yielding is dominant, but
at higher stresses crazing makes an increasing contribution as the
strain and strain rate rise (18). These effects are due to
differences in kinetics, which will be discussed in more detail
later. Under similar test conditions, ABS made by the mass-
suspension process gives a slope close to 1.0 in the volume-
elongation plot, as shown in Fig. 4. The difference is probably
due to a larger particle size, but may also be connected with the
large sub-inclusions in the cellular particles. The presence of
the polar acrylonitrile group in the styrene-acrylonitrile
copolymer matrix of ABS raises its crazing resistance (19), so
that a combination of shear and craze deformation is observed at
strain rates that produce only crazing in HIPS; another
consequence of the crazing resistance of SAN copolymer is that
higher stresses are needed to cause creep or yielding in ABS. Hot
drawing introduces orientation in the SAN matrix, which further
increases its resistance to crazing when stressed along the
orientation direction (20).

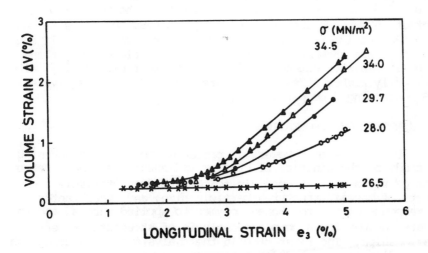

Fig. 3   Volume-elongation plots for ABS emulsion polymer

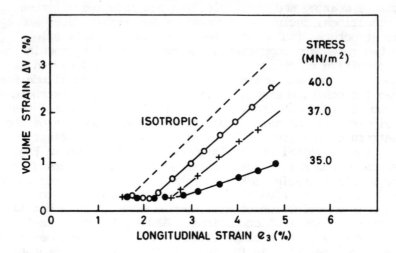

Fig. 4   Volume-elongation plots for isotropic and hot-drawn ABS
mass-suspension polymer.   Draw ratio 1.47.

Resistance to shear yielding is increased to a smaller
extent, with the result that the hot-drawn mass-suspension ABS
shows a combination of the two mechanisms of deformation, at
stresses higher than those needed to produce the same strain rate
in the isotropic polymer.

It has been suggested that particles less than $1\mu m$ in
diameter are incapable of initiating crazes in HIPS, and that
smaller particles can only initiate shear bands (22).   As
discussed above, there is some evidence that the mechanism of
deformation depends upon particle size in ABS, but the same cannot
be said about isotropic HIPS at room temperature.   Bucknall and
Partridge have shown that HIPS containing particles less than
0.6 μm in diameter extends exclusively by multiple crazing, albeit
at slightly higher stresses than are required for an equivalent
HIPS containing 3 μm particles (23).

3   KINETICS OF DEFORMATION

One of the most powerful approaches to rubber toughening is
the study of the kinetics of crazing and shear yielding.   Measurements of rates make it possible to compare quantitatively the
effects of matrix structure, particle size, and other factors upon
the response of the toughened polymer to applied stress, and thus
prepare the way for a complete analysis of structure-property
relationships.   The first step in the analysis is to distinguish
between dilatational and constant volume contributions to total

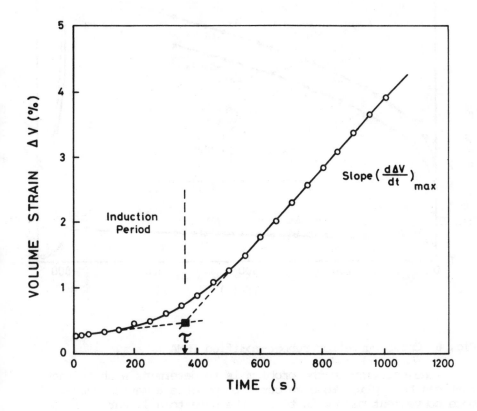

Fig. 5 Typical creep curve for multiple crazing in HIPS.

extension, which can be done easily using the methods mentioned in the preceding section. The second step is to correlate data obtained over a range of stresses, strains, and times under load. The task is simplified if measurements are made at constant applied stress.

Figure 5 is a typical curve of volume against time at constant stress for HIPS. A similar pattern of behaviour is observed in ABS polymers: there is a slow increase in volume during an initial 'induction period', after which the rate of volume change dV/dt becomes constant. Both the length τ of the induction period and the maximum value of dV/dt are useful parameters for comparing deformation behaviour.

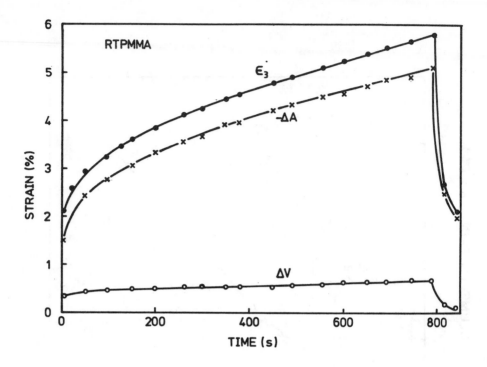

Fig. 6 Creep curve for rubber-modified PMMA

One solution to the problem is to determine a shift factor that can be applied to each curve to produce a master curve. A more convenient method is to fit the data to a linear relationship. The Andrade equation provides a satisfactory basis for this procedure for a number of toughened plastics and their parent matrix polymers (9, 24). Andrade found that, for a wide range of metals, polymers and ceramics at constant true stress the strain $\varepsilon(t)$ at time t was given by:

$$\varepsilon(t) = \varepsilon(0) + b\,t^{1/3} \qquad (1)$$

where $\varepsilon(0)$ is the instantaneous strain following the application of load (25, 26). Figure 7 presents data on toughened PMMA plotted on this basis. Although not all of the points from each test lie exactly on a single line, the fit is good enough to define the parameter b quite precisely.

Deformation due to crazing or shear yielding can be treated as the conversion of flow units from an initial to a strained state by passing over an energy barrier defined by the free energy

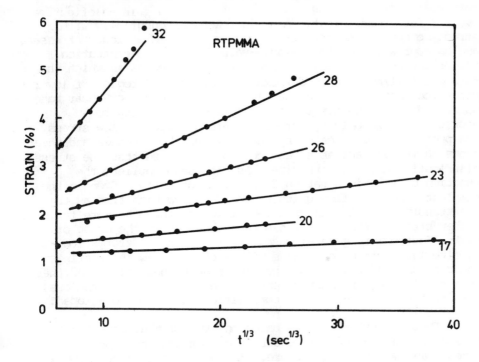

Fig. 7 Andrade plots for toughened PMMA (9)

of activation ΔG. Most of the energy needed to surmount this barrier comes from thermal vibrations; the effect of the applied stress is to lower the free energy of activation in the 'forward' direction by an amount $\gamma v^* \sigma_\infty$ where σ_∞ is the applied stress, v^* is the activation volume, and γ is a stress concentration factor which takes account of the fact that the local stress in the flow unit is not in general equal to σ_∞ . The rate of conversion of the flow units can then be expressed in terms of a rate coefficient which is defined by the Eyring equation (27):

$$\kappa = 2 \, A \, \exp \, (\, \Delta G/kT) \, \sinh \, (\, \gamma v^* \, \sigma_\infty \, /kT)$$

$$\approx \quad A \, \exp \, (\, \Delta G/kT) \, \exp \, (\, \gamma v^* \, \sigma_\infty \, /kT) \tag{2}$$

where A is a constant, k is Boltzmann's constant, and T is temperature. The overall rate is then given by:

$$\text{Rate} = \sum \kappa_i \, N_i \approx \quad \kappa_e \, f(t) \tag{3}$$

where N_i is the number of flow units in stress state i, κ_i is the rate coefficient for that stress state, and κ_e is an effective overall rate coefficient. As κ increases exponentially with stress concentration the series converges rapid at lower stresses, and an approximation to a single-valued stress concentration factor can be justified. However, there are cases in which the rate controlling step in deformation occurs in a region of lower stress concentration. An example would be a craze or shear band nucleated at a high stress concentration but growing to appreciable size only by extending into a region of low stress. The problem is complicated by the fact that both crazes and shear bands themselves act as stress concentrators, so that the stress distribution changes with time. As discussed earlier, the function of time in equation 3 is often not unity. Even in the simplest case, the flow units become depleted with time as they are transformed into the extended state, so that the rate falls at higher strains; this effect has been observed in the creep of HIPS. In the more general case, the kinetics of deformation involve interactions of crazes with each other, with existing shear bands, and with rubber particles near their path, and there is a similar set of interactions of shear bands. Nevertheless, despite these problems, some correlations have been developed.

These correlations are based on experimental observations of rate as a function of time under load, temperature, applied stress, and materials structure. The determination of γ for a series of polymers based on the same matrix but having different rubber contents or particle sizes is of particular interest. In HIPS, for example, it can be assumed that v^* of craze formation does not change, since it is a property of polystyrene, and that variations in τ or dV/dt are due to differences in stress concentrations. Values of γv^* can be obtained by plotting τ, dV/dt or b (in the case of shear yielding) on a logarithmic scale against applied stress. Absolute values of γ can be calculated if comparable measurements can be made on the untoughened matrix polymer. However, this is not always possible for crazing in relatively brittle materials such as polystyrene.

Figure 8 shows the effects of 16 volume percent of ethylene-propylene copolymer rubber particles upon the kinetics of deformation in polypropylene (PP). The rubber not only accelerates creep, but increases the stress concentration factor by 1.56. The curves for toughened and untoughened PP coincide if log b is plotted against $\gamma\sigma_\infty$; in other words, the pre-exponential factor A and the number of available sites for flow remain unaffected by the addition of rubber. The value of γ in this case is interesting: calculations based on Goodier's equations indicate that stress concentrations of approximately 2.0 are to be expected at the equator of an isolated spherical hole or particle of low modulus (28, 29). Results for PP and for a number of other

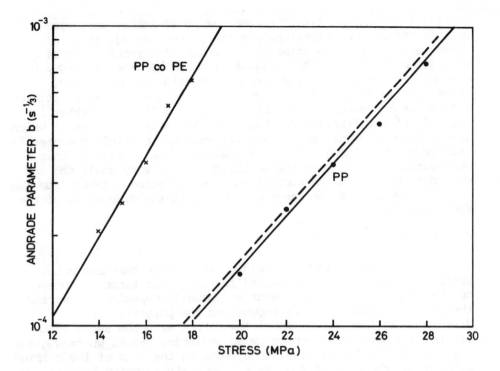

Fig. 8 Eyring plot of Andrade parameter b against applied stress
for polypropylene containing 0% and 16% by volume of
rubber particles (24).

materials show that high stress concentrations of this magnitude
do not always control the overall rate of deformation (9, 24): the
main effect of the rubber particles is not to produce high local
stress concentrations but to increase the average stress in the
matrix simply by reducing the volume fraction of the matrix. For
materials in this category, the relationship between γ and the
volume fraction of particles ∅ can be fitted approximately to the
Ishai—Cohen equation, which estimates stresses in the matrix on
the principle that the load—bearing plane passes through the
equators of the (soft) particles in its path (30):

$$\gamma = [1 - \pi (3 \emptyset/4 \pi)^{2/3}]^{-1} \qquad (4)$$

Recent experiments on HIPS show that γ for crazing in this
material can have much larger values than are predicted by
equation 4, and may reach as high as 8 (23). These results can be
interpreted in terms of three separate but connected effects: (a)
a reduced volume fraction of load—bearing matrix, as discussed in
connection with equation 4; (b) stress concentrations near the
equators of rubber particles of the type calculated by Goodier;
and (c) overlap of stress fields of neighbouring particles in

materials containing a high concentration of rubber particles. Maximum stress concentrations are the product of all three of these terms. The same experiments showed that rates of crazing were strongly affected by particle size in the range 0.3 to 3.0 microns, but that γ obtained from dV/dt data was unaffected. These results suggest that at any given applied stress the stress fields are the same in two HIPS materials of the same rubber particle volume fraction $\emptyset$, but that the number of sites at which crazes form is significantly smaller when the particle diameter is well below one micron. The best available explanation of this observation is that when the particle size is very small the distance over which the high stress concentration extends from the particle equator is not large compared with the dimensions of the flow unit (28).

4   FRACTURE RESISTANCE

Accelerated rates of yielding are a necessary but not sufficient condition for enhanced fracture resistance in rubber toughened plastics. Glass beads are equally capable of lowering the yield stress but not of toughening the polymer. The difference is that rubber particles deform with the surrounding matrix, and have to be broken by an advancing crack, whereas glass beads debond at low strains, thus reducing the area of loadbearing material in the path of the crack. Adequate adhesion between rubber particle and matrix and some degree of strength within the rubber particle are necessary for the same reason.

Because of associated yielding around the crack tip, it is difficult to measure the work done in separating the two opposite surface of a crack. This problem is especially acute in rubber-toughened plastics. However, information about energy absorption at the crack surface can be deduced from fracture mechanics measurements made under plane strain conditions, which limit the extent of plastic deformation very severely. In order to achieve plane strain conditions in toughened plastics, it is necessary to work at low temperatures, preferably below -60°C. Williams and co-workers have shown that under conditions of maximum constraint the fracture toughness of HIPS and of toughened PP becomes equal to, or less than the toughness of the parent polymer (31-33).

These observations help to put the phenomenon of rubber toughening in perspective. The rubber particles do not actually strengthen the crack, but on the other hand they do not weaken it if the material is properly made. The function of the rubber is to lower the yield stress in a controlled manner, so that the stresses transmitted to the tip of the crack are greatly reduced. Energy absorption is therefore much greater than in the parent polymer under the same test conditions. Because of the reduced yield stress the extent of the yielded zone around the crack tip is greatly increased.

References

1. Wagner, E.R. and Robeson, L.M., Rubber Chem.Technol. 43 (1970) 1129.

2. Turley, S.G. and Keskkula, H. Polymer 21 (1980) 466.

3. Echte, A., Angew. Makromol. Chem. 58/59 (1977) 175.

4. Reiss, G., Schlienger, M., and Marti, S., J.Macromol Sci.Phys (1980) 355.

5. Bucknall, C.B., and Yoshii, T., Br.Polym.J. 10 (1978) 53.

6. Bucknall, C.B., and Clayton, D., J.Mater.Sci. 7 (1972) 202.

7. Bucknall, C.B. Toughened Plastics (London, Applied Science Publisher, 1977).

8. Hooley, C.J., Moore, D.R., Whale, M. and Williams, M.J., Plast.Rubber Process Appl. 1 (1981) 345.

9. Bucknall, C.B., Partridge, I.K., Ward, M.V., J.Mater.Sci. 19 (1984) 2064.

10. Bucknall, C.B., Clayton, D., and Keast, W.E., J.Mater.Sci. 7 (1972) 1443.

11. Coumans, W.J., Heikens, D., and Sjoerdsma, S.D., Polymer 21 (1980) 103.

12. Heikens, D., Sjoerdsma, S.D., and Coumans, W.J., J.Mater.Sci. 16 (1981) 429.

13. Beahan, P., Thomas, A., and Bevis, M., J.Mater.Sci., 11 (1976) 1207.

14. Haaf, F., Breuer, H., and Stabenow, J., J.Macromol Sci.Phys., B14 (1977) 387.

15. Bascom, W.D., and Cottington, R. L. J.Adhesion 7 (1976) 333.

16. Yee, A.F., and Pearson, R.A., NASA Contractor Report 3718 (1983).

17. Kambour, R.P., Polymer Communications 24 (1983) 292.

18. Bucknall, C.B., and Drinkwater, I.C., J.Mater.Sci. 8 (1973) 1800.

362

19. Kambour, R.P., and Gruner, C.L., J.Poly.Sci.Phys.Ed. 16 (1978) 703.

20. Bucknall, C.B., Page, C.J., and Young, V.O., ACS Advances in Chemistry Series, 154 (1976) 179.

21. Donald, A.M., and Kramer, E.J., J.Mater.Sci. 17 (1982) 2351.

22. Donald, A.M. and Kramer, E.J., J.Appl.Poly.Sci. 27 (1982) 3729.

23. Bucknall, C.B., Cote, F., and Partridge, I.K., to be published.

24. Bucknall, C.B., and Page, C.J., J.Mater.Sci. 17 (1982) 808.

25. Andrade, E.N. da C., Proc.Roy.Soc. A84 (1910)1.

26. Idem in 'Creep and Recovery' (Metals Park Ohio, American Society for Metals, 1957).

27. Eyring, H., J.Chem.Phys. 4(1936) 283.

28. Goodier, J.N., Trans.Am.Soc.Mech.Eng. 55 (1933) 39.

29. Bucknall, C.B., J.Materials 4 (1969) 214.

30. Ishai, O., and Cohen, L.J., J.Composite Mater. 2 (1968) 302.

31. Parvin, M., and Williams, J.G., J.Mater.Sci. 11 (1976) 2045.

32. Mai, Y.W., and Williams, J.G., J.Mater.Sci. 12 (1977) 1376.

33. Fernando, P.L. and Williams, J.G., Poly.Eng.Sci. 20 (1980) 215.

FATIGUE OF HIGH-IMPACT POLYMERS

Clive B. Bucknall

Cranfield Institute of Technology, Cranfield, Bedford MK43 OAL

1. APPROACHES TO FATIGUE

The fatigue properties of plastics have attracted an increasing level of interest in recent years, as plastics have become more widely used in engineering applications. However, the problems of impact failure and solvent cracking are still regarded in the industry as being of far greater importance, and the amount of information available about the fatigue properties of specific polymer grades is at present rather limited. This is in marked contrast to the situation in metallurgical engineering, where it is recognised that dynamic fatigue is by far the commonest cause of failure. Fatigue testing requires the commitment of expensive equipment for long periods of time, and is undertaken only when there is a demonstrable need for the information. The number of applications of thermoplastics (as opposed to fibre composites) that have generated a demand for comprehensive fatigue data has so far been limited, but is likely to increase. At present, many fatigue failures of plastics probably go unrecognised.

The approach of manufacturers to fatigue of rubber-toughened plastics should be viewed against this general background. The fracture resistance of a novel rubber-modified material is assessed mainly on the basis of impact tests, and in most cases, little if any attention is given to dynamic fatigue behaviour. It is usually assumed that improvements in impact strength are a guarantee of enhanced resistance to other forms of fracture. The validity of this assumption is questionable.

364

The traditional method for studying fatigue is to subject an unnotched bar to repeated flexural or tensile loading at a fixed amplitude of stress or strain (1). The data are usually presented in the form of an S-N curve, a plot of stress amplitude against the logarithm of the number of cycles to failure. Figure 1 is an example, which shows some of the typical features. Decreasing the stress amplitude results in an increase in fatigue lifetime, until at the endurance limit the curve flattens out, to give lifetimes in excess of ten million cycles. For design purposes, the endurance limit is probably the most useful piece of information obtained from fatigue tests. In plastics, the endurance limit typically corresponds to a strain amplitude of about 0.4 to 0.5%, and designers can avoid fatigue failure by ensuring that the component never reaches these strains. However, this approach can lead to overdesign where the component is not expected to experience a very large number of loading cycles during normal service. In these cases the data obtained at high stress amplitudes will be more relevant. A practical problem is that significant temperature rises occur in plastics because of their high mechanical hysteresis and low thermal conductivity (2,3). Even if the material does not reach Tg and soften, fracture resistance is affected. The amount of heating is of course dependent upon geometry and testing frequency. The need to limit frequency is one of the great drawbacks in fatigue testing of plastics.

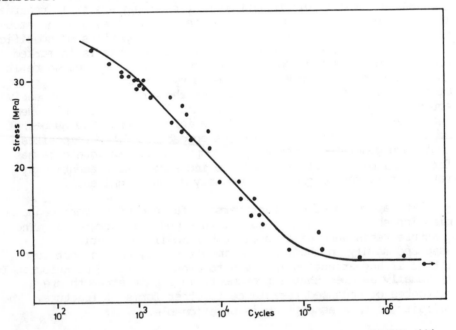

Fig. 1 Fatigue endurance curve for a rubber-toughened PMMA (4)

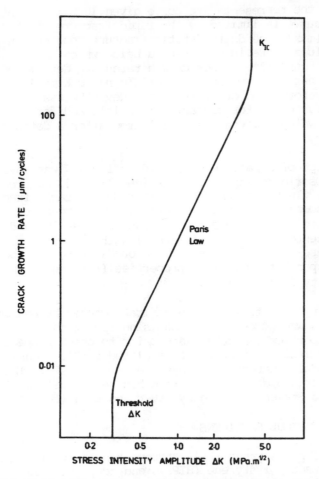

Fig. 2 Schematic diagram of relationship between fatigue crack
growth rate and amplitude of stress intensity factor.

Most current research on fatigue of materials is based firmly
upon the fracture mechanics approach, in which the rate of crack
growth per cycle, da/dn is measured as a function of upper and
lower values of stress intensity factor K_I at the tip of a growing
crack. The factors affecting crack growth rates in plastics are
reviewed by Hertzberg and Manson (2). Over a restricted range of
conditions, fatigue data can be fitted to the Paris law (5):

$$da/dn = A (\Delta K)^N \qquad (1)$$

where A is a constant, ΔK is the difference between maximum and
minimum values of K_I, and N is a constant which is typically
about 4. A theoretical explanation of this fourth power
relationship has been advanced by Izumi et al (6). A Paris plot

forms part of the representative curve given in Fig. 2. At high
K values the curve is limited by the rapid fracture of the
specimen as K(max) exceeds the static fracture toughness K_{IC}, and
at low ΔK values there is a threshold below which no fatigue
crack growth occurs (7). Crack growth rates are determined not
simply by ΔK but also by the ratio of K(min) to K(max), and by
frequency, waveform, and other factors. Nevertheless, the Paris
plot is a useful basis for comparing materials and for calculating
the length to which a given crack will grow after a defined
loading history.

The concept of a fatigue threshold at low ΔK was advanced by
Williams and Osorio to explain the fatigue limit in unnotched
specimens (7). In the absence of an introduced crack, fracture
initiates at some defect in the material. The largest of these
'intrinsic flaws' is equivalent to a crack between 100 and 200μm
in most components, but the equivalent length may be as much as
1mm in some cases. The fatigue limit should, therefore, be seen
as depending upon both materials properties (threshold ΔK) and
specimen quality (defect size).

A third approach to fatigue was first adopted by Beardmore
and Rabinowitz, who demonstrated that significant changes in
mechanical properties can take place prior to crack propagation in
plastics, especially at higher load amplitudes (8). Tensile and
compressive moduli decrease, and hysteresis increases. This
technique of fatigue damage monitoring has been developed further
by Bucknall and Stevens (9) and by Sauer and Chen (10).

2. UNNOTCHED FATIGUE SPECIMENS

Fatigue of polystyrene (PS), styrene-acrylonitrile copolymer
(SAN), high-impact polystyrene (HIPS), and acrylonitrile-
butadiene-styrene polymer (ABS) has been studied in unnotched
specimens by several groups (8 - 13). The comparison between
toughened and untoughened plastics is striking, because rubber
particles actually reduce both the fatigue endurance and the
endurance limit, as illustrated in Fig. 3 (10,12). The essential
function of rubber particles in toughened styrene polymers and
copolymers is to accelerate crazing without weakening the material
in the crack tip, and the reduced fatigue life is a consequence of
this response: crazes form more rapidly, at lower stress amplitude
than in the parent polymer, eventually breaking down under the
cyclic loading to form cracks. The endurance limit marks the
point at which the critical conditions for craze formation are
reached. In this context, the intrinsic defect should be regarded
not as a true crack, but as a region of polymer which can form a
small craze as a precursor to a crack.

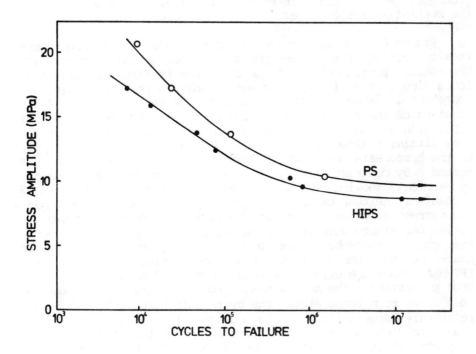

Fig. 3 Fatigue endurance curves for PS and HIPS (11).

Experiments on styrene-acrylonitrile copolymers show that the craze initiation stress increases with acrylonitrile content (10). The crazing stress for a typical SAN polymer containing about 25 wt % of acrylonitrile is about 50% higher than that of PS. This difference is reflected in the higher fatigue endurance of ABS in comparison with HIPS. Another factor affecting the craze initiation stress is the molecular weight of the matrix polymer, which plays an important part in determining the fatigue endurance (10).

The endurance limit is usually well below the yield stress of the polymer. Consequently, there is little evidence of yielding in the bulk of the specimen when the material fails at low stress amplitudes. On the other hand, at stresses much closer to the short-term yield stress there can be a significant amount of multiple crazing or shear yielding before final fracture of an unnotched specimen. Heating effects are also more severe at these higher stress amplitudes. The overall result is that the mechanical properties of the material change as the fatigue test proceeds. By monitoring these changes, it is possible to identify the mechanisms responsible.

Figure 4 compares hysteresis loops obtained in tension-compression testing of HIPS mass polymer and ABS emulsion polymer (9). The tensile portion of the loops developed by HIPS resemble those for a single craze (15), and clearly indicate the progressive formation of large numbers of crazes as the test proceeds. The kinetics of the process are similar to those observed in creep of HIPS, with an induction period during which the properties change very little, followed by more rapid crazing. The compressive part of the hysteresis loop simply shows the effect of compressing the crazes. By contrast, the hysteresis loops for the ABS are more symmetrical, showing that shear yielding is the dominant mechanism. Neither of these toughened polymers showed more than a few degrees temperature rise at the very low frequency of two cycles per minute used for the tests, and the changes in properties can therefore be related almost entirely to the formation of crazes or shear bands. Rubber-toughened poly(methylmethacrylate) (RTPMMA) shows the effects of both thermal and mechanical changes upon properties. The matrix has a prominent secondary relaxation just above room temperature; the hysteresis loop is therefore relatively large initially, and increases in size as a result of shear yielding and of a modest rise in temperature (16). By comparing widths or areas of tensile and compressive portions of the hysteresis loop over a fatigue test, it is possible to determine the number of cycles for the initiation of detectable crazing or shear yielding as a function of stress (10). This measurement again shows that ABS has a higher resistance than HIPS to crazing and fatigue fracture. Analysis of hysteresis loops can also be used to study mechanisms and extent of deformation in rubber-toughened plastics that have previously been strained in a completely different test (14); it is even possible to study impact specimens in this way (17).

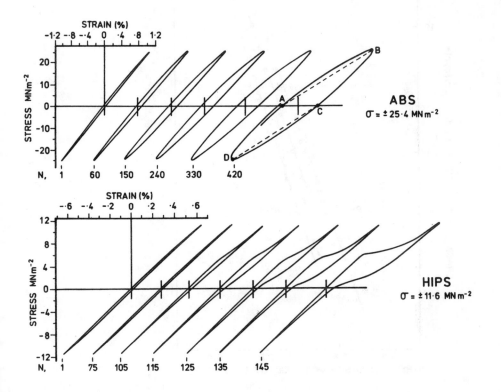

Fig. 4 Hysteresis loops for ABS and HIPS after N cycles under
     reversed tension–compression loading (9).

     Rubber–toughened plastics, like other materials, will fail
after prolonged periods under static loading. The effects of
dynamic loading can be demonstrated by comparing times to fracture
under static load (creep rupture) with cumulative times under peak
tensile load in a square–wave fatigue test. Experiments of this
type on ABS show very similar times to failure at relatively high
stresses close to the yield stress, but much shorter lifetimes
under dynamic loading as the stress level is decreased (18).

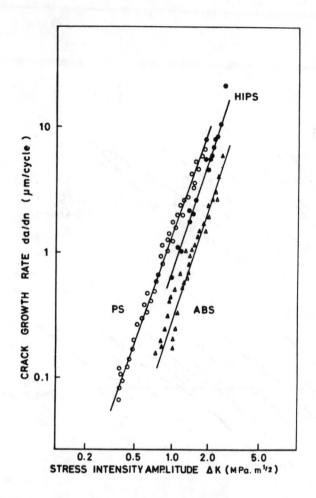

Fig. 5   Paris law plots for PS, HIPS and ABS

3.   FRACTURE MECHANICS TESTS

Fracture mechanics experiments show that rubber particles do reduce crack growth rates, although the improvement may not be very large. Paris law plots for PS, HIPS and ABS are shown in Fig. 5 (19). The stress intensity amplitude ΔK needed to produce a given growth rate in HIPS in this experiment is only 25% higher than that needed for PS, whereas much greater increases are observed in monotonic fracture mechanics tests at room temperature (20). The limited improvement in fatigue resistance may be an example of restricted yielding in toughened plastics under plane strain conditions (21). The rubber toughening effect is most apparent when the material is able to develop an extensive yield zone around the crack tip or notch.

Skibo and co-workers have made a systematic study of fatigue crack growth in a series of blends of rigid PVC with methacrylate-butadiene-styrene (MBS) polymer, which is basically a grafted polybutadiene toughening agent (21). They found that the MBS produced a marked reduction in fatigue crack propagation (FCP) rates in low molecular weight PVC (Mw = 67,000), and a smaller effect in a high molecular weight polymer (Mw = 208,000). Interestingly, addition of more than 6% of MBS produced no observable improvement in FCP resistance in either material, whereas 10% or more of MBS was needed to raise impact strength significantly. The data do not extend sufficiently far to define a threshold value of ΔK with any precision, but curvature in the Paris plots indicates that the threshold ΔK is probably about a factor of two higher in the higher molecular weight PVC (0.6 as compared with 0.3 MPa$\sqrt{m}$); the MBS appears to have a relatively small effect on the threshold. At the other end of the Paris plot, K_{IC} for low molecular weight PVC is 1.0 MPa$\sqrt{m}$ for low molecular weight PVC, and 1.8 MPA$\sqrt{m}$ for the high molecular weight grade. Addition of rubber has a major effect upon K_{IC}, especially in the low molecular weight polymer. Shifts in the curve of FCP rate against ΔK should be seen in the context of these effects of molecular weight and rubber content upon the threshold ΔK and K_{IC}.

Experiments on nylon 66 also show a reduction in FCP rate on addition of rubber (22). However, in a recent study of epoxy resins it was found that the rubber-modified material had a similar FCP rate to the parent epoxy resin (23), although the rubber did increase fracture toughness and therefore the maximum attainable ΔK.

4. CLOSING COMMENTS

This brief review has drawn attention to some important issues concerning the engineering properties of rubber-toughened plastics. The fatigue endurance limit is a property that has so far received little attention from manufacturers or academic research laboratories. A more detailed investigation of the relationship between polymer structure and endurance limit would be fruitful in enabling this rapidly expanding group of high-performance polymers to be used more effectively in load-bearing applications.

References

1.   ASTM - D671

2.   Hertzberg, R.W., and Manson, J.A., Fatigue of Engineering

Polymers (New York, Academic Press, 1980).

3. Kinloch, A.J., and Young, R.J., Fracture Behaviour of Polymers (London, Applied Science Publishers, 1983).

4. Bucknall, C.B., and Marchetti, A., J.Appl.Poly.Sci. 28 (1983) 2689.

5. Paris, P.C. and Erdogan, F., Trans., ASME D85 (1963) 528.

6. Izumi, Y., Fine, Y., and Mura, T., Int.J.Fract. 17 (1981) 15.

7. Williams, J.G., and Osorio, A.M.B.A., Conference on Fatigue Thresholds, Stockholm, June 1982. (See also Osorio, A.M.B.A., PhD. Thesis, Imperial College).

8. Beardmore, P. and Rabinowitz, S., Appl.Poly.Symp. 24 (1974) 25.

9. Bucknall, C.B., and Stevens, W.W., J.Mater.Sci. 15 (1980) 2950.

10. Sauer, J.A. and Chen, C.C., in Crazing in Polymers, ed. Kausch, H.H., Heidelberg, Springer-Verlag, 1983).

11. Chen, C.C., Chheda, M., and Chen, C.J., Macromol Sci.Phys., B19 (1981) 565.

12. Sauer, J.A., Habibullah, M., and Sauer, J.A., Polymer 22 (1981) 699.

13. Woan Der-jin, Habibullah, M., and Sauer, J.A., J.Appl.Phys. 52 (1981) 5970.

14. Bartesch, H., and Williams, D.R.G., J.Appl.Poly.Sci., 22 (1978) 467.

15. Kambour, R.P., and Kopp, R.W., J.Poly.Sci., 7 (1969) 115.

16. Bucknall, C.B., and Marchetti, A.J., Appl.Poly.Sci. 28 (1983) 2689.

17. Idem, Poly.Eng.Sci., 24 (1984) 535.

18. Bucknall, C.B., and Stevens, W.W., in Toughening of Plastics (London, Plastics & Rubber Institute, 1978).

19. Hertzberg, R.W., Manson, J.A., and Wu, W.C., ASTM STP 536 (1973) 391.

20. Parvin, M., and Williams, J.G., J.Mater.Sci. 11 (1976) 2045.

21. Skibo, M.D., Manson, J.A., Webler, S.M. and Hertzberg, R.W., in Durability of Macromolecular Materials, ed. Eby, R.K. (Washington, D.C., Am.Chem.Soc., 1979). ACS Symposium Series 95 (1979) 311.

22. Skibo, M.D., Manson, J.A. and Hertzberg, R.W., Conference on Deformation Yield & Fracture (London, Plastics & Rubber Institute, 1979).

23. Shah, D.N., Attalla, G., Manson, J.A., Connelly, G.M. and Hertzberg, R.W., ACS Adv.Chem.Ser. 1983, in the press.

YIELDING AND FAILURE CRITERIA FOR RUBBER MODIFIED POLYMERS, PART 1

Dr. Hans Breuer

BASF AG, Anwendungstechnik Thermoplaste, KTE/WMM,
D-6700 Ludwigshafen/Rhein, FRG

1. INTRODUCTION

According to current theories, rubber modifiers in
polymer blends with brittle and rigid matrices act as
follows:
an applied tensile load induces local stress concentrations. As a consequence, zones of plastic deformation are formed at the rubber interfaces in the
matrix. The strain energy involved is high in the
case of shear or crazing. The deformation mechanism
depends on the nature of the matrix and the stress
concentration, i.e. the "multiaxiality" of the stress.

Two principles underlie the toughening mechanism in a
two-phase system, viz.

(1) the behaviour of the matrix under different combi-
    nations of multiaxial stress. These are the so-
    called failure criteria that apply to the onset of
    shear flow and crazing.

(2) the nature of the stress in the matrix and the
    dependence of its spatial distribution on the
    rubber morphology and the applied load.

2. GENERAL SHAPE OF FAILURE DIAGRAMS FOR ISOTROPIC MATERIALS

An isotropic material will fail if the stress exceeds a threshold, which is given by the six independent components of the local stress tensor in the arbitrary, orthogonal system of x, y and z coordinates. By rotation of the axes it can be transformed into an orthogonal coordinate system, in which the local load is represented by only three normal stresses, the so- called principal stresses: σ_1, σ_2 and σ_3.

In the following, we are not interested in the direction of failure (shear band or craze direction) but only in the limiting stress conditions. Thus, we can neglect the angles of rotation α_x, α_y and α_z, which transform the x, y, z coordinates into the principal ones 1,2 and 3. The boundaries set by the failure criteria form a closed surface in the principal stress space, which includes the origin. Any stress system within this space is safe, and any outside it is unsafe.

With isotropic materials the surfaces representing the failure boundaries are symmetrical about the axis of rotation formed by the diagonal in the principal stress space (1). Therefore the failure criteria can be defined by two cylindrical coordinates instead of the three variables σ_1, σ_2 and σ_3. The one is the length of the space diagonal $\sqrt{3} \cdot \sigma$, and the other is the radial distance $\sqrt{3} \cdot \tau$ from the axis. σ is the octahedral normal stress and represents the hydrostatic portion of the stress system (the portion of negative hydrostatic pressure $-P = \sigma$); in the following, it is referred to as the "dilational stress". Likewise, τ is the octahedral shear stress. It is equal to the mean square of the shear components and can be derived from the second principal invariant II_D of the deviator $D = \sigma_{ik} - \sigma \cdot \delta_{ik}$: $II_D = (3/2) \cdot \tau^2$. In the following, it is referred to as "deviatoric stress".

$$3\,\sigma \quad = \sigma_{xx} + \sigma_{yy} + \sigma_{zz} \tag{1}$$

$$(3\,\tau)^2 = (\sigma_{xx} - \sigma_{yy})^2 + (\sigma_{yy} - \sigma_{zz})^2 + (\sigma_{zz} - \sigma_{xx})^2 +$$
$$+6 \cdot (\sigma_{xy}^2 + \sigma_{yz}^2 + \sigma_{zx}^2) \tag{2}$$

$$w \quad = \pm\,3\,\sigma^2/(\sigma^2 + \tau^2) \quad (\text{sign as } \sigma) \tag{3}$$

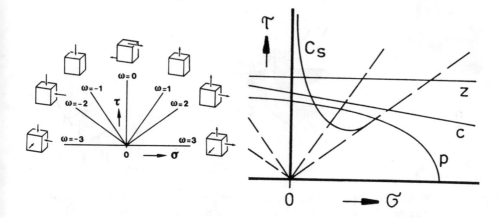

Fig. 1 Left: Simple stress systems in the deviatoric
(τ)/dilatoric (σ) stress diagram , in which
w denotes the type of stress, e.g. w = 1 monoaxial stress, w = 2 biaxially symmetrical stress,
w = -3 hydrostatic pressure
Right: types of failure criteria: z cylindrical,
c conic, p rotational-parabolic, and c_s surface crazing, see chapter 4

Various simple stress systems in the octahedral stress
diagram are shown on the left of Figure 1; the type
of stress w is derived from Eqn. (3). Examples of
failure curves are shown on the right of Figure 1.
They were obtained from the first three constants in
the Tschoegl power series expansion (2) for the octahedral stresses, i.e.

$$1 = a_1\tau + a_2\sigma + b_1\tau^2 + b_2\sigma^2 + b_3\tau\sigma \tag{4}$$

z cylindrical $a_1 \neq 0$, c conic a_1 and $a_2 \neq 0$, and p
rotational-parabolic a_1 and $b_1 \neq 0$.

The corresponding failure boundaries in the principal stress space are shown in Figure 2 together with the lines of intersection with the principal stress planes (areas of plane stress).

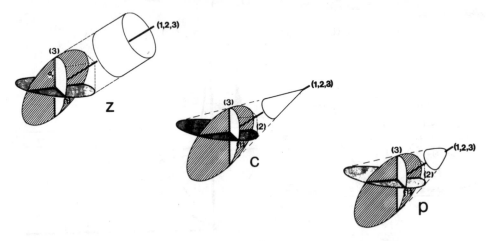

Fig. 2 Failure boundaries in the principal stress space determined from the failure curves on the right of Fig. 1. (1), (2) and (3) are the principal stress axes, (1,2,3) is the space diagonal and (1,2), (2,3) and (1,3) are the plane diagonals.

The second power terms give rise to a great variety of shapes, because Eqn. (4) embraces all conic sections as contours of the rotationally symmetric surfaces (1).

3. CRITERIA FOR THE ONSET OF SHEAR YIELD

The onset of shear yield under multiaxial stress has been measured by several authors on a number of brittle, rigid polymers. We can sum up by saying that the criteria for the onset undoubtedly correspond to curves c and p in Figure 1.
In most cases, a difference between the shapes of these curves can hardly be distinguished, because the curvature is not very pronounced within the range of dilatoric stress measured. Usually the results are presented in plane stress diagrams (section curves on the right of Fig. 1). It can be seen that the differences in shape between the section curves of the conical and the rotational parabolic boundaries are too small to allow a definite differentiation. However, this uncertainty is of no significance for rubber-modified systems, in which case linear (conic) relationships can

be assumed as an approximation and the slope between
w = -1 and w = +1 can be estimated from the asymmetry
between tensile and pressure tests.

Criteria for the onset of crazing are much more difficult to define.

4. SPATIAL FORM OF THE STERNSTEIN PLANE STRESS CRAZE CRITERION

The results of the experiments performed by Sternstein
et al (3) on PMMA are shown on the left of Figure 3.
If σ_3 = 0, they can be described by

$$|\sigma_1 - \sigma_2| = B/(\sigma_1 + \sigma_2) - A \tag{5}$$

where A and B are constants for the material that
depend on time and temperature.

Eqn. (5) is valid for plane stress systems and does
not apply direct to the onset of crazing in rubber-
modified polymers with triaxial stress concentrations.
Several attempts have been made to transform it into
a three-dimensional criterion. Sternstein (3) succeeded
in doing so by adding σ_3 or the hydrostatic pressure
0 to the sum of $\sigma_1 + \sigma_2$ in the right-hand side. But
this results in noninterchangeable principal stress
axes (1).

Matsuo et al (4,5) proposed a form with interchangeable
axes, viz. $|\sigma_i - \sigma_k|_{max} = B/3\,\sigma - A$ $\qquad$ (6)

The subscript "max" for the term on the left-hand side
of Eqn. (6), however, is responsible for discontinuities
in the plane stress curves (1) and thus for discrepancies compared to the results of Sternstein's
measurements. Oxborough and Bowden (6) proposed the
following notation:

$$\sigma_i - \nu(\sigma_k + \sigma_l) = C - D/(\sigma_i + \sigma_k + \sigma_l) \tag{7}$$

where ν is Poisson's ratio (approx. 0.4).

Bucknall (7) showed that the left-hand side of Eqn. (7)
is equal to $E\varepsilon_i$, where E = Young's modulus and
ε_i = major principal strain. Thus Eqn. (7) represents
a critical tensile strain criterion. However, it can
be shown that critical tensile stress or strain criteria
generally cannot be enveloped by a unique three-

dimensional failure boundary (1).

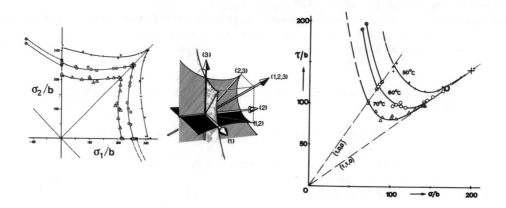

Fig. 3 Plane stress (σ_3 = O) surface crazing according
to Sternstein et al (3). Left: in one plane of
principal stress. Middle: in three planes of
principal stress. Right: correctly transformed
to the octahedral stress diagram.

If symmetry about an axis of rotation is assumed, the
results of measurements obtained from biaxial plane
stress systems can be correctly transformed into a
spatial (triaxial) criterion (cf. Fig. 3, right).

$$6\tau^2 = (B/3\sigma - A)^2 + 3\sigma^2 \qquad\qquad (8)$$

This curve can hardly be interpreted. As the dilational
stress increases, minima occur in the threshold of the
deviatoric stress , and there is no threshold in
purely hydrostatic stress systems.

5. SURFACE AND BULK CRAZING

Sternstein et al (3) studied those stress systems
"that produce visible surface crazing". There are
strong indications, however, to the effect that surface crazing generally commences at a markedly lower
stress than that of bulk crazing. Argon found that the
threshold stress of crazing distinctly depends on
surface quality, rising almost to the shear yield
stress on carefully polished samples (8).

Rehage and Goldbach (9) realized the ultimate in
surface quality on PMMA samples. They found a polymeric plasticizer that, after acting at elevated
temperatures, yields a thin superficial layer of
plasticized polymer with a stable concentration
profile and a glass temperature below that at which
the mechanical tests are performed.
The transition between the plasticized and unplasticized layers of polymer can be regarded as a
flawless and scratch-free interface. The threshold
stress of crazing in specimens for the tensile test
treated in this manner thus increased by about
35 % at room temperature and coincided with the
(brittle) fracture stress of untreated specimens.
Bulk crazing was uniformly distributed throughout
treated specimens but not preferentially at the surface.

Injection-moulded specimens displayed, in contrast
to those obtained by compression moulding, no signs
of surface crazing in the tensile test. This effect
was exploited in measuring the threshold stress for
bulk crazing in polystyrene: increasing the mould
temperature until it approaches the glass transition
results in specimens with an almost unoriented,
crazing central zone. The highly oriented surface
zone in the specimen provides protection from surface crazing. Hence, the threshold of bulk crazing
for isotropic polystyrene can be reliably determined
by extrapolation. The results thus obtained agreed
with Argon's. Similar to PMMA, bulk crazing in
polystyrene commences near the threshold stress of
shear yield.
It is known that for high impact strength the bonds
at the rubber/matrix interfaces in modified polymers
must be of a chemical nature, and this is effected
by graft polymerization. The graft polymerization
results in transition zones that, on a smaller scale,
correspond to those that occur in single-phase
vitreous polymers treated with plasticizer, as
mentioned above. Hence the rubber/matrix interfaces
in toughened polymers must be regarded as flawless
and scratch-free surfaces, and it must be assumed
that the prevention of surface crazing is an
essential function of graft polymerization. Furthermore, the failure criteria in bulk crazing must be
adopted to explain crazing in rubber-modified
polymers. The only one that we know at present relies on a onepoint estimation, viz. the threshold
at monoaxial tensile stress.

The second part of this paper is devoted to stress patterns in two-phase systems and a reevaluation of model experiments carried out by Matsuo et al. We can then see what information can be derived from the results of light-scattering experiments on stress-whitened zones.

(1) Breuer, H., Angew. Makromol. Chem. 78 (1979) 45
(2) Tschoegel, N.W., J. Polym. Sci. C 32 (1971) 239
(3) Sternstein, S.S., and F.A. Myers, J. Macromol. Sci.-Phys. B8 (1973) 539
(4) Wang, Tsuey T., M. Matsuo and T.K. Kwei, J. Applied Phys. 42 (1971) 4188
(5) Matsuo, M., Tsuey T. Wang, and T.K. Kwei, J. Pol. Sci. A-2, 10 (1972) 1085
(6) Oxborough, R.J. and P.B. Bowden, Phil. Mag. 28 (1973) 547
(7) Bucknall, C.B., Toughened Plastics, Appl. Sci. Publ., LTD, London 1977
(8) Argon, A.S., paper presented at the 4th Discussion Conference on Macromolecules "Heterogeneity in Polymers", 1974, Marianske Lazne, Czechoslovakia
(9) Rehage, G., and G. Goldbach, Angew. Makromol. Chem. 1 (1967) 125

YIELDING AND FAILURE CRITERIA FOR RUBBER MODIFIED POLYMERS, PART 2

Dr. Hans Breuer

BASF AG, Anwendungstechnik Thermoplaste, KTE/WMM,
D-6700 Ludwigshafen/Rhein, FRG

6. INTRODUCTION

Part 1 of this paper closed with the dismal fact that
there is very little knowledge on the form of the
failure criteria for bulk crazing, although it is
essential for understanding the behaviour of rubber-
modified, craze-yielding matrix materials, e.g.
polystyrene. It was demonstrated that the Sternstein
plane stress craze criterion does not yield a significant failure surface in the principal stress space
or a significant failure curve in the octahedral stress
diagram. Its validity is restricted to plane stress
systems (cf. Figure 3, centre). It is certainly a
good criterion for describing typical surface crazing,
i.e. multiaxial environmental crazing.

The failure curve for the bulk criterion is defined
by only the one point in the octahedral stress diagram, viz. that for monoaxial tensile stress (w = 1).
It is derived from measurements on specimens whose
surface is protected from crazing by swelling with a
plasticizer (formerly on PMMA only) or by orientation
in the direction of tensile stress. In view of the
necessity of this surface protection, there is not
much hope of finding useful methods for measuring
bulk crazing in pure matrix materials subjected to
multiaxial stress systems.
In Part 2 of this paper, the stress systems that occur
in rubber-modified polymers will be discussed. Reevaluation of former, macroscopic model experiments

performed by Matsuo et al gives an idea on the bulk
craze limit for w values (types of stress systems)
that actually occur. Finally, it will be described
how light scattering in zones of stress whitening can
yield information on the yield mechanisms that pre-
dominate in rubber-modified polymers.

7. TYPES OF STRESS SYSTEM AROUND A RUBBER INCLUSION

Let us regard the stress fields in a rigid matrix
around a single spherical rubber particle in the
light of a simple model. The fundamental equations
arising from the theory of linear elasticity are
given in the literature, e.g. Timoshenko and Goodier
(10) for thermal prestresses and Goodier (11) for
monoaxial applied stress. However, too much discussion
was devoted in these works to unsuitable elastic
constants for the rubbery phase and the resulting
stress fields, which comprise four components, have
not been discussed in the octahedral stress diagram.
A good review of the state of the art in 1977 was
presented by Bucknall (12).
The first problem arises in the discussion on the
thermal prestresses that occur when the two-phase
system cools in the range between the glass temperature
of the matrix and the temperature of mechanical test.
The original Timoshenko and Goodier equations (10)
were compiled for the case of foreign bodies in steel.
They contain the Poisson constant ν_R of the particle
in a form of a factor $(1-2\nu_R)$ that cannot be adopted
for the case of a rubber particle, because ν_R for
rubber is of the order of 0.5 within the limits of
experimental accuracy, i.e. $1-2\nu_R$ is indeterminate.
Matsuo et al (4,5) assumed a numerical value of
$\nu_R = 0.49$ in order to evaluate their model experiments.
Böhn (13) and subsequently Morbitzer et al (14) were
forced to interpret measured values for the reduction
in glass temperature of rubber particles by assuming
$\nu_R = 0.499$. Booij was the only author to use a suitable
expression, viz. the well-known relationship for
converting elastic constants:
$$(1-2\nu) = E \, \kappa /3$$
where E is Young's modulus and κ is the compressibility.
This relationship allows the above factor to be reliably
estimated resulting in $\nu_R = 0.4996$. Accordingly the
thermal prestresses calculated by Matsuo et al are too
small by a factor of 14.

The tensor fields for the matrix can be more easily
understood if one takes into account the following
facts.

All components of stress attenuate as the distance
from the surfaces of the particles increases. The
rate of attenuation is so high that only the loads
at the interfaces are of interest for the nucleation
of yield.

The four components of stress that occur can be
reduced to the two octahedral magnitudes σ_M and τ_M .

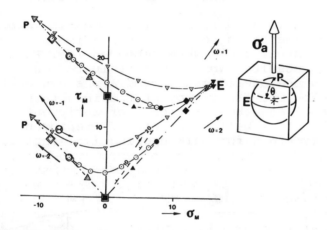

Fig. 4 Octahedral stresses σ_M and τ_M for a polystyrene
matrix at the interfaces to butadiene rubber
particles as a function of the polar coordinate
angle between the pole P and the equator E for
various values of applied stress σ_a.
Lower set of curves: without thermal prestressing
Upper set of curves: with thermal prestressing
All stresses in N/mm^2

Polystyrene modified with butadiene rubber was taken
as the example to demonstrate the matrix loads acting
at the interfaces. The results are plotted in Figure 4.
Thermal prestressing preferentially has an effect
on the load at the pole P, whereas practically identical
values occur at the equator E for applied stresses of
$\sigma_a \geqq 20$ N/mm^2. The stress system at the equator includes positive dilatoric stress σ_M, which, according
to general opinion, favours craze yield. As opposed
to this, hydrostatic pressure $P = -\sigma_M$ occurs at the
pole. Consequently all that can be expected here is

386

shear yield or brittle fracture and not yield with
void formation. Thus the stress level at the pole
should be kept low for predominantly craze-yielding
matrix materials such as polystyrene: in this case,
thermal stresses are injurious. This explains why
preference is given to modifying matrix materials of
this nature by means of particles with low difference
in thermal expansion to the matrix, known as cellular
or capsule particles, "filled" particles in general.

8. BULK CRAZING IN RUBBER-MODIFIED POLYSTYRENE

Matsuo et al (4,5) carried out interesting macroscopic
model experiments on polystyrene specimens in which
rubber and steel balls were embedded (cf. Figure 5,
left). The results, which were evaluated in a manner
similar to that described above, are shown on the right
of Figure 5. The failure point C_{st} is obtained from the
stress level and the position Θ of incipient crazing
around the steel ball. The point C_A represents that
measured by Argon for bulk crazing (cf. Part 1, section
5).

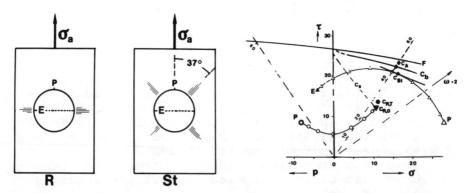

Fig. 5 Left: model experiments performed by Matsuo et
al (4,5): incipient crazing in polystyrene at
the equator E around a rubber ball R for an
applied stress $\sigma_{a,R}$ = 14 N/mm^2 and around a
steel ball St at $\sigma_{a,St}$ = 20 N/mm^2.
right: evaluation of model experiments.Δ stress
system around the steel ball and $\odot$ around the
rubber ball.

If allowance is made for the differences in the nature
of the experiments and the specimen materials, C_{st} and
C_A agree well with C_b, a value that we extrapolated from
measurements on injection-moulded specimens. All three
values occur under stress systems that are roughly
equivalent to monoaxial tension and thus do not yield
any fresh information direct on the shape of the
failure curve for bulk crazing.

The steel ball model experiment, however, allows the
slope of the failure curve to be obtained at the point
C_{st}, which is on the common tangent to the curve for
the stress systems around the steel ball. The failure
curve F above this point represents brittle fracture
as determined in the monoaxial tension and compression
tests (for W = +1 and w = -1). As far as a distinction
can be made, the section representing the shear yield
curve commences immediately below this boundary F.
presumably, therefore, the failure curve for bulk
crazing closely follows or coincides with that for
shear yield. This agrees with Argon's theory (8) to
the effect that both processes have a common nucleation
mechanism.
A major discrepancy can be discerned on the right of
Figure 5: the craze point obtained in the rubber ball
experiment lies far below the bulk crazing discussed
above, regardless of whether the experiment is evaluated with thermal prestressing $C_{R,T}$ or without it
$C_{R,O}$.
On the other hand, these values lie on the failure
curve for surface crazing C_S, measured by Retting (15)
and shown in dotted lines. From this fact, it can be
concluded that there was inadequate adhesion between the
ball and the matrix in the rubber ball experiment and
that surface crazing thus ensued. It can be calculated
that a normal stress of 24 N/mm^2 acted on the interface to the rubber ball at the commencement of crazing,
whereas a shear stress of only (21 N/mm^2), which less
affects coupling, acted on the steel ball interface.

9. RESULTS OF STRESS-WHITENING ANALYSES

Now that we have dealt with the theory and the evaluation of model experiments, we can turn our attention
to the study of real systems. For this purpose,
interesting information of the predominant yield
mechanism in matrices can be obtained from stress-
whitening analyses. The term "stress whitening"

describes the blushing observed when thermoplastics
and, especially, rubber-modified polymers are subjected to high stresses. This effect has been discussed
extensively in literature. The pronounced light
scattering in the zones of stress whitening only can
be explained by the formation of voids in material
that has undergone plastic deformation.

A special case of voids or zones of much lower density
is represented by crazes. Studies on rubber-modified
PVC, however, have revealed that practically all the
light scattering from stress-whitened zones in this
system originates from ruptured rubber particles. It
was not until all the results from electron and light
microscopy, density measurements, and turbidity
spectra were taken into consideration that the un-
expected light scattering in these systems could be
explained (16,17). The difference in light scattering
between the stress-whitened zones of rubber-modified
polystyrene and those of modified PVC is shown on the
left of Figure 6. The samples with the polystyrene
matrix show light scattering parallel to the direction
of applied stress (monoaxial tensile stress in dumbbell
specimens). This can easily be explained by light
scattering at crazes, which are propagated perpen-
dicular to the direction of the highest principal
stress.

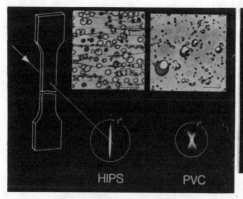 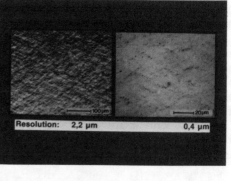

Fig. 6 Left: electron micrographs (above) and light
        scattering patterns (below) for stress-whitened
        zones in a rubber-modified PS (HIPS) and a
        modified PVC.
        Right: transmitted-light micrographs of the
        stress-whitened zones in PVC modified by large
        rubber particles (0.2-2 μm) at different
        optical resolutions.

In the stress-whitened zones of modified PVC, however, the scattered light forms a cross-shaped pattern. The photographs shown on the right of figure 6 are for a modified PVC whose particle size extends to the limit of resolution for light microscopy. At the poor resolution of 2.2 μm, a rhombic network can be seen consisting of bands of reduced density whose direction agrees with that of the light scattering. At higher resolution, viz. 0.4 μm, the individual ruptured particles can be recognized as dark spots. They occur preferentially in rows and thus form the bands mentioned above. Modified PVC displays hardly any crazes but shows rupturing of rubber particles as a voiding mechanism: here shear deformation of the matrix is combined with rupturing of the particles in front of the preceding shear bands.

The azimuthal angle μ in the light scattering diagram which is taken to be zero at crazes and should be of the order of μ = 45 $^{\circ}$ for shear bands in unmodified materials, decreases with increasing particle size from 35 $^{\circ}$ for particles of 80 nm size to 25 $^{\circ}$ for the large particles mentioned above.

Since the stress-whitening intensity also increases with particle size, the angle μ indicates how easily the rubber particles fail. The smallest numerical value for μ was measured on PVC blended with an Ethylene-Vinylacetate copolymer;milled at temperatures below 170°C the resulting structure was a coherent network consisting of a rubber-rich phase (see Figure 7, left). From the study of electronmicrographs alone, processes of this nature have been referred to as "rubber-crazing". This interpretation is misleading if crazing is taken to mean the formation of a highly oriented fibrillar structure that cannot occur in the rubbery state of a polymer.

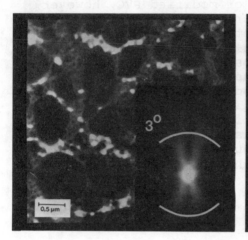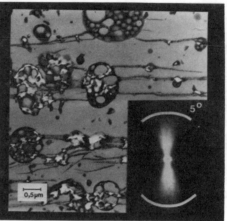

Fig. 7 Left: electronmicrograph and light scattering
pattern for a stress-whitened zone of PVC
modified with an Ethylene-Vinylacetate-copolymer.
Right: idem for solution-polymerized ABS with
large-cell particles

Of all the matrix polymers tested, the one with the
most complicated relationships for rubber morphology
and rate of straining is PSAN, a copolymer that usually consists of 75 % wt of styrene and acrylonitrile.
The electronmicrograph for PSAN with large-cell
particles (solution-polymerized ABS) strained at a
low rate shows partially ruptured particles connected
by crazes (figure 7, right), similar to the modified
polystyrene in Figure 6, left. If the polymer is modified by smaller particles, viz. of 400 nm, the pattern
of light scattering in the stress-whitened areas indicates a transition from predominant crazing to prevailing shear yield as the rate of straining increases
(Figure 8, left).

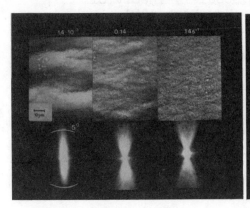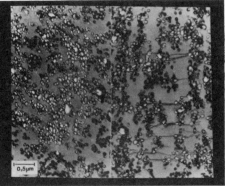

Fig. 8 Left: transmitted-light micrographs and light
scattering in the stress-whitened areas of PSAN
modified with 400 nm particles strained at the
rates indicated.
Right: electronmicrographs of stress-whitened
areas in PSAN modified with 70 nm particles and
strained at a rate of 1.4×10^{-3} s^{-1}; with well-
dispersed particles on the left and with lightly
agglomerated particles on the right.

Finally, it can be seen from the right of Figure 8,
which is for material with 70 nm particles, that the
dominant yield mechanism does not depend so much on
the particle size as on the spacing between particles
in the matrix. The large distances between lightly
agglomerated regions are bridged by crazes, and the
small distances between well-dispersed particles by
rupturing accompanied by shear yield. The two electronmicrographs shown on the right of Figure 8 were obtained
at different points on the same specimen. Thus, photographs of light scattering taken on large specimens
often show areas of diffuse mixed processes. Therefore,
electronmicrographs and light scattering patterns used
for clarification must be obtained on micropreparations
taken from identical parts of the specimen: all the
light scattering diagrams shown here were taken direct
under the light microscope.

(10) Timoshenko, S. and J.N. Goodier, Theory of Elasticity, McGraw-Hill, New York 1951.

(11) Goodier, J.N., Trans ASME (J. Appl. Mech.) 55 (1933) A 39

(12) Bucknall, C.B., Toughened Plastics, Applied Science Publishers LTD, London 1977

(13) Bohn, L., Angew. Makromol. Chem. 20 (1971) 129

(14) Morbitzer, L., K.H. Ott, H. Schuster and D. Kranz, Angew. Makromol. Chem. 7 (1972) 57

(15) Retting, W. unpublished data

(16) Haaf, F., H. Breuer, and J. Stabenow, Angew. Makromol. Chem. 58/59 (1977) 95

(17) Breuer, H., F. Haaf and J. Stabenow, J. Macromol. Sci.-Phys. B14 (3) (1977) 387

MULTIPHASE THERMOSETTING POLYMERS

A J Kinloch

Imperial College, Dept of Mechanical Engineering,
Exhibition Road, London, SW7 2BX, UK

ABSTRACT

Thermosetting polymers are very widely employed as structural
adhesives and in glass-, polyamide - and carbon-fibre composites.
They are amorphous, highly-crosslinked polymers and this structure
results in these materials possessing a high modulus and fracture
strength, low creep and good performance at elevated temperatures.
However, their structure also leads to low toughness and poor
crack resistance. One of the most successful methods of increasing
their toughness, without significantly decreasing other important
properties, is to incorporate a second phase of dispersed rubbery
particles into the crosslinked polymer. The present review will
consider the inter-relationships between the chemistry, microstructure and mechanical behaviour of multiphase thermosetting
polymers and discuss some of the applications of such materials.

1. INTRODUCTION

The use of structural adhesives and fibre composites in
aircraft, guided weapons, ships and vehicle construction has
increased markedly in the last decade and this dramatic growth
rate shows every sign of continuing in the future. At present
these materials employ thermosetting polymers almost exclusively:
either as the basis for the adhesive compositions or as the matrix
for glass-, polyamide- and carbon-fibre composites.

The current dominance of thermosetting polymers in these
applications may be ascribed to two main aspects. Firstly, they are
usually prepared from low molar mass resins, which are polymerised

via the addition of a curing agent. The resin/curing agent mixture generally exhibits a relatively low viscosity, particularly at elevated temperatures (assuming rapid curing has not commenced) and/or upon solvent dilution - a low viscosity at some stage is obviously a necessary requirement for attaining good wetting and adhesion at the polymer substrate or polymer/fibre interface. Secondly, when cured, thermosetting polymers are amorphous, highly-crosslinked polymers. This structure results in many useful properties for structural engineering applications - such as high modulus and fracture strength, low creep and good performance at elevated temperatures.

However, the structure of thermosetting polymers also leads to one highly undesirable property - they are relatively brittle materials with poor resistance to crack growth. Several methods have been proposed (1) to increase the toughness of thermosetting polymers and one of the most successful is to incorporate a second phase of dispersed rubbery particles into the crosslinked polymer. Such a multiphase microstructure of a rubber-modified epoxy material is clearly shown in Figure 1 and the data given in Table 1 reveals that this microstructure leads to major improvements in toughness, without significant losses in other important properties.

The present review will consider the inter-relationships between the chemistry, microstructure and mechanical behaviour of multiphase thermosetting polymers and discuss some of the applications of such materials.

2.   CHEMISTRY/MICROSTRUCTURE RELATIONSHIPS

2.1 Introduction

The microstructure of a multiphase thermosetting polymer is obviously dependent upon the detailed polymerisation chemistry of the starting products of resin/curing agent/rubber. The microstructure, in turn, controls the mechanical behaviour of the cured polymer.

TABLE 1

Properties of an Unmodified and Rubber-Modified
Epoxy Thermosetting Polymer

Property	Unmodified Epoxy	Rubber-Modified Epoxy
Molar mass between crosslinks in epoxy phase	~500	~500
T_g of epoxy phase (°C)	100	100
T_β of epoxy phase (°C)	– 74	– 74
Volume fraction of dispersed rubbery phase	–	0.18
Mean diameter of dispersed rubbery phase (μm)	–	1.6
T_g of rubbery phase (°C)	–	– 55
Tensile fracture stress (MPa)	63	58
Tensile fracture strain (%)	5	9
Young's Modulus (GPa)	3.2	2.8
Fracture energy, G_{Ic}, (kJ/m^2)	0.2	4 → 6
Izod impact strength (J/m of notch)	0.7	3 → 5

For several reasons (eg ease of manufacture, stability of reactants, uniformity of end-product) the reactants are usually selected to give an initially homogeneous solution of resin/curing agent/rubber and cure of this mixture generally involves the sequential processes of phase separation, gelation and vitrification (2-4). During cure, phase separation of the rubber largely occurs due to the molar mass of the rubber increasing, often via copolymerisation with resin monomer or oligomers. This copolymerisation may result in the volume fraction of dispersed rubbery phase being greater than the volume fraction of added rubbery and/or the rubbery phase not simply consisting of homogeneous particles but the particles themselves possessing a separated-phase microstructure.

The main factors which affect the microstructure are the chemical and physical properties of the rubber and resin/curing agent and the cure conditions (eg temperature/time of cure).

2.2 Chemical and Physical Properties of Rubber

2.2.1 Molecular structure. The optimum molecular structure of the rubber is determined by two main factors.

Firstly there is the compatibility requirement: as discussed above the rubber is usually initially dissolved in the resin but must phase separate upon cure. This requirement is met by using polar, low molar-mass (liquid) rubbers, such as carboxy-terminated butadiene-acrylonitrile rubber (CTBN) which may be represented by:

$$HOOC \left[(CH_2-CH=CH-CH_2)_x-(CH_2-\underset{\underset{CN}{|}}{CH})_y \right]_m COOH$$

where typically $x = 5$, $y = 1$ and $m = 10$ giving a molar mass of 3500 g/mol. Miscibility and phase separation are governed by the thermodynamics of the system and when the free energy of isothermal mixing, ΔG_m, is negative then miscibility will result. ΔG_m is given by:

$$\Delta G_m = \Delta H_m - T\Delta S_m \qquad (1)$$

where ΔH_m and ΔS_m are the molar enthalpy and entropy of mixing and T is the temperature. ΔG_m may be approximately expressed by (5, 6):

$$\Delta G_m = V(\delta_1 - \delta_2)^2 \phi_1 \phi_2 + RT(n_1 \ln \phi_1 + n_2 \ln \phi_2) \qquad (2)$$

where n_1 and n_2 are mole fractions of rubber and resin system, ϕ_1 and ϕ_2 are volume fractions, δ_1 and δ_2 are the solubility parameters, R is the gas constant and V is the molar volume of the liquid mixture. The second term in equation (2) represents $- T\Delta S_m$ and is usually negative (ie the entropy, ΔS_m, is usually positive, since mixing involves an increase in the disorder of the system). Since equation (2) predicts that the first term is always positive (this represents the enthalpy ΔH_m, ie endothermic mixing) then this contribution counterbalances the effect of the second term and the deciding factor is the relative magnitude of the two terms. This is where the molar mass plays an important role, since as the molar mass of the components increases the first term in equation (2) becomes increasingly positive (due to V increasing), whilst the second term is not significantly affected. Thus in the rubber/resin systems the aim is to match the respective values of δ_{resin} and δ_{rubber} fairly closely so that ΔG_m is initially negative and the rubber dissolves in the resin. However, the values of δ_{resin} and δ_{rubber} need to be sufficiently different so that upon the rubber polymerising and increasing its molar mass the first term becomes sufficiently positive to make ΔG_m positive - and phase separation results. The preferred liquid rubbers for toughening thermoset resins usually have a solubility parameter lower than the resin by about 0.5 $(\text{cal/cm}^3)^{\frac{1}{2}}$. This difference meets the above requirements without the rubber possessing an excessively polar structure; a polar structure leads to high inter-chain attractions and a high T_g (rubbery phase) which impairs toughness.

The second factor is a chemical requirement since the rubber needs to react with the resin, not only to increase its molar mass and so lead to phase separation as discussed above, but also to ensure that there are intrinsically strong chemical bonds across the rubbery phase/resin matrix interface. Weak interfacial bonding leads to poor toughness as shown later in Section 3.2.4. Indeed, to ensure "end-capping" of the liquid rubber by resin monomer units, the rubber and resin are often pre-reacted and the end-capped rubber added to the resin/curing agent mixture to give the initial homogeneous system.

2.2.2 Concentration of rubber. As discussed previously there is no simple relation between the concentration of initially added rubber and the volume fraction v_f, of precipitated rubbery phase. If phase separation does not occur, but the rubber remains completely in solution in the thermosetting polymer (possible as a random copolymer of short rubber/resin units), then v_f is obviously zero. On the other hand, if phase separation is very efficient and the rubbery phase contains copolymerised resin, as it usually does, then v_f may be appreciably greater than the volume fraction of initially added rubber. (However, T_g (rubbery phase) may now be greater than T_g (homopolymer rubber).) Obviously, any situation between these two extremes is possible and will be largely influenced by the detailed chemical composition of the reaction products, especially the length of the blocks of rubber and resin in the copolymers formed.

2.3 Chemical and Physical Properties of Resin/Curing Agent

The chemical and physical properties of the resin/curing agent obviously have a major effect on such factors as the overall compatibility and extent of phase separation, the type of reactive groups the rubber needs to possess, etc. However, the choice of resin/curing agent is usually determined by the end-application and the consideration of factors such as pot-life, cure temperature, desired thermal resistance, etc. Thus, it is the chemical and physical properties of the rubber which are usually "tailored" in order to achieve a multiphase microstructure.

Two resin systems which have been successfully toughened by the addition of rubbers to give multiphase thermosetting polymers are indicated in Figures 2 and 3. Figure 2 shows a typical epoxy

resin based upon the diglycidyl ether of bisphenol A, while
Figure 3 indicates two basic types of imide resins currently
available commerically. As might be expected from their relative
structures, the polyimides typically possess higher T_g's and
better thermal stability than the epoxy polymers.

2.4 Cure Conditions

As mentioned earlier the curing of the resin/curing agent/rubber
mixture generally involves the sequential process of phase
separation, gelation and vitrification. Since these factors are
temperature-dependent, the cure conditions have a profound effect
on the microstructure of the cured material.

The process of phase separation, gelation and vitrification
in an epoxy resin/amine curing agent/CTBN rubber system are
illustrated in Figure 4 in the form of a time-temperature-
transformation (TTT) isothermal cure diagram (4). By consideration
of a particular isothermal cure temperature this diagram indicates,
for example, the time at which phase separation initiates and is
completed, gelation occurs and vitrification occurs. As may be
seen, the process of phase separation is virtually completed
before gelation of the epoxy phase. If this was not the case
then the high viscosity accompanying gelation would inhibit any
subsequent phase separation and result in a very low volume
fraction. In the system illustrated a cure temperature of 200°C
for three hours resulted in a v_f of 0.11 (volume fraction of
initially added CTBN was about 0.9). The T_g (matrix phase) was
161°C whilst the T_g (pure epoxy polymer) was 166°C - indicating a
little rubber still remained in solution in the matrix phase.

3   MICROSTRUCTURE/MECHANICAL PROPERTY RELATIONSHIPS

The microstructural features of multiphase thermosetting
polymers which may affect the mechanical properties are:

(a) Thermosetting matrix

   (i) crosslink density

   (ii) concentration of non-phase separated rubber

(b) Rubbery dispersed phase

    (i) volume fraction

   (ii) particle diameter

  (iii) distribution of particle size

    (iv) intrinsic adhesion across particle/matrix interface

(v)   morphology of dispersed phase

(vi)   glass transition temperature of dispersed phase.

However, before considering the influence of the above features on the mechanical properties it is of interest to consider the mechanisms whereby the presence of rubbery particles increase the toughness of a thermosetting polymer, since it is the property of most interest.

3.1   Toughening Mechanisms

Many different mechanisms have been proposed (1) to explain the greatly improved toughness that may result when a thermosetting polymer possesses a multiphase microstructure of dispersed rubbery particles. The recent mechanism proposed (7,8) by the present author and colleagues best explains the experimental evidence and has received independent verification from the work of Yee and Pearson (9,10). This mechanism proposes that the greater crack resistance in the rubber-modified thermosetting polymers arises from a greater extent of energy-dissipating deformations occurring in the material in the vicinity of the crack tip. The deformation processes are (i) cavitation in the rubber, or at the particle/matrix interface, and (ii) multiple, but localised, plastic shear yielding in the matrix initiated by the rubbery particles. The localised cavitation of the rubber gives rise to the stress whitening that often accompanies crack growth, especially at high temperatures. Evidence for cavitation is clearly visible in Figure 5 with no indication of cavitation in the matrix, as would occur if multiple crazing was a toughening mechanism. The shear yielding is the main source of energy dissipation and increased toughness and therefore occurs to a far greater extent in the matrix of the rubber toughened polymers, compared to the unmodified material, due to interactions between the stress field ahead of the crack and the rubbery particles. Recently Yee and Pearson (10) have obtained direct evidence for such shear yielding from optical microscopy studies of thin sections viewed between cross polars. The birefringent localised shear bands were clearly visible running between the rubber particles at an angle of approximately 45° to the principal tensile stress, ie in the direction of the maximum shear stress.

3.2   Microstructural Features of the Matrix Phase

The toughening mechanism outlined above highlights the role of the inherent ductility of the matrix in influencing the toughness of the multiphase polymer. A decrease in the stress needed for localised shear yielding should obviously assist in increasing the toughness. This may be achieved, for example, by (i) increasing the molar mass between crosslinks (9), (ii) increasing the test temperature or decreasing the test rate (7, 8, 11, 12) or

400

(iii) decreasing the glass transition temperature, T_g, of the matrix (4). A corollary to this argument is that thermosetting polymers which are extremely tightly crosslinked (ie possess a very low molar mass between crosslinks) and possess a high T_g are difficult to toughen to a high absolute level. This may be seen from the data (13, 14) shown in Figure 6 for a rubber-modified polyimide (based upon an addition-polymerised bismaleimide) which has a tightly crosslinked structure and a T_g of about 300°C: comparison of the fracture energy, G_{Ic}, values to those for a rubber-modified epoxy (T_g(epoxy phase) = 100°C) shown in Table 1 clearly reveal the much lower values of G_{Ic} for the rubber-modified polyimide.

3.3 Microstructural Features of the Dispersed Rubbery Phase

3.3.1 Volume fraction. The toughness generally increases as the volume fraction of dispersed rubbery phase increases, but the modulus and yield strength will generally decrease slightly. The relationship between G_{Ic} and v_f for a series of rubber-modified epoxies is shown (15) in Figure 7 where the epoxy matrix phase is the same in all cases (T_g(epoxy phase) = 100°C). However, other work (4, 16) has demonstrated that in other multiphase thermosetting polymers a limiting value of G_{Ic} may be attained and further increases in v_f produces little, if any, further improvements in toughness. Finally, it is usually found that the maximum value of v_f that can be attained is about 0.3 and attempts to produce higher volume fractions have resulted in phase inversion (17).

3.3.2 Particle diameter. In order to study the effect of the particle size then the value of v_f must be measured, and ideally held constant. In the few studies (18, 19) where this has been achieved then the particle size does not appear to have a significant influence on the mechanical properties. However, the average particle diameter was only varied between about 0.5 and 5 μm and no materials have been compared which contain very small particles.

3.3.3 Particle size distribution. Again, until recently, the evidence has been very contradictory concerning the effect of the distribution of particle sizes upon the measured toughness. However, recent work (11, 15) has clearly shown that, for a given value of v_f, a bimodal distribution of particle sizes can result in the highest values of G_{Ic}, but only over certain test temperature/rate conditions.

3.3.4 Intrinsic adhesion across particle/matrix interface. Nearly all of the studies reported in the literature have well bonded particles as a consequence of the chemical reactivity of the rubber. But poor interfacial bonding has been specifically engineered, by using an unreactive system, as is evident in the fracture surface shown in Figure 8. This multiphase material showed virtually no increase in toughness (18).

3.3.5 Morphology of dispersed phase. In only a few studies (10, 19) has the presence of a phase-separated morphology in the rubbery particles, as opposed to homogeneous particles as shown in Figure 1, been reported. There is no evidence to indicate whether a complex morphology in the particle is desirable or not - but it should at least give a relatively high value of volume fraction for the dispersed particles.

3.3.6 Glass transition temperature of dispersed phase. Kinloch et al (7) have demonstrated that it is not necessary for the particles to be rubbery (ie test temperature, T_{test}, above T_g (dispersed phase)) for a multiphase thermosetting polymer to be appreciably tougher than the unmodified material. Nevertheless the greatest increases in toughness are observed when the particles are rubbery ($T_{test} > T_g$ (dispersed phase)). This probably arises from several causes. Firstly, the bulk modulus of rubbery particles under the triaxial tensile stress state at a crack tip is approximately the same as that of the cured resin matrix. So the particles will carry their full share of the local loads. However, the shear modulus of rubbery particles is several orders of magnitude lower than that of the cured resin and so rubbery particles will have a stress concentration located at their equator of value about 1.6. This local stress concentration is the initiation point for shear yielding in the matrix and the rubbery particles will also be able to accommodate the necessary shear strains imposed by the plastic deformation in the matrix. Secondly, rubbery particles will eventually cavitate due to the dilatation produced by the triaxial stress state and the plastic deformation in the matrix (which occurs at essentially constant volume). Cavitation of the particles will reduce the triaxiality of the stress in the matrix, lower the stress necessary for shear yielding to occur and so promote further shear yielding and energy dissipation in the matrix.

4   APPLICATIONS

4.1 Adhesive Joints

Rubber-modified epoxies form the basis for the current generation of tough structural adhesives for advanced engineering applications. Their toughness is usually assessed via peel or fracture mechanics tests and examples of such tests, and others referred to later in Table 2, are shown in Figure 9.

Some typical properties of a commercial rubber-modified epoxy adhesive, measured in bulk and in various adhesive joints, are listed (2) in Table 2. It is noteworthy that the effect of the multiphase microstructure is not only reflected in high peel and G_{Ic} values but also in the relatively extensive plastic shear strain the adhesive may undergo - all these improvements in physical properties lead to stronger and tougher structural adhesive joints compared to unmodified epoxy adhesives.

TABLE 2

Physical properties of rubber-modified epoxy adhesive

measured in bulk and in various adhesive joints

Test [a]	Substrates	Property	
Bulk Adhesive			
Flexure	–	Modulus	2.8 GPa
	–	Failure stress	74 MPa
	–	Failure strain	2.7%
Simple shear	–	Modulus	1.1 GPa
	–	Shear yield stress	54 MPa
	–	Shear failure strain	50%
[b,c]	–	Glass transition temperature, T_g	120°C
Adhesive joints			
Torsional shear	Aluminium alloy	Failure stress	61 MPa
Axially-loaded butt joint	Steel	Failure stress	58 MPa
Single-lap joint, loaded in tension	Aluminium Alloy/Steel	Failure stress Failure stress	28 MPa 38 MPa
Double-lap joint, loaded in tension	Cfrp-steel (Tapered substrates)	Failure load (per 25.4 mm width)	80 kN
Peel tests			
90° peel angle	Aluminium alloy	Peel strength	5 kN/m
135° peel angle	"	Peel strength	4 kN/m
Precracked TDCB [d]	Steel	Fracture energy, G_{Ic}	0.9 kJ/m^2
Precracked compact shear	Steel	Fracture energy, G_{IIc}	2.2 kJ/m^2

a. Tests conducted at 23°C and a moderate rate.

b. See Figure 9.

c. All joints failed by cohesive fracture through the adhesive,
    except for the cfrp-steel double-lap joints which failed by
    interlaminar fracture of the cfrp.

d. TDCB: tapered-double-cantilever-beam specimen.

4.2 Fibre Composites

Aligned fibre-composite materials possess low interlaminar
strengths in comparison to their strengths parallel to the fibre.
Usually, interlaminar failure initiates from stress concentrations
such as holes or resin cracks and proceeds by delamination, indeed
most fatigue damage occurs in this fashion. As Bascom et al (23)
have commented it seems intuitively reasonable that an increase in
the toughness of the thermosetting polymer used should improve the
interlaminar fracture resistance of the composite. However, as
discussed above, the high fracture energy of multiphase thermosetting polymer is associated with a relatively large volume of
plastic deformation at the crack tip. Thus, if there is
"insufficient polymer" to undergo such deformation the expected
increase in toughness is severely curtailed. Indeed, this effect
has been observed when tough multiphase thermosetting polymers
have been employed as very thin layers, either in adhesive joints
(24, 25) or fibre composites (26, 27). In the former case the
metallic substrates restrict the extent of plastic deformation
at the crack tip whilst in the latter the high-modulus glass- or
carbon-fibres exert the same restrictive influence.

However, Bascom and co-workers (23) have shown that the
enhanced toughness of multiphase thermosetting polymers may indeed
be translated into greatly improved interlaminar fracture
resistance when they are used for woven-fabric composites.
Essentially the woven fabric of glass or carbon fibres leads to
polymer-rich regions at the strand crossover points and in these
regions the extensive energy-dissipating plastic deformations,
arising from the multiphase microstructure as discussed previously,
can now occur. Thus, the interlaminar fracture energy is
considerably higher than when a low-toughness unmodified thermoset
is employed.

Bascom et al (23) have described a novel fracture-mechanics
test specimen to determine the interlaminar fracture energy and
it is sketched in Figure 10. The width is tapered to give a
constant rate of charge compliance with crack length and the
interlaminar fracture energy, G_{Ic} (interlaminar) is given by:

$$G_{Ic} \text{ (interlaminar)} = \frac{12 \, P_c^2 \, m}{E_b h^3} \qquad (3)$$

P_c = load for crack growth

E_b = bending modulus of composite

h = beam height

m = geometry constant $(=a^2/b^2)$

Hence G_{Ic} (interlaminar) can be calculated independently of the crack length, a, which can be difficult to determine, and the crack may be driven at a constant load and an average P_c value measured.

The results obtained for both an unmodified and a rubber-modified epoxy-novalac polymer are shown in Table 3. The improvements in both G_{Ic} (bulk polymer) and G_{Ic} (interlaminar) that have occurred due to the muliphase microstructure of the material are clearly evident and are extremely encouraging for future developments in this area.

TABLE 3

Effect of multiphase microstructure on both

polymer and composite toughness (23)

Polymer	Bulk Properties		Woven-Fibre Composite Properties			
			Woven glass cloth		Woven carbon cloth	
	E	G_{Ic} (bulk)	E	G_{Ic} (interlaminar)	E	G_{Ic} (interlaminar)
Unmodified	2.8	0.3	31.6	1.0	54.1	0.6
Rubber-modified	2.2	5.1	24.0	4.4	42.1	4.6

a. E is flexural modulus (GPa)

b. Fracture energy values in kJ/m^2

c. Fibre volume in composites was ~60%

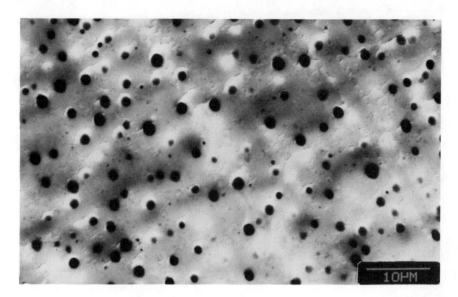

Fig 1

Transmission electron micrograph of osmium-tetroxide stained
section of a typical rubber-modified epoxy thermosetting polymer.

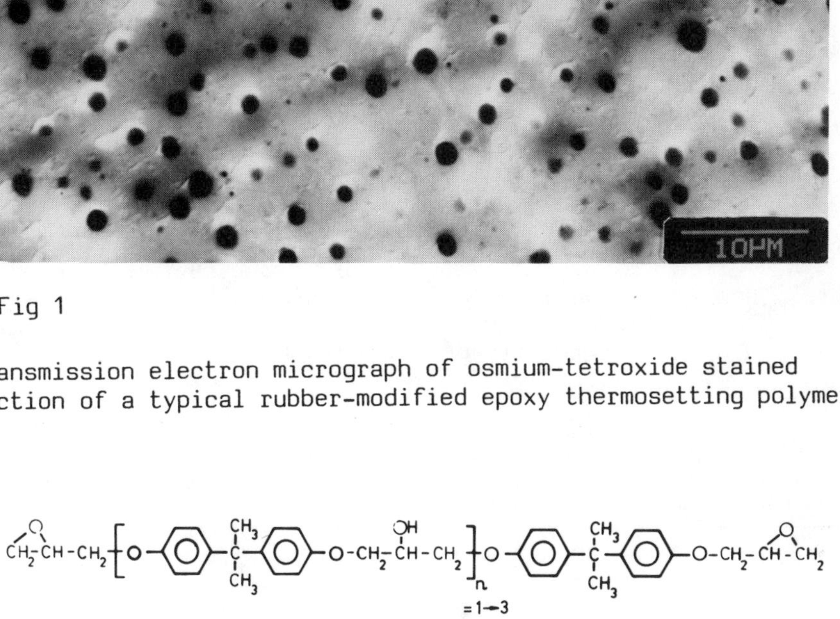

diglycidyl ether of bisphenol A

Cross linked with wide range of curing agents, e.g.,

1.   amines e.g. $NH_2-R-NH-R-NH_2$

2   polysulphides $HS-R-SH$

(long R aliphatic chain makes adhesive more flexible but
  decreases modulus, strength, heat resistance).

Fig 2   Chemistry of epoxy resins.

Condensation _Type_

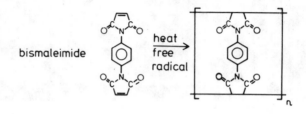

+ oxidative crosslinking

Addition _Type_

bismaleimide

Fig 3   Chemistry of imide resins.

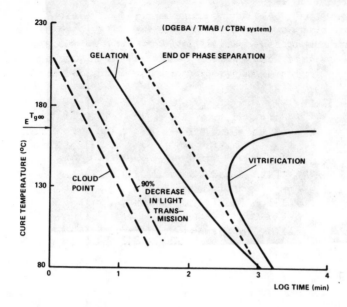

Fig 4

Time-temperature-transformation cure diagram for an epoxy resin/
amine curing agent/CTBN rubber (4).

(phase separation determined from light transmission studies)

(TMAB = trimethylene glycol di-p-aminobenzoate)

408

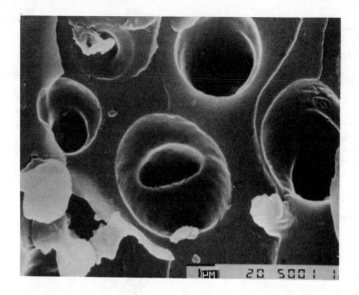

Fig 5

Scanning electron micrograph of fracture surface of rubber-modified epoxy showing cavitated rubber particles.

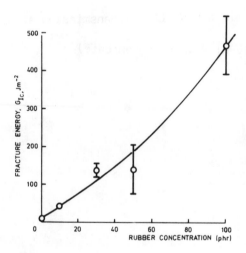

Fig 6

Fracture energy, G_{Ic}, as a function of added rubber concentration for a rubber-modified polyimide (13, 14).

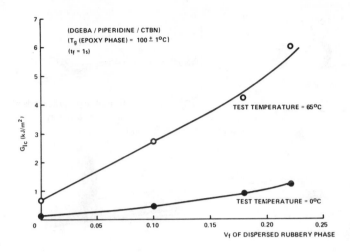

Fig 7

Fracture energy, G_{Ic}, as a function of volume fracture, v_f, of dispersed rubbery phase in a series of rubber-modified epoxy polymers (T_g(epoxy phase) ≈ 100°C). (Note dependence of relation upon test temperature.)

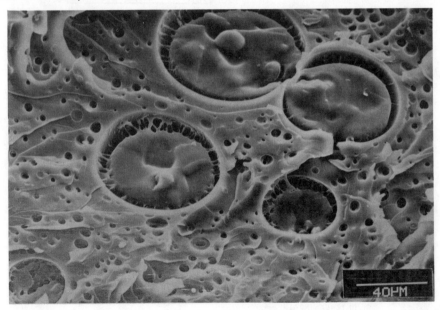

Fig 8

Scanning electron micrograph of a fracture surface of a rubber-modified epoxy where poor adhesion across the particle/matrix interface was present (18).

410

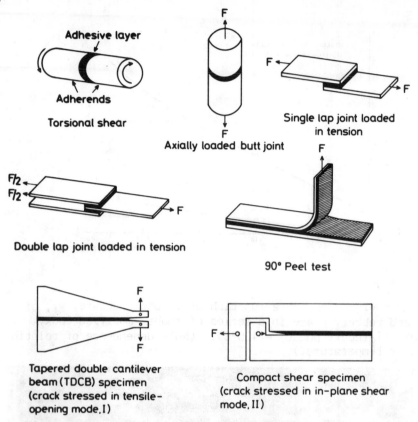

Torsional shear

Axially loaded butt joint

Single lap joint loaded in tension

Double lap joint loaded in tension

90° Peel test

Tapered double cantilever beam (TDCB) specimen (crack stressed in tensile-opening mode, I)

Compact shear specimen (crack stressed in in-plane shear mode, II)

Fig 9 Sketches of joint geometries employed to obtain the data shown in Table 2. (20-22).

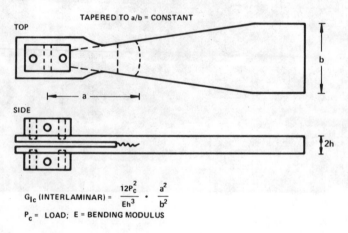

$$G_{Ic} \text{ (INTERLAMINAR)} = \frac{12P_c^2}{Eh^3} \cdot \frac{a^2}{b^2}$$

P_c = LOAD; E = BENDING MODULUS

Fig 10 Sketch of the width-tapered double cantilever beam test specimen for determining interlaminar fracture energy (23).

REFERENCES

1. Kinloch, A.J and R.J. Young. Fracture Behaviour of Polymers
    (Applied Science, London 1983).

2. Riew, C.K., E.H. Rowe and A.R. Siebert. In Toughness and
    Brittleness of Plastics, Ed. R Deanin and A Crugnola, Adv.
    Chemistry Series 154 (ACS, Washington D.C., 1976) p326.

3. Manzione, L.T., J.K. Gillham and C.A. McPherson. J Appl.
    Polym. Sci. 26 1981 907.

4. Chan, L.C., J.K. Gillham, A.J. Kinloch and S.J. Shaw. In
    Rubber-Modified Thermoset Resins Ed. C.K Riew and J.K Gillham,
    Adv. Chemistry Series 209 (ACS, Washington DC, 1984).

5. Flory, P.J. Principles of Polymer Chemistry (Cornell
    University Press, Cornell, 1953).

6. Bucknall, C.B. Toughened Plastics (Applied Science, London,
    1977).

7. Kinloch, A.J., S.J. Shaw, D.A.Tod and D.L. Hunston. Polymer
    24 1983 1341.

8. Kinloch, A.J., S.J. Shaw and D.L. Hunston. Polymer 24 1983 1355.

9. Pearson, R.A and A.F. Yee. Polym. Mat, Sci. Engng. Preprints
    49 1983 316.

10. Yee, A.F and R.A. Pearson. NASA Contractor Report 3718, 1983.

11. Hunston, D.L., A.J. Kinloch, S.J. Shaw and S.S. Wang. Proc.
    of Amer. Chem. Soc. Meeting, Kansas City, 1982 (Plenum Press,
    New York, to be published).

12. Hunston, D.L et al. J. Adhesion 13 1981 3.

13. Kinloch, A.J and S.J. Shaw. Polym. Mat. Sci. Engng. Preprints
    49 1983 307.

14. Kinloch, A.J. and S.J. Shaw. In Rubber-Modified Thermoset
    Resins Ed. C.K Riew and J.K Gillham, Adv. Chemistry Series
    209 (ACS, Washington DC, 1984).

15. Kinloch, A.J., D.L. Hunston and S.J. Shaw. Unpublished work.

16. Kunz, S.C., J.A. Sayre and R.A.Assink. Polymer 23 1982 1897.

17. Bascom, W.D et al. J. Appl. Polym. Sci. 19 1975 2545.

412

18. Chan, L.C., J.K Gillham, A.J. Kinloch and S.J. Shaw. In Rubber-Modified Thermoser Resins Ed. C.K. Riew and J.K. Gillham, Adv. Chemistry Series 209 (ACS, Washington D.C. 1984).

19. Bucknall, C.B. and T. Yoshii. British Polym. J. 10 1978 53.

20. Kinloch, A.J. J. Mater. Sci. 17 1982 617.

21. Kinloch, A.J (Ed). Developments in Adhesives - 2 (Applied Science, London, 1981).

22. Kinloch, A.J (Ed). Durability of Strucutral Adhesives (Applied Science, London, 1983).

23. Bascom, W.D et al. Composites 11 1980 9.

24. Bascom, W.D and R.L. Cottingdon. J. Adhesion 7 1976 333.

25. Kinloch, A.J and S.J Shaw. J Adhesion 12 1981 59.

26. McKenna, G.B and F.F Mardell and F.J McGarry. Soc. Plastics Industry, Ann. Tech. Conf (RPD) 1974, Section 13-C.

27. Scott, J.M and D.C Phillips. J. Mater. Sci, 10 1975 551.

PROCESSING AND PHASE MORPHOLOGY OF INCOMPATIBLE POLYMER BLENDS

James L. White
Kyonsuku Min
Polymer Engineering Center
The University of Akron
Akron, Ohio, USA

ABSTRACT

A critical review of experimental research of development of phase morphology in stratified and disperse two phase flow of polymer melts. It is shown that in stratified flow, phase morphology variations are primarily caused by viscosity differences between the two phases. For disperse two phase flow, the phase morphology is strongly influenced by the interfacial tension as well as the viscosity ratio. This appears to be through the capillarity or Taylor number. Large interfacial tensions and Taylor numbers lead to coarse phase morphologies and phase growth under quiescent conditions. Compatibilizing agents may be interpreted in terms of reduction of interfacial tensions. Applications to polymer processing operations and interpretation of rheological properties are described.

INTRODUCTION

Polymer blends play an increasingly important role in commerce. This is because of the ability to produce new products with a wide range of properties of interest to commerce with minimal investment. The development of compounds and blends of polymers dates back over a century to the earliest rubber and plastics industry when rubber was mixed with substances ranging from pitch (1) to gutta percha (2). As each new plastic has been developed its blends with previously existing materials have been explored. Thus synthetic rubbers were in an early period mixed into natural rubber and found to produce superior performance in tire components. Polystyrene was blended with natural and synthetic rubbers after its commercialization and this led to high impact polystyrenes (HIPS) which now hold a stronger position in the market place than the base plastic. Attention has been especially focussed on blends since the commercialization of General Electric's Noryl®, a blend of HIPS with poly(2,6 dimethyl-1,4 phenylene oxide). We are now in a period of enormous expansion of both new blend and related compound systems and the scientific principles underlying blend characteristics.

Blend systems may generally be classified as compatible and incompatible. The former materials are homogeneous on a molecular level and their characteristics are similar to those of normal homopolymer and random copolymers themselves being compatible mixtures of macromolecules. The situation with incompatible systems is quite different. They possess complex phase morphologies which are dependent upon their deformation and thermal histories. The manner of this dependence varies with the nature of the polymers involved. Clearly factors such as the chemical character of the materials and their individual rheological properties must also be of importance in determining the processes by which the phase morphology evolves. The multi-phase nature of polymer blends and the variation of this morphology implies that flow responses may well be quite complex in character. The scientist and engineer concerned with processing blends and fabrication of products from blends must be concerned not only with the molecular weights and distributions and thermal/crystallization/stress histories of importance in controlling single polymer systems but with the phase morphology and its evolution.

In this paper we will develop the perspectives described in the above paragraph. We first consider stratified bulk two phase flow of polymer melts, and the material characteristics determining the distributions of the two melt phases. This leads into a discussion of phase morphology in both incompatible low molecular weight liquids and polymer melt blends and its variation with thermal and deformation history. We then turn to consideration of two applications (i) forming very small diameter fibers using blend technology and (ii) the rheological properties of polymer melts and their dependence upon blend morphology.

STRATIFIED TWO PHASE FLOW

General

Before considering the flow of polymer blends we will briefly consider the bulk-stratified two phase flow of polymer melts and the factors governing the distribution of phases in the melt system. We may divide the types of flow of the sort considered here into (i) sequential and (ii) simultaneous side-by-side flows. In the former two volumes of polymer melt are injected into the apparatus in question and the relative motions of the two phases in a pressure induced flow followed. This type of melt flow is practiced in the sandwich injection molding process (3). In simultaneous side-by-side flow one co-extrudes the two melt phases. This is a common industrial practice in fabricating multilayer film (4) or producing bicomponent fibers (5). While these are important process technologies there is little in the associated patent literature which describes the influence of material on the flow of the two melts.

Sequential Injection

The sequential flow of melts into a mold has been studied by Young, White, Clark and Oyanagi (6). They observe that the shear viscosity plays the predominant role in influencing the mode of penetration into the mold (see Figure 1). If the two melts have about the same viscosity, the second melt uniformly enters the mold following the first melt. A uniform skin of the first polymer injected covers a core of the second polymer. If however the second polymer injected has a much higher shear viscosity than the first material the second polymer injected does not penetrate uniformly from the gate but remains concentrated in that region. On the other hand if the second polymer has a much lower viscosity than the first material injected, breakthrough of the second melt is generally observed.

Side-by-Side Flow

Studies of side-by-side simultaneous extrusion of two polymer melts has been studied by Southern and Ballman (7), Han (8) and Lee and White (9). Striking observations of encapsulation by one of the melts during flow were made by each of the research terms (Figure 2). The study of the latter authors critically consider the relative influence of (i) shear viscosity (ii) first and second normal stress difference and (iii) interfacial tension. It is clear from these studies that the mechanism of encapsulation is due to differences in shear viscosity. The dependence of extent of encapsulation on viscosity ratio is shown in Figure 3.

Summary

It seems clear that the distribution of phases in stratified two phase flow of polymer melts, whether involving sequential or simultaneous (side-by-side) flow through a conduit is dominated by the relative shear viscosities of the two melts. Other rheological properties play a minor role as does interfacial tension.

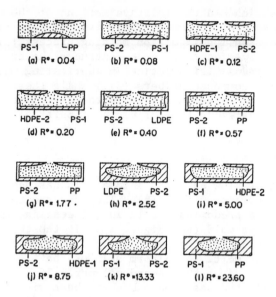

Figure 1. Sequential injection of power melts with differing viscosity levels into a mold.

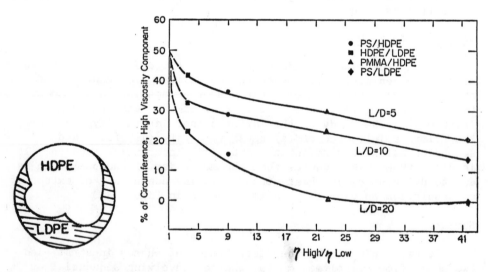

Figure 2. Encapsulation for stratified side-by-side two phase flow through a capillary die.

Figure 3. Encapsulation as a function of viscosity ratio for stratified side-by-side two phase flow through a capillary die.

PHASE MORPHOLOGY IN INCOMPATIBLE POLYMER BLEND SYSTEMS - QUALITATIVE

The level of dispersion in incompatible polymer melt may vary
considerably depending upon the nature of the polymer pair and
deformation and temperature history. There have been few studies
of this subject.

Generally polymers of similar viscosity and chemical character
seem to blend easily. Indeed there is usually little difficulty in
mixing together materials of similar character of quite different
molecular weight. Blending polar and hydrocarbon polymers of
greatly differing viscosity level, e.g. polyisobutylene and nylon-6,
is extremely difficult. Min, White and Fellers (10) have described
the blending of polyethylene/polystyrene, polyethylene/polycarbonate
and polyethylene/nylon-6 using a screw extruder plus static mixer.
The nylon-6 blend gives the coarsest phase morphology and the polystyrene blend the finest. These differences continue through
subsequent melt processing.

We now turn to quiescent blends. The basic study of these
characteristics was given by Liang et al (11) who observed the
phase morphology variations in molten polypropylene/nylon-6
blend in a quiescent reservoir as a function of time. When nylon-6
was the continuous phase, rapid phase growth was observed. This
was interpreted in terms of the large interfacial tensions of this
blend system. Similar observations have been made by Endo et al
(12) with polyethylene/polycarbonate. A broader perspective is
given in a paper by Min et al (10) which contrasts blend morphology
in a range of systems. Before proceeding it would seem useful to
critically review the literature for low viscosity liquids.

PHASE MORPHOLOGY OF NEWTONIAN FLUID EMULSIONS

The formation of phase morphology in Newtonian liquid systems
is based on the interaction of interfacial and viscous forces.
These forces have magnitudes in a shear flow of

$$\frac{\kappa}{a} \quad \text{and} \quad \eta \dot{\gamma}$$

respectively where κ is the interfacial tension, η the viscosity, a,
the droplet radius and $\dot{\gamma}$ the shear rate. The relative magnitude of
these forces defines the capillarity or Taylor number

$$\frac{\kappa/a}{\eta \dot{\gamma}} = \frac{\kappa}{\eta a \dot{\gamma}} = \frac{\kappa}{\eta v} \tag{1}$$

There is a long history of theoretical and experimental invest--
of the deformation of droplets and the formation of emulsions (13)
which date to a classical investigation by G.I. Taylor (13) in 1934.
Taylor (13) analyzed the deformation of a droplet in various shear
flows using an earlier analysis of the viscosity of an emulsion, (14),
he showed that for simple shear flow:

$$v_1 = \dot{\gamma} \, x_2 \qquad\qquad v_2 = 0 \qquad\qquad (2)$$

The droplet deformation is given by

$$v_1 = c \, x_1 \qquad\qquad v_2 = -c \, x_2 \qquad\qquad (3)$$

while for

$$\frac{L - B}{L + B} = \frac{\eta_A a}{\kappa} \, \dot{\gamma} \; \left(\frac{19\eta_B + 16\eta_A)}{16\eta_B + 16\eta_A}\right) \qquad (4)$$

one obtains

$$\frac{L - B}{L + B} = \frac{\eta_A a}{\kappa} \, 2c \; \left(\frac{19\eta_B + 16\eta}{16\eta_B + 16\eta_A}\right) \qquad (5)$$

where L is the length direction, B the breadth of the drop, η_B the
viscosity of the suspended droplet and η_A the viscosity of the matrix.
The quantity in parenthesis in Eqs (3) and (5) only ranges between
1.0 and 1.18 as the ratio of η_B/η_A goes from zero to infinity. Thus
the droplet deformation may be represented as

$$\frac{L - B}{L + B} = \frac{1}{N_{Ta}} \qquad\qquad (6)$$

where N_{Ta} is the dimensionless Taylor number defined by Eq (1).
The larger the interfacial forces relative to the viscous forces,
the less the droplet will deform during shear flow.

Experimental studies of the deformation of droplets in the
flows described by Eqs (2) and (4) have been described by Taylor
(13). A parallel band instrument was used to carry out the flow
described by Eq (2) and a four roller instrument to carry out the
kinematics of Eq (4). Droplets of CCl_4-paraffin mixture, various
oils and a tar-pitch mixture were suspended in a concentrated sugar
solution. These gave viscosity ratios η_B/η_A ranging from 0.003 to
20. Generally Eqs. (3) and (5) were found to be valid at small
droplet deformations. In the case of droplets with the lower

viscosity ratio, the linearity remained valid until (L-B)/(L+B) reached a value of 0.5, i.e. a ratio of the major to the minor principal axis, L/B, or 3/2. At high values of $1/N_{Ta}$, the droplet deformation occurs at a less rapid than linear manner.

Droplets with viscosity ratios of 0.9 and 20 were observed by Taylor (13) to become sinuous and burst, the former at a value of $1/N_{Ta}$ of 0.4, and the latter at a ratio of 0.3. This suggests

$$\left(\frac{L}{B}\right)_{crit} = F \left[N_{Ta}, \frac{\eta_B}{\eta_A}\right] \qquad (7)$$

where $(L/B)_{crit}$ is a decreasing function of increasing capillarity number and viscosity ratio.

The observations of Taylor (13) have been extended in succeeding papers by Cerf (15), Bartok and Mason (16), Rumscheidt and Mason (17, 18), Karam and Bellinger (19) and by Torza, Cox and Mason (20). Their observations generally confirm Eqs (5) and (7). Rumscheidt and Mason (18) described three modes of deformation behavior which depend upon viscosity ratio. When the viscosity ratio η_B/η_A is small, the droplet is drawn out to great lengths but does not burst. At intermediate viscosity ratios, the drawn out filaments break up into a number of droplets. When the viscosity ratio is of order unity the extended droplets break up into smaller droplets. At very high viscosity ratios, the droplets undergo only very limited deformations. Karam and Bellinger (19) conclude that the breakup of droplets occurs most readily when the interfacial tension is low, the viscosity of the continued phase is high and the radius of the droplet is large. Droplets could only be broken when the viscosity ratio was in the range 0.005 to 4. With lower viscosity ratios, the ends of the drops stretched out. At higher viscosity ratios, the droplets did not deform. Torza et al (20) find droplet bursting most easily in a viscosity ratio range of 0.3 to 0.6 and not to occur above a value of 3. They concluded, contrary to earlier researchers, that the mode of burst, when it occurred, was independent of viscosity ratio, but depended upon the rate of increase of velocity gradient.

Various views of the mechanism of breakup of droplets have been described. Taylor (13) and Karam and Bellinger (19) have argued that when the shear flow induced flow stresses overcome the pressure difference across the interface due to interfacial tension i.e.

$$f\left(\frac{\eta_B}{\eta_A}\right) \eta_A \dot{\gamma} > \frac{2\kappa}{a} \qquad (8a)$$

or

$$\frac{\eta_A a}{\kappa} \; \dot{\gamma} \; f \cdot (\frac{\eta_B}{\eta_A}) = N_{Ta} \; f \; (\frac{\eta_B}{\eta_A}) > 1 \qquad (8b)$$

Tomotika (21, 22) had considered the stability of a viscous filament surrounded by a continuum of a liquid with a different viscosity. The instability is considered to be due to interfacial tension in a manner analogous to Rayleigh's capillarity instability (23).

QUANTITATIVE STUDIES OF THE DEVELOPMENT OF PHASE MORPHOLOGY DURING FLOW OF POLYMER MELT BLENDS

Transverse Morphology in a Capillary

In investigations of the cross-sectional phase morphology Liang et al (11) and Min et al (10, 24) conclude that the scale of the cross-sectional morphology was determined by the dimensionless group of Eq. (7). Coarse phase morphologies were caused by large values of the group and thus implicitly by large interfacial tensions. The systems studied included polyethylene (PE)/ polystyrene (PS), polyethylene/polycarbonate (PC), polyethylene/ nylon-6 (N6) and polypropylene (PP)/nylon 6. The blends were all prepared in the same manner with a single screw extruder and a Koch static mixer. The finest phase morphology was found with the PE/PS and the coarsest with the PE/N6 and PP/N6. They associated the highest interfacial tensions with the latter materials and the smallest with the former. The scale of the phase morphology found with PE/PS system proved to be independent of mixing temperature from 180° to 240°C and polyethylene molecular weight. The temperature and polyethylene molecular weight variables considerably changed the viscosity ratios. Indeed the value of ratios of the zero shear viscosity η_o (PE)/η_o(PS) varied from 0.13 to 3.8 in the experiments. Min et al (24) also noted that the scale of dimensionless of the PE/PS phase morphologies are the same as reported in the literature by other research groups. The scale of the phase morphology was made much finer by shearing at high stresses. PE/N6 extrudates generally showed coarse morphologies along their axis and finer grained morphologies near the outer wall in capillary die experiments (Figure 4). This serves to strengthen the views of this section in terms of local capillarity numbers.

Min et al (10) were able to quantitatively correlate the variation of scale of phase morphology with Eq. (2) by choosing suitable values of interfacial tensions. This is shown in Figure 5. They took κ for PE/PS to be 0.8 dyne/cm, PE/PC to be 15 dyne/cm and PE/N6 to be 30 dyne/cm.

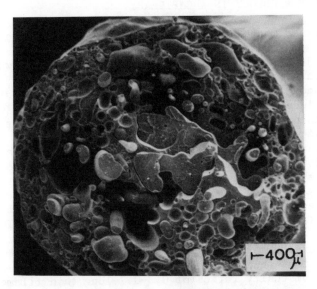

Figure 4. SEM photomicrograph of PE/N6 –50/50 blend extrudate at die wall shear stress 8.3 x 10^4 dyne/cm^2.

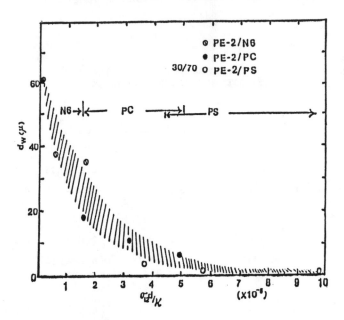

Figure 5. Variation of dispersed globule size as a function of inverse Taylor number $1/N_{Ta}$ with κ of 0.8 dyne/cm for PE/PS, 15 dyne/cm for PE/PC and 30 dyne/cm for PE/N6.

LONGITUDINAL MORPHOLOGY IN A CAPILLARY

Basic experimental studies of phase morphology development in incompatible polymer melt blends were first reported in papers summarizing joint research by M.V. Tsebrenko and his coworkers at the Kiev Technological Institute of Light Industry and G.V. Vinogradov and his colleagues at the USSR Academy of Science (25-29) conclude using extraction and scanning electron microscopy that ultra thin long filaments were formed during extrusion of blends through long capillary dies when the viscosities of the two phases are about the same. For lower viscosity ratio, η_B/η_A, films were to be formed together with fibers. At higher viscosity ratios, only droplets or short rods were obtained.

Other research groups have also explored this problem including Liang et al (11) and Min, White and Fellers (11, 24). The latter authors have explored both cross-sectional and longitudinal phase morphology development during shear flow. They found studying PE/PS blends that when the continuous phase had a lower viscosity specifically viscosity ratios of 0.3-0.7 long thin filaments were formed, at viscosity ratios of 0.8-1.3 undulating threads were obtained, at viscosity ratios of 1.4-1.9 ellipsoids were obtained and at values over 2.2 the dispersed phase is spherical (see Table 1).

TABLE 1

LONGITUDINAL PHASE MORPHOLOGY
OF POLYETHYLENE/POLYSYTRENE BLENDS

PE/PS Composition	Fibrillation	η_d/η_m (dispersed/matrix)	$\kappa/\eta_m\gamma a$	a (μm)
PE-1/PS				
10/90	undulating thread	1.2 - 1.3	0.37~0.76	0.4 - 0.8
90/10	undulating thread	0.8 - 1.3	0.64~1.12	0.7 - 1.8
PE-2/PS				
10/90	thin thread	0.5 - 0.7	0.31~0.92	0.5 - 1.7
90/10	ellipsoids	1.4 - 1.7	0.47~0.54	1.3 - 2.1
PE-3/PS				
10/90	thin thread	0.3 - 0.5	0.09~0.66	0.3 - 1.3
90/10	spheres	2.2 - 3.7	1.0~2.0	6.5 - 6.9

They may also be an influence of the Taylor number of Eq. (1) on the longitudinal morphology. We list its values in Table 1 also. Tsebrenko, Vinogradov et al dealt with lower viscosity continuous phase systems than PE or PS, and their transition to films may represent a capillarity number effect.

INFLUENCE OF COMPATIBILIZING AGENTS

Compatibilizing agents would be expected to reduce interfacial tension κ and reduce the capillarity number. Heikens and his co-workers (30) introduced styrene-ethylene graft copolymers into polyethylene/polystyrene and showed it led to finer grained morphologies.

A second quantitative study of the influence of compatibilizing agents on phase morphology is that of Endo et al (12). They added styrene-ethylene-styrene block copolymer to polyethylene-polycarbonate blends and described its influence on phase morphology variation in annealing. Blends containing compatibilizing agents exhibited much reduced phase growth with time. The results of Cimmino et al (31) with maleic anhydride grafted EPDM in EPDM/N6 blends may be similarily interpreted. Compatibilizing agents also produce much more uniform dispersions, i.e. smaller ranges of dispersed phase globule size (12, 30, 31, 32).

APPLICATIONS OF PHASE MORPHOLOGY ANALYSIS

Small Diameter (Mini) Fibers

The formation of small diameter (or mini) fibers from polymer blends has a long history (10, 11, 24, 32-35). This involves extrusion through a capillary die followed by drawdown of the emerging extrudate onto the take-up roll. The continuous phase is subsequently extracted leaving small diameter mini fibers (Figure 6). Fibers with diameter or order 0.1μm have been so produced.

However it is clear from the discussion of the previous sections that the development of such small diameter continuous filaments has certain preconditions. It seems clear from the studies of Tsebrenko and Vinogradov (29) as well as Min et al (11) that the viscosity of the dispersed phase must be of the same order or smaller than that of the continuous phase for success. Furthermore the Taylor number should be as small as possible to ensure a small diameter dispersed phase in the extrudate.

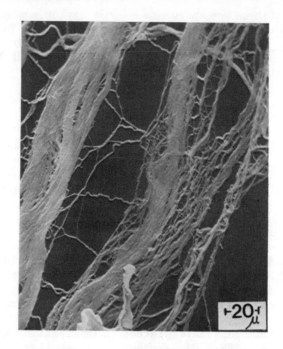

Figure 6. Small diameter polyethylene fibers produced by
extrusion of a 10% polyethylene PE/PS blend and
subsequent extraction.

Rheological Properties

There have been extensive investigations of the rheological
properties of polymer blends since about 1970 ($\underline{10}$, $\underline{11}$, $\underline{36-46}$). It
has long been realized in the polymer community that such rheological
properties are associated with the phase morphology. However there
has been relatively little effort to follow through on this.

Theoretically studies of the rheological properties begin with
the work of Lees ($\underline{47}$) who developed various simple models of two
phase flow. The simplest of these involve the viscosity of parallel
lamellae oriented in the direction of flow. The mean shear rate in
a volume element will be according to the volume fractions

$$\dot{\gamma} = \phi_A \, \dot{\gamma}_A + \phi_B \, \dot{\gamma}_B \tag{9a}$$

The stresses will be the same in all layers so that

$$\frac{\dot{\gamma}}{\sigma_{12}} = \phi_A \frac{\dot{\gamma}_A}{\sigma_{12}} + \phi_B \frac{\dot{\gamma}_B}{\sigma_{12}} \tag{9b}$$

$$\frac{1}{\eta} = \frac{\phi_A}{\eta_A} + \frac{\phi_B}{\eta_B} \tag{10}$$

This predicts a monotonic variation of viscosity with volume fraction. The above theory has frequently been rediscovered by more recent authors considering polymer melts (36, 37) and attempts made to apply it across the concentration range.

A more sophisticated approach to the viscosity of emulsions was given by Taylor (14) who modified Einstein's theory of viscosity of dilute suspensions to include internal circulations in the drop due to viscous shearing in the matrix. Taylor concludes that the viscosity of a dilute emulsion is increased from the matrix according to

$$\eta = \eta_A \left[1 + 2.5 \left(\frac{\eta_B + \frac{2}{5} \eta_A}{\eta_B + \eta_A} \right) \phi \right] \tag{11}$$

This always predicts an increase in viscosity with addition of a second polymer. It neglects however the deformation of the dispersion phases and presumes it to be spheres. Min et al (24, 32) attempt to rationalize viscosity-composition plots in terms of applying Eq (11) to dilute systems and Eq (10) to intermediate compositions (see Figure 7). This is successful in some cases but not in others. It would appear most successful when the dispersed phase is more viscous and shows little deformation and least when it is much less viscous and readily deforms to produce a continuous filament.

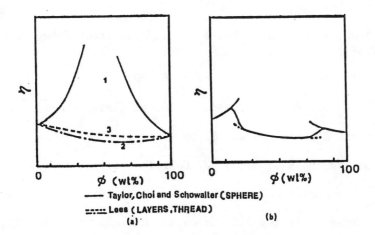

Figure 7. Prediction of shear viscosity - composition behavior on the basis of phase morphology.

REFERENCES

1. T. Hancock, English Patent 4768 (1823)

2. C. Hancock, English Patent 11,147 (1846)

3. D.F. Oxley and D.J. H. Sandifor, Plast, Polym. 39, 288 (1971).

4. W.J. Schrenk, K.J. Cleereman and T. Alfrey, SPE Trans. 19, 192 (1963).

5. J.H. Saunders, J.A. Burroughs, L.P. Williams, D.H. Martin, J.H. Souther, R.L. Ballman and K.R. Lea, J. Appl. Polym. Sci. 19 1387 (1975).

6. S.S. Young, J.L. White, E.S. Clark and Y. Oyanagi, Polym. Eng. Sci, 20, 79B (1980).

7. J.H. Southern and R.L. Ballman, Appl. Polym. Symp. 20, 175 (1973).

8. C.D. Han, J. Appl. Polym. Sci. 17, 1289 (1973).

9. B.L. Lee and J.L. White, Trans. Soc, Rheology 18, 467 (1984).

10. K. Min, J.L. White and J. F. Fellers, Polym. Eng. Sci. (in press).

11. B.R. Liang, J.L. White, J.E. Spruiell and B.C. Goswami, J. Appl. Polym. Sci. 28, 2011 (1983).

12. S. Endo, K. Min, J.L. White and T. Kyu, unpublished research (1983-4).

13. G.I. Taylor, Proc, Roy. Soc. A146, 501 (1934).

14. G.I. Taylor, Proc, Roy. Soc, A138, 41 (1932).

15. R. Cerf, J. Chim. Phys. 48, 59 (1959).

16. W. Bartok and S.G. Mason, J. Colloid Sci. 14, 13 (1959).

17. F.D. Rumscheidt and S.G. Mason, J. Colloid Sci. 16, 210 (1961).

18. F.D. Rumscheidt and S.G. Mason, J. Colloid Sci, 16, 238 (1961).

19. H. J. Karam and J. C. Bellinger, IEC Fund 7, 576 (1968).

20. S. Torza, R. G. Cox and S. G. Mason, J. Coll, Interf. Sci. 38, 395 (1972).

21. S. Tomotika, Proc, Roy. Soc A150, 322 (1935).

22. S. Tomotika, Proc, Roy. Soc A153, 302 (1936).

23. Lord Rayleigh, Proc, London Math Soc. 10, 4 (1879).
                    Phil. Mag. 34, 145 (1892).

24. K. Min, J.L. White and J. F. Fellers, J. Appl. Polym. Sci. 29, 2117 (1984).

25. M.V. Tsebrenko, M. Jakob, M. Yu. Kuchinica, A.V. Udin and G.V. Vinogradov, Int. J. Polym. Mat. 3, 99 (1974).

26. T. I. Ablazova, M.V. Tsebrenko. A.V. Yudin, G.V. Vinogradov and B.V. Yarlykov, J. Appl. Polym. Sci. 19, 1781 (1975).

27. G.V. Vinogradov, B.V. Yarlykov, M.V. Tsebrenko. A.V. Yudin and T.I. Ablazova, Polym. 16, 609 (1975).

28. M.V. Tsebrenko, A.V. Yudin, T. I. Ablazova and G. V. Vinogradov Polym. 17, 831 (1976).

29. M.V. Tsebrenko, N.M. Rezanova and G. V. Vinogradov, Polym. Eng. Sci. 20, 1023 (1980).

30. W.M. Barentsen, D. Heikens and P. Piet, Polymer 15, 119 (1974).
                    Ibid. 18, 69 (1977).

31. S. Cimmino, L. D'Orazio, R. Greco, G. Maglio, M. Malinconico, C. Mancarella, E. Martuscelli, R. Palumbo and G. Ragosta. Polym. Eng. Sci, 24, 48 (1984).

32. K. Min, Ph. D. Dissertation in Polymer Engineering, University of Tennessee (1984).

33. W.A. Miller and C.N. Merriam, U.S. Patent 3,097,991 (1963).

34. A.L. Breen, U.S. Patent 3,382,305 (1968).

35. J. Shimizu, N. Okui, T. Yamamoto, M. Isah and A. Takaku, Sent. Cakkaishi 38, T-1 (1982).

36. R.F. Heitmiller, R.Z. Naar and H.H. Zabusky, J. Appl. Polym. Sci. 8, 873 (1964).

37. K. Hayashida, J. Takahashi and M. Matsui, Proc. 5th Int. Cong. Rheol. 4, 525 (1970).

38. C.D. Han and T.C. Yu, J. Appl. Polym. Sci. 15, 1163 (1971).

428

39. C.D. Han, J. Appl. Polym. Sci. 15, 2579 (1971).

40. H.L. Doppert and W.S. Overdiep, ACS Advances Chem. Ser. 99, 53 (1971).

41. Y. Shoji, M. Sato, T. Yahata and F. Komatsu, Res. Repts, Muroran Inst. Technol 7, (3), 131 (1972).

42. A. Plochocki, J. Appl. Polym. Sci. 16, 987 (1972).

43. C.D. Han and Y.W. Kin, Trans. Soc. Rheol. 19, 245 (1975).

44. B.L. Lee and J. L. White, Trans. Soc. Rheol. 19, 481 (1975).

45. K. Iwakura and T. Fujimura, J. Appl. Poly. Sci. 19, 1427 (1975).

46. N. Alle and J. Lyngaae-Jorgensen, Rheol. Acta. 19, 94 (1981).

47. C.H. Lees, Proc. Phys. Soc. 17, 460 (1900).

PULSE-INDUCED CRITICAL SCATTERING

Manfred Gordon

Statistical Laboratory, University of Cambridge, 16 Mill Lane, Cambridge, England

ABSTRACT

The instrumental technique of Pulse-induced Critical Scattering (PICS), and its use in determining spinodals for the construction of phase diagrams, is briefly described. Like a cloud-chamber or bubble-chamber in particle physics, PICS exploits the principle of bringing a sample of a mixture into a highly sensitive labile (metastable) condition for a short time only, so as to keep the sample from demixing; but long enough to measure the very special properties which become manifest in such a labile system.

1. TYPICAL PHASE DIAGRAM

The nature and significance of phase diagrams, especially in the temperature-concentration plane, is once more summarised by the **schematic figure 1. Dr. Koningsveld has explained elsewhere in this** book the classical notion of the spinodal, and its general significance.

2. THE PICS TECHNIQUE

The PICS technique (1,2) was developed at Essex University with much cooperation from Koningsveld and coworkers; see especially the contribution by Koningsveld and Kleintjens (3) in this book. The instrument measures intensities I_{30} of scattered laser light at a low angle (30°). The sample, ca. 10 mg of the mixture, is contained in a capillary cell traversed axially by the incident laser beam. The scattered beam intensity is sampled repeatedly while

430

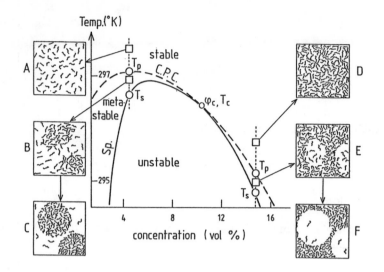

Fig. 1. Typical phase diagram of temperature T against volume % ϕ,
here for polystyrene in solution with cyclohexane. The side
panels A-F show the molecular states schematically, see text.
T_p cloud point temperature; T_s spinodal temperature; ϕ_c
critical volume % of polystyrene; T_c critical temperature.
PICS scans the low angle scattering intensity after fast
thermal steps along the broken vertical lines. Cf. fig. 2.

successive thermal pulses step the capillary and sample from a fixed
temperature in the stable temperature region of the phase diagram
(fig. 1), to temperatures which approach by steps of ca. $0.01°$ ever
more closely to the spinodal (and hence to instability).
    In fig. 1, panels A-F, the difference between molecular states in
polymer solutions is shown schematically, the solvent being implied
by the white background, while the polymer molecules are represented
by the black lines (random coils). In polymer mixtures, fluctuations
in concentration occur in the metastable part of the phase diagram,
quite analogously to those in panels B and E for solutions. In this
metastable region, the fluctuations give rise to intense laser
scattering. When the pulses approach sufficiently closely to the
spinodal temperature, the nucleation of the new emulsion phase
(panels C and F) begins to cut off the scattered light which is
being sampled some way along the cell. The shape of the observed
scattered-light pulse indicates when this happens, so that the
existence of true scattering from fluctuations ('critical scattering')
can be unambiguously tested (4). In the Debije-Scholte plot (fig. 2),
the reciprocal intensities $1/I_{30}$ of successive critical scattering
pulses are plotted against the temperature. The spinodal is found by

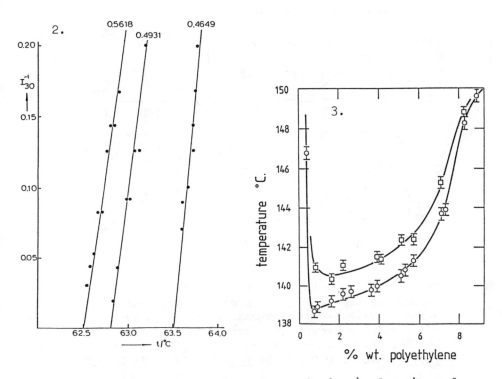

Fig. 2. Debije–Scholte plots (by L.A. Kleintjens) of reciprocal scattering intensity I_{30} (arbitrary units) against temperature for three mixtures of low molecular weight polystyrene (M_w = 2.10 kg/mole) and polyisoprene (M_w = 0.82 kg/mole), at the weight fractions indicated.

Fig. 3. PICS measurements by L.A. Kleintjens of spinodals (squares) and cloud points (circles) in the LCST region, for linear polyethylene (M_w = 177 kg/mole) in n-hexane. The measurements were taken under the vapour pressure (ca. 5 atm.) of the solutions prevailing in the PICS capillaries.

a short extrapolation to $1/I_{30}$=0. By collecting plots for mixtures of different concentration ratios, the spinodal curve can be constructed.

The PICS instrument is fully automated, collects and collates data from four scattering cells in one run, and produces the Debije plots using its dedicated computer. Samples of polymer mixtures are homogenised in their capillary cells before use in PICS by means of the Centrifugal Homogeniser (3).

432

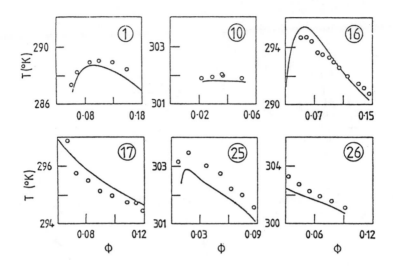

Fig. 4. Six typical spinodal temperature plots (circles) against
volume fraction φ of polystyrene in cyclohexane solutions.
Number-, weight- and z-average molecular masses (in kg/mole)
of mixed polystyrene fractions; Plot 1: 51, 55, 57; plot 10:
86, 86, 87; plot 16: 113, 324, 411; plot 17: 166, 395, 860;
plot 25: 1210, 1675, 1900; plot 26: 1147, 1690, 1900.

3. THE STUDY OF COMPLEX SPINODALS BY PICS

   The study of spinodals in the construction of theoretical models
has been described by Koningsveld and Kleintjens (3), who also stress
the complexity of shapes observed. The occurrence of bimodal spinodal
curves, as in their fig. 5, provides one illustration of such
complexity, and two further examples may be quoted from the field of
polymer solutions.

a) The whole shape of the familiar (single-maximum) spinodal in
fig. 1 may of course be inverted by passing from upper to lower
critical solution temperature regions (UCST to LCST). The PICS
technique is equally adapted to both kinds of extrapolations, i.e.
by pulses stepping to progressively lower or progressively higher
temperatures in the UCST and LCST regions respectively. Proceeding
in the latter direction, Kleintjens observed the phase diagram for
solutions of linear polyethylene in n-hexane (fig. 3). The cloud
points shown were also obtained at the same time; for the emulsion-
seeding technique of measuring cloud points in the PICS instrument,
without supercooling or superheating errors, see (4).

b) An example of inverting the sense of the curvature of the
familiar spinodal plot, while remaining in the UCST region (!) rather
than passing to the LCST domain, occurred in the study of rather
pathologically polydisperse mixtures of polystyrene fractions in
cyclohexane solution (5). This example is included as the panel
marked 17, in fig. 4, which reproduces fig. 4 of Gordon and
Torkington (6). The other panels illustrate both the wide variety
of shapes of spinodals observed as a function of varying polydispersity of the mixtures, and the ability of a theoretical model (5)
to capture these shape effects (continuous lines). Koningsveld and
coworkers have more recently improved such models by means of more
powerful computer optimisation programmes - see for instance ref (44)
of Koningsveld and Kleintjens' contribution (3).

The complexities observable with polymer mixtures in absence of
solvents is expected and found to be even greater, and PICS may be
helpful in refining statistical-mechanical models for these also (3).
Koningsveld and coworkers have been able to study mixtures in PICS
capillary cells using constituents in the range of molecular weight
of about 10^5.

REFERENCES

1) Gordon, M., Cowie, I.M.G., Goldsbrough, J., Ready, B.W.,
British Pat. 1,377,478 (1971); US Patent 3,807,865 (1974)
2) Galina, H., Gordon, M.,Irvine, P., Kleintjens, L.A., Pure Appl.
Chem., 54 (1982) 365
3) Koningsveld, R., Kleintjens, L.A., this Volume
4) Derham, K.W., Goldsbrough, J., Gordon, M., Pure Appl.Chem., 38
(1974) 97
5) Gordon, M., Irvine, P., Macromolecules, 13 (1980) 761
6) Gordon, M., Torkington, J.A., Ferroelectrics, 30 (1980) 237

PHASE SEPARATION IN POLYMER BLENDS

R.G. Hill, P.E. Tomlins and J.S. Higgins

Department of Chemical Engineering and Chemical Technology,
Imperial College,
London SW7 2BY, England.

In recent years the thermodynamics of polymer-polymer miscibility has been studied extensively in contrast to the rather scant attention received by the kinetics of phase separation.

Phase separation proceeds essentially by either nucleation and growth or by spinodal decomposition depending on the temperature and composition of the sample (the principle features of these processes are discussed in Chapter 5). This abstract is concerned only with the location and kinetics of the latter process which is the dominant mechanism within the unstable region of the phase diagram.

The Cahn-Hilliard theory (1) of spinodal decomposition predicts that the scattering pattern from a sample during the early stages of the process will appear as a (fairly well defined) peak. The intensity of this peak increases exponentially with time although its position remains constant. When phase separation occurs in the metastable region the initial domain sizes are much larger than those formed by spinodal decomposition, the scattering is therefore confined to much smaller angles. In this case a scattering maximum is generally not observed. This difference in scattering behaviour can be used to locate the position of the spinodal.

Cloud point curves have been determined for an oligomeric system of polystyrene/polybutadiene (PS/PB) (2) and a high molecular weight blend of ethylene vinyl acetate copolymer/solution chlorinated polyethylene (EVA/SCPE) (3). The apparatus used in these experiments consists of a laser light source and a photodiode array which is used to monitor the scattered intensity over

an operational range of 0-90° (3). Only one composition out of a
range of PS/PB blends when phase separated was found to show a
scattering maximum which suggests that the cloud point curve lies
close to the binodal for this composition. The composition which
shows a scattering maximum is thought to correspond to the 'critical' mixture; this result has been corroborated by pulse induced
critical scattering measurements (4). With the EVA/SCPE blends a
scattering maximum was observed at all compositions implying that
for this system the cloud point curve lies close to the spinodal
locus. In both systems where scattering maxima were observed
plots of ℓnI against time were linear during the early stages in
accordance with the Cahn-Hilliard theory. These results agree
with Binder's theoretical predictions (5) that insignificant nucleation and growth will occur in high molecular weight entangled
blends.

The wavelength, λ_m of the dominant concentration fluctuation
in a blend can be found from the position of the scattering maximum using Bragg's law. van Aartsen (6) using Cahn-Hilliard and
Flory-Huggins (7,8) theory derived the following expression to
calculate λ_m for a LCST system:

$$\lambda_m = 2\pi\ell \left[3(\frac{T-T_s}{T_s}) \right]^{-\frac{1}{2}}$$

where ℓ is the Debye interaction length (9). T_s and T are the
temperatures of the spinodal and measurement respectively. ℓ is
approximately the same size as the radius of gyration (R_g). For
the PS/PB blend λ_m would be c.a. 500Å and unobservable by small
angle light scattering. For high molecular weight blends such as
EVA/SCPE or poly(methyl methacrylate)/SCPE (PMMA/SCPE) λ_m would
be c.a. 10,000Å and observable by light scattering. This value is
consistent with our experimental data on the EVA/SCPE system but
not with the PMMA/SCPE blends (10). In the latter system the
initial phase size is c.a. 300Å and only observable by small angle
X-ray or neutron scattering. Although this result is contrary to
van Aartsen's predictions it is consistent with the theories of
Pincus (11), de Gennes (12) and Binder (5), which are based on a
reptation model and predict the initial phase size to be near to
the R_g of the polymer.

The discrepancies between our limited experimentally determined λ_m values and those predicted suggest that the present theories are not sufficiently comprehensive to be applicable to all
systems.

References

1. Cahn, J.W. and J.E. Hilliard, J.Chem.Phys. 31, (1959) 688.

2. Hill, R.G., P.E. Tomlins and J.S. Higgins, submitted to Polymer.

3. Hill, R.G., Ph.D. Thesis, Imperial College (In prep.).

4. Atkin, E.L., L.A. Kleintjens, R. Koningsveld and L.J. Fetters, Makromol. Chem. 185 (1984) 377.

5. Binder, K., J. Chem. Phys. 79 (1983) 6387.

6. van Aartsen, J.J., Eur. Polym. J. 6 (1970) 919.

7. Flory, P.J., J. Chem. Phys. 9 (1941) 660 and 10 (1942) 51.

8. Huggins, M.L., J. Phys. Chem. 46 (1942) 151
                    J. Amer. Chem. Soc. 64 (1942) 1712.

9. Debye, P., J. Chem. Phys. 31 (1959) 680.

10. Hill, R.G., P.E. Tomlins and J.S. Higgins, submitted to Polymer.

11. Pincus, P., J. Chem. Phys. 75(4) (1981) 1996.

THERMODYNAMICS OF COMPATIBILITY IN BINARY POLYMERIC MIXTURES

Ivet Bahar

Istanbul Technical University
Chemistry Department
Istanbul - Turkey

According to the conventional Flory-Huggins theory, the change in chemical potential of the solvent in solution is given by,

$$(\mu_1 - \mu_1^0)/RT = \ln(1 - v_2) + (1 - 1/r)v_2 + \chi v_2^2 \tag{1}$$

where v_2 is the volume fraction of the solute, r is the ratio of the molar volumes of solute to solvent. χ is the heat interaction parameter associated with a van Laar type enthalpy of mixing. The dependence of χ on concentration, from activity measurements, and the observed changes of volume on mixing lead to the development of the equation of state theory (1), that takes into account the contribution of the characteristic properties of the components in solution. A major implication of this approach is that χ in Eq. (1) is referred to as the reduced residual chemical potential and may be expressed as:

$$\chi = \chi_1 + \chi_2 v_2 + \chi_3 v_2^2 + \ldots \approx \chi_1 + \chi_2 v_2 \tag{2}$$

where the coefficients χ_1, χ_2, etc. depend only on specific liquid state properties of components in solution, and on temperature. Explicit expressions for χ_1 and χ_2 are given in literature (1). For a particular binary system, the parameters χ_1 and χ_2 may be calculated as a function of temperature provided that sufficient data (2,3,4) on the variables involved in the expressions for χ_1 and χ_2 are available. The $\chi_1\chi_2$ curve, so obtained, is characteristic of the system considered.

Combining Eqs. (1) and (2), and applying the two conditions for critical miscibility yields the critical curve for transition from homogeneous to heterogeneous state (Fig. 1). Also, as a further

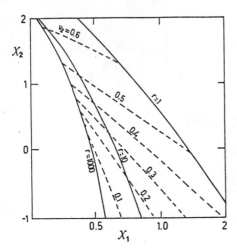
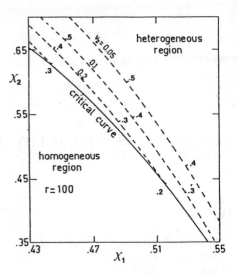

Fig. 1 Critical curves for various r. Each pt along each curve corresponds to a particular critical compositions. The dashed lines join pts of constant critical composition.

Fig. 2 Phase equilibrium curves. Each curve corresponds to the indicated constant composition for the phase more dilute in comp. 2. Figures along each curve represent the volume fraction of component 2 in the more conc. phase.

step, for a selected r, the composition of the two phases in the heterogeneous region can be computed from simultaneous solution of the phase equilibrium equations (Fig. 2). The family of curves, so obtained, is applicable to any binary mixture by appropriate selection of the variable r. The intersection of the above mentioned X_1X_2 curve, characteristic of the specific system chosen, with this family of curves gives the UCST (and/or LCST which is more frequent in polymer-polymer systems) and the compositions of the respective phases if phase splitting occurs.

As a simple application a mixture of monodisperse polystyrene-cyclohexane may be considered. The intersection of the characteristic X_1X_2 curve with the critical curves and phase equilibrium curves, shown in Figure 3, gives the binodals in Figure 4, for two different chain lengths. The binodals constructed, by the above described procedure are in close agreement with experimental results (5). However in general, a satisfactory agreement of the predicted values with experimental results requires a correct knowledge of the mixture properties present in the equation of state theory, such as X_{12} and Q_{12} representing the energy and entropy of interaction, respectively. A rational determination of these quantities, considering

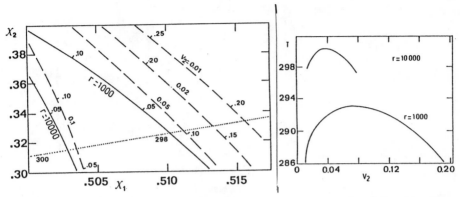

Fig. 3 C_6H_{12} PS system. $\chi_1\chi_2$ curve may be calculated from references (3) and (4) as a function of temperature, each point corresponding to a seperate temperature. The intersection with critical curves occurs at T = 293°K for r = 1000 and T = 299°K for r = 10000.

Fig. 4   Binodals for PS C_6H_{12}

the contribution of both dispersion forces and polar interactions, on molecular level, has not yet been achieved.

Acknowledgement

The author is indebted to Dr. Burak Erman for helpful suggestions.

References

1.   Flory, P.J. J. Amer. Chem. Soc., 87, 1833 (1965).
2.   Flory, P.J. Disc. Faraday Soc., 49, 7 (1970).
3.   Shih, H. and P.J. Flory, Macromolecules, 5, 758 (1972).
4.   Höcker, H., G.J. Blake and P.J. Flory, Trans. Faraday Soc., 67, 2251 (1971).
5.   Shultz, A.R.. and P.J. Flory. J. Amer. Chem. Soc., 75, 3888 (1953).

P V C   B L E N D I N G   R E S I N S

P R O P E R T I E S   A N D   A P P L I C A T I O N S

Jean M. LEHMANN

Electrochemical Industries (Frutarom) Ltd.

Introduction

The emulsion polymerization of vinyl chloride yields a PVC
latex which, after drying and deagglomeration consists in spheri-
cally shaped particles of very low porosity ranging from 0.5 to a
few microns in size. They are called dispersion resins. When
mixed under moderate shear into an oil (plasticizer), they form
pastes, usually called plastisols. Stringent control of the
particle size of the latex at the polymerization stage is necessary
in order to control the rheological properties of the plastisol
essential for subsequent conversion into finished articles. The
incidence of the particle size distribution on the viscosity of the
plastisols has been reported in the literature (1,2,3); it is well
admitted that the viscosity of dispersed systems depends on the
packing efficiency of the polymer particles and hence on the amount
of plasticizer remaining to dilute the dispersion once the inter-
particles' voids have been filed (4). Generally, finer particle
size resins yield higher viscosity and dilatancy (shear rate thickening). In order to reduce these effects mixture of dispersion
resins and blending resins are used.

Blending resins are special PVC grades made by suspension polymerization processes. They are discrete, spherical and dense
particles, intermediate in particle size between dispersion and conventional suspension resins. For large particle resins the range
is 90 to 130 microns; for the fine particle resins, 25 to 50 microns.
Also called extender resins, they are employed to achieve a
plastisol of high polymer content but adequate fluidity and they are
utilized as partial replacement of dispersion resins for paste viscosity reduction, viscosity stability, air release enhancement,
gelation temperature adjustment capabilities and gloss reduction if
desired. Cost savings can also be made by the fact that the lower

viscosity achieved allowes the addition of higher filler contents.
Also, the processes used for their preparation are less expensive
than those for making dispersion resins.

The viscosity reduction effect is achieved primarily in two ways:
(i) a denser particle packing achieved by combination of smaller
PVC paste and larger blending resin particles; (ii) a better sur-
face-volume ratio of the blending resins. Both these effects
result either in a lower plasticizer content for the same viscosity or, with the same plasticizer content, a lower viscosity is
achieved provided that the blending resins are spherical, smooth
and non-porous particles.

Effect of particle size: changes in the average particle size of
blending resins, provided that the total range of particle size
distribution is maintained, have not shown any effects on the viscosities of plastisols (5). These results have been further
evidenced in our laboratories where we have studied the rheological
behaviour of paste blending resins over a very wide shear rate
range. Other authors have found that the coarser the particle size
of the blending resin, the lower the plastisol viscosity (6); this
can only be understandable if the change in the maximum size of the
particle is combined with a change in the particle size distribution.

Effect of ratio of blending resin to dispersion resin: in theory
the greater this ratio, the greater the viscosity reduction. In
practice, however, there are some limitations: a blending resin,
because of its large particle size will not form a paste by itself;
also, on excess of blending resin will cause settling tendencies.
The lowest viscosities (both at low and high shear rates) are usually
observed at a 45-55% level.

Various commercial grades of blending resins, used at the same
level can give different rheological properties although the particle characteristics (shape, size) are very close. It is supposed
that the interaction dispersion resin - blending resin - plasticizer
depends also on the type of surfactant used for the preparation of
the resins (the diffusion rate of plasticizer depends on the emulsifiers, surfactants present in the system).

Effect on the properties of the finished articles: The use of
blending resins effects, amongst other things, the surface, transparency and the mechanical properties. With the same plasticizer
content, the tensile strength and break elongation are somewhat
reduced. However, the use of blending resins makes it possible to
have a harder material as a result in an increase in the gelling
temperature or time. The water absorption is also significantly
reduced and consequently, paste blending resins give a dryer feel
to most finished articles. Blending resins, especially the higher
molecular weights tend to maintain a uniform level of gloss control
over a wide range of processing conditions, since the particles do

not readily flux in the plastic matrix.

Applications for blending resins: there are many areas where the use of paste blending resin can assist in providing an economical solution and/or improves the properties of finished articles. For example in flooring, the base layer is made for economic reasons with a high filler content: as high as 50% of blending resins are needed in order to achieve loadings of 400 parts by weight of filler and 100 parts of plasticizer; in the foam layer, blending resins act primarily as viscosity reducers which thus make the foam production possible; the cover layer requires high abrasion resistance and hardness which means that a low plasticizer content must be used: however such a paste would not be spreadable without the use of a blending resin. For similar reasons, hard synthetic leather for the travel goods industry can only be produced by the addition of paste blending resins. Other application areas include cast articles with high self-supporting rigidity combined with low viscosity in the plastisol in order to ensure that complicated mouldings are properly filled; dipped articles (PVC gloves with a textile carrier) where low viscosities are required to allow a thin layer to be coated without striking through the fabric; vinyl wall papers, produced at very high speeds can lead to dilatancy problems, even with low viscosity dispersion resins.

(1) Collins E.A., Hoffman D.J. and Soni P.L. Rubber Chem. Technol. 52 (3) (1979) 676.
(2) Nakajima N., Daniels C.A. Journal Appl. Pol. Sci. 25 (1980) 2019.
(3) Tester D.A. Manufacture and Processing of PVC. Burgess R.H. Ed. (Macmillan Publishing Co., New York, 1982), p. 261.
(4) Portingell G.C. Particulate Nature of PVC. Butters G. Ed. (Applied Science Publishers Ltd., London, 1982), p. 188.
(5) Weinhold G.,Paul K.P. Kunstoffe 71 (1981) 436.
(6) Skiest E.N. Mod. Plast. 47 (1970) 132.

I am grateful to Electrochemical Industries (Frutarom) Ltd. for the permission to submit this Seminar and I acknowledge with thanks Dr. F. Lerner for his active contribution in the field of PVC paste resins, M. Hesse, S. Nemet and our Quality Control Laboratory for the experimental part in this work.

BLOCK COPOLYMERS AS HOMOGENIZING AGENTS IN BLENDS OF AMORPHOUS AND SEMICRYSTALLINE POLYMERS

Michael A. Drzewinski

Massachusetts Institute of Technology
Cambridge, Massachusetts

Previous research in our laboratory has focused on amorphous rubbery-rubbery diblock copolymers and the corresponding homopolymer blend. This work can be represented by a unified "phase diagram" consisting of three regions:

   (i)   A low molecular weight region in which diblock copolymers
         and their corresponding blends are homogeneous materials.
   (ii) A high molecular weight region in which diblock co-
         polymers and blends are heterogeneous.
   (iii) A region of intermediate molecular weight, spanning
         regions (i) and (ii) by a factor of at least three,
         in which the diblock copolymers are homogeneous but the
         corresponding blend is heterogeneous

It is in this largely unexplored region (iii) that we have acquired data verifying that diblock copolymers can be used as homogenizing agents in amorphous homopolymer blends.

The move towards studying simple, semicrystalline systems was motivated especially because of the potential of applying the results from the amorphous analog systems. Thus, this newer area of research is partly based on the assumption that a semicrystalline diblock copolymer will organize itself at the interface between two homopolymer and thereby act to improve adhesion between the components. This in turn should lead to favorable morphological and mechanical property changes.

Considerable work has been underway in an effort to find appropriate synthetic procedures for generating model semicrystalline diblock copolymers to perform such studies. A procedure which has shown promise is a transformation reaction from an anionic to coordinative or Ziegler-Natta type polymerization. By this method we have been able to produce some fairly well controlled, novel diblock copolymer structures. Initial tests have produced some interesting morphological and mechanical property profiles. Therefore, this seminar will briefly review some of our previous findings with purely amorphous systems and introduce some of our more recent results in the area of semicrystalline materials.

STUDY OF COPOLYMER-HOMOPOLYMER BLENDS

P. MARIE[2], J. SELB[1], A. RAMEAU[1], R. DUPLESSIX[1] and Y. GALLOT[1]

1) C.R.M., 6, rue Boussingault, 67083 Strasbourg-Cedex, France
2) I.L.L., 156X, 38042 Grenoble-Cedex, France

Some years ago, we started studying the structure and thermodynamics of binary and ternary copolymer-homopolymer blends. Two
aspects have been developped. The first one concerns the investigation of demixing in such systems ; the second one focuses on the
formation of block copolymer micelles in the case of binary blends
of a block copolymer with one of the parent homopolymers. In the
present paper are given some of the results we have obtained for
polystyrene-polybutadiene (PS-PB) block copolymer-homopolymer systems.

1 PHASE SEPARATION BEHAVIOUR

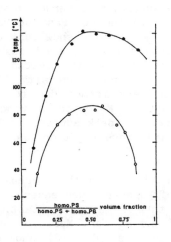

Fig. 1 Cloud point curves :
($\bullet$) binary mixtures of
PS (M_W = 2 400) and
PB (M_W = 3 300) ;
(o) ternary mixtures with
20 wt % of a block
copolymer (M_W = 6300,
%PS : 47.6)

450

Phase separation temperatures (cloud points) were determined from
the measurement of the intensity of light transmitted through the
polymer mixtures. Experimental phase diagrams (fig. 1) show that
the coexistence curve of ternary blends of a block copolymer with
the two corresponding homopolymers is located in a temperature
range much lower than in the case of the binary homopolymer
mixture. Moreover, in the particular case of the ternary blends
investigated, the temperature of demixing decreases linearly with
increasing copolymer content (fig. 2). These results are in good
agreement with theory (1) and confirm the compatibilizing effect
of a block copolymer towards its parent homopolymers.

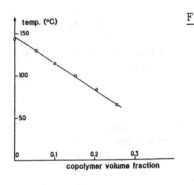

Fig. 2 Cloud point temperatures
versus block copolymer
content for ternary blends
with a constant PS/PB homopolymers ratio (46/54 in
volume)

2 BLOCK COPOLYMER MICELLE FORMATION

Small angle neutron scattering (SANS) experiments have been
carried out to characterize the structure of "solid solutions" of
PS-PB diblock copolymers embedded in a matrix of polybutadiene. In
these systems, only the PS block of the copolymer is perdeuterated
; then the scattering function results from the contrast between
PS and PB parts of the blends. Experimental results show that the
scattered intensity exhibits several subsidiary, smeared maxima
and minima (fig. 3) which are characteristic of the scattering
function of spherical particles with a low polydispersity (2).
More precisely, spherical PS microdomains are formed due to the
micellar association of copolymer molecules. Moreover, beyond a
given copolymer concentration, an interference peak appears at low
q values. The main conclusions of this investigation can be
summarized as follows :

i)   Influence of copolymer concentration : with increasing
copolymer concentration, the radius of spheres decreases to a
copolymer concentration of about 5%, then remains constant. No
quite satisfying explanation has been found so far. Moreover, the
intensity of the interference peak increases (Fig. 3) and its
location is shifted towards greater q values. Therefore, the

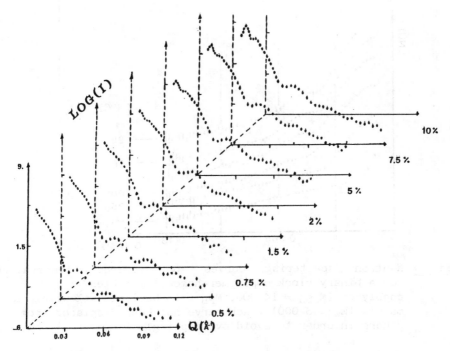

<u>Fig. 3</u> Neutron scattering curves for binary block copolymer-homo-
polymer blends at different copolymer concentrations ;
cop. : M_{PSD} = 35 000, M_{PB} = 75 000
PB matrix : M_w = 3 000

higher the copolymer content, the greater the interparticle
interactions are.

ii)   Influence of temperature : with increasing temperature,
a continuous decrease of the scattering intensity is observed
and the subsidiary maxima gradually disappear (Fig. 4). This
means that the equilibrium between micelles and single molecules
is shifted towards molecularely dispersed copolymer chains.

iii) Swelling of the shell of micelles : for a given copolymer at
a given concentration, the radius of micelle core increases with
the molecular weight of the matrix, but the interparticle
interactions decrease. This result has been explained as follows :
the lower the molecular weight of the matrix, the higher the
degree of swelling of the shell by homopolybutadiene chains and,
therefore, the higher the overall dimensions of the micelle will
be, although the degree of association decreases. This feature
confirms theoretical predictions (3).

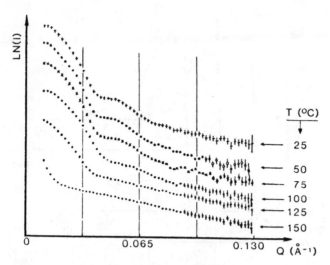

Fig. 4 Neutron scattering curves at different temperatures
for a binary block copolymer-homopolymer blend :
copolymer (M_{PSD} = 14 000, M_{PB} = 15 000) at 5wt% within a PB
matrix (M_W = 3 000) (each curve has been displaced from
others in order to avoid overlapping)

1. L. Leibler, Makromol.Chem.Rapid Commun. 2, 293 (1981)
2. J. Selb, P. Marie, A. Rameau, R. Duplessix and Y. Gallot,
   Polym.Bull. 10, 444 (1983).
3. L. Leibler, H. Orland, J.C. Wheeler, J.Chem.Phys., 79, 3550
   (1983).

RUBBER-PLASTICS BLENDS

P J Corish

Dunlop Technology Division, Birmingham

To complement the papers in the course describing Plastics-Rubber blends, an example of practical significance in the Rubber-Plastics area is described.

The objective was to develop a compound with low specific gravity as the play ball product had a weight restriction, as well as meeting high resilience, strength and modulus requirements.

Initial studies using a low Tg rubber (natural rubber) and a high Tg plastics as organic filler, showed high resilience in the room temperature range. However, strength properties were not sufficiently high due to poor dispersion of the high Tg plastics.

Higher mixing temperatures gave better dispersion and higher levels of properties up to a point, but, at very high mixing temperatures, degradation of the natural rubber occurred leading to a lower property level (c.f. Table 1). Use of unmasticated natural rubber (higher molecular weight) gave some improvement.

The rubber and plastics phases were obviously incompatible as the rubber/plastics blends had two distinct Tgs as shown by loss peaks obtained on a Rolling Ball Spectrometer. In order to obtain better general properties, several ways of improving compatibility, to a limited extent, were tried.

Use of a very high styrene SBR, in place of polystyrene, gave reasonably good properties at lower mixing temperature, attributed to interphase crosslinking between the polybutadiene part of the SBR copolymer and the natural rubber (c.f. Table 2).

TABLE 1

NATURAL RUBBER/POLYSTYRENE BLENDS (100/38)

Mixing Temp (^{o}C)	Tensile Strength (Kg/cm^2)	Modulus at 300% (Kg/cm^2)	Tear Strength (Kg/tp)	RT Resilience (%)
130	113	53	4	81
160	133	66	5	81.5
190	103	33	5	70

TABLE 2

	Tensile Strength (Kg/cm^2)	Modulus @ 300% (Kg/cm^2)	Tear Strength (Kg/tp)	RT Resilience (%)
NR/V. High Styrene SBR (100/38)	149	105	8.8	76

It was found essential to use a plastics phase with a high Tg or the resilience level fell to an unacceptable value.

Use of a compatibilising agent was tried, utilising a 50/50 block copolymer of butadiene and styrene. General properties of an unmasticated natural rubber/polystyrene blend were improved without too much loss in resilience (c.f. Table 3).

TABLE 3

	Tensile Strength (Kg/cm^2)	Modulus @ 300% (Kg/cm^2)	Tear Strength (Kg/tp)	RT Resilience (%)
Unmasticated NR/Polystyrene (100/38)	166	52	5.3	80
Unmasticated NR/Polystyrene + BS Block Copolymer (93/38/7)	247	128	11	76

Co-crosslinking of rubber and plastic phases was tried, viz. peroxide curing of natural rubber/polyethylene blends, but the general level of properties was not satisfactory, being appreciably lower than for sulphur-based cures.

To summarise, by blending rubbers and plastics of varying Tgs, some control of physical properties for end-use applications can be achieved. The ability to achieve a modicum of co-crosslinking appears to be important and compatibilising agents could provide useful benefits.

ISOTACTIC POLYPROPYLENE/RUBBER BLENDS: EFFECT OF CRYSTALLIZATION
CONDITIONS AND COMPOSITION ON PROPERTIES

R. Greco

Istituto di Ricerche su Tecnologia dei Polimeri e Reologia del
C. N. R. - Arco Felice, Napoli - Italy

A generalized discussion on the influence of crystallization
conditions and of the addition of a rubbery component on a polypropylene matrix is presented in this paper. Such a discussion is
based on the results obtained in our institute for isotactic polypropylene (iPP)/polyisobutylene (PIB) blends (1).

The mixtures (containing 0,10,20,30% in weight of PIB) have
been obtained by melt-mixing in a Brabender-like apparatus. Then
the materials, remelted between two plates in a hot press, have
been immersed in a thermostatic bath kept at prefixed temperatures
(T_c=25÷132°C). In such a way different degrees of undercoolings
and cooling rates could be achieved. Differential scanning calorimetry, wide and low angle X-ray diffraction, optical transmission
and electron scanning microscopy and mechanical tensile tests have
been the used techniques.

The results can be briefly summarized as follows:
a) At constant PIB percentage, the overall crystallinity
content, X_c, the modulus, E, the average spherulite diameter, D,
and the lamellar thickness, L_c, increase with increasing T_c, whereas
the ductility, d, and the strength, σ_r, decrease.
b) At constant crystallization temperature T_c; X_c, E, L_c, d,
and σ_r decrease with enhancing %PIB whereas D stays nearly constant.
For further details on these findings see previous work (1).

Now we refer to the scheme of Fig. 1, where the effect of the
crystallization conditions on the superstructure and the morphology
of a semicrystalline polymer is briefly illustrated. On the
vertical axis the temperature is reported to show the state of
the polymer:

1) At $T > T°m$ it is a pure melt without nucleation seeds;

2) At $Tm < T < T°m$ the melt contains a certain number of self-seeded nuclei. From the temperature T various undercoolings and rates of coolings can be achieved.

   a) In principle, using an infinite rate of cooling, the melt can be frozen into an amorphous glass with a local order range (noduli or paracrystals).

   b) At higher Tc, a fringed micelle type of morphology can result since only local crystallizations can occur. The tiny crystallites are interconnected by very many tie-chains coming from the original network existing in the melt.

   c) At Tc=0°C (iced water quenching) a microspherulitic structure is able to develop and the system acts as a multiple cross-linked rubber or leather (depending on the amount of crystallinity). The lamellar thickness Lc is very low and the crystallites are linked together by a considerable number of tie-molecules. Therefore the material exhibits a ductile behaviour at $T > Tg$.

   d) At low undercoolings (Tc equal to some ten below Tg) more perfect crystals and thicker lamellae are formed since most of the macromolecules are able to disentangle from each other, leaving very few tie-chains among the crystallites. The material is brittle in nature since the crystallinity is high, the lamellae are very thick and the interconnecting strands (carrying the applied load) very few.

Now if to such semicrystalline system a rubber is added before the crystallization takes place its presence influences the final morphology. In fact the PIB addition induces a Lc reduction but leaves the spherulite diameter unaltered since there is no nucleation effect as in other cases. Perhaps also the number of tie-chains among different spherulites can be reduced since the matrix must crystallize working against the PIB particles and therefore exhibits a more opened structure. Furthermore, when the blend is stretched and the deformation becomes sufficiently high, the PIB particles hamper the cold-drawing and hence the elongation to break decreases with increasing the % PIB.

Concluding this short discussion one can say that the ultimate properties of such blends depend mainly on the number of tie-chains carrying the applied load and on the lamellar thickness offering the resistance to breakage, sliding during the deformation. These structural features depend in their turn on:

   1) the rate of cooling;

   2) the degree of undercooling;

   3) the molecular characteristics of the matrix;

   4) the nature and amount of rubber.

459

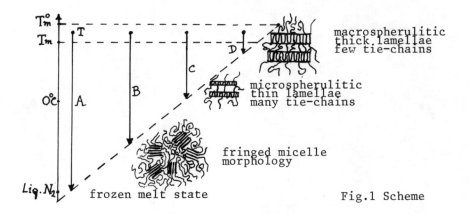

Fig.1 Scheme

REFERENCE

1. Bianchi, L., S. Cimmino, A. Forte, R. Greco, E. Martuscelli, F.
   F. Riva and C. Silvestre. Isotactic Polypropylene: Polyiso-
   butilene Blends: Morphology-Structure-Mechanical Properties
   Relationships. J.Mat.Sci., in print.

ETHYLENE-PROPYLENE RUBBER AND POLYOLEFIN POLYMER BLENDS: PRESENT SITUATION AND FUTURE TRENDS

P. Galli, G. Foschini and A. Moro

MONTEDISON/DUTRAL S.p.A. (Italy)

INTRODUCTION

Ethylene-propylene random copolymers (EPM), which sometimes contain a third diene (EPDM), are the most widely used rubbers in compound with crystalline polyolefins: the advantage of having fully polyolefinic systems is obvious, so as not to affect the inherent performance of these matrixes, that is: low specific weight, chemical inertia, low cost of raw materials and processing, easy transformation and recycling of scraps.

As this work mainly deals with those systems of considerable interest in terms of application, the various types of rubber-modified crystalline polyolefins can be classified into two main classes: a) blends with prevailing plastomer content, to obtain low temperature impact resistant grades; and b) blends with prevailing rubber content, with the characteristics of the thermo-plastic elastomer.

IMPACT RESISTANT BLENDS

In the first class of blends, a considerable role is played by the modification of PP, whose brittleness at low temperatures is a great handicap. In order to overcome this trouble, research efforts were directed towards two different areas: ethylene (or butene)/ propylene copolymerization, or mechanical blending of the homopolymer with elastomer. Even if in theory the first approach looks more attractive from an economical point of view, it however requires a great sophistication of the Ziegler-Natta catalyst system (1). The second approach is the one generally followed up to now, as it allows the most favourable balance, in terms of minimum stiffness loss and greater impact resistance increase, to be achieved. Investigations on the PP deformation behaviour, state that at room temperature the main way of dissipating the failure energy is of the "shear yielding" type, while at low temperatures a crazing-type dissipative mechanism prevails (2).

To take the best advantage of the reinforcing effect of an elastomeric phase dispersed within a matrix which undergoes deformation for shear yield, a very fine dispersion of the rubbery phase should be achieved. On the other hand, should the matrix deform by crazing, it is important to focus attention on a good quality rubber (Tg lower than the minimum working temperature) adhering well to the matrix (3).

POLYOLEFINIC THERMOPLASTIC ELASTOMER BLENDS

Thermoplastic elastomers have been available on the market for several years now, filling the gap between traditional plastic materials and vulcanized elastomers. Among this product family, thermoplastic rubbers produced by blending EPM or EPDM with PP or PE, have quickly grown thanks to the excellent physico-mechanical properties/cost balance.

By a proper choice of both elastomeric and plastomeric components a wide range of properties can be achieved. Grades based on a dynamic curing method (4), with considerably improved thermal elasticity performance, are also commercially available.

NEW TRENDS

Following the increasing demand for polyolefin elastomers to be used as plastomer modifiers, research efforts were directed towards new product ranges. As an example, elastomeric-polyolefinic copolymers with "controlled blocking", spherical-form, non-sticking rubbers obtainable directly from the polymerization reactor, particularly suitable to dry-blending and melt-extrusion are hypothetically possible. Finally, an additional improvement in grafting techniques will make it possible to further widen their use in non-polyolefinic matrices.

REFERENCES

1. Galli, P., T. Simonazzi and P.C. Barbe. VI Convegno AIM, Pisa, 1983.
2. Ramsteiner, F., G. Kanig, W. Heckmann and W. Gruber. Polym. 24 (1983) 365.
3. Bucknall, C.B.   Toughened Plastics (London, Appl. Science, 1977).
4. Coran, A.Y. and R. Patel. Rubber Chem. Technol. 53 (1980) 141.

REFERENCES

Smith, T. ... Cohousos & J.M. ...
1951.

Ramachandra, ... O. Kanik, W. Hackman and W. Cooper. *Polym.*
... 1955, ...

Sacchdil, C.S., *Doughnut* ... *Compos.* Instn. Science

...na, R.L. and L. Patel ... *Chem. Techn* ..., 55 1980, No.
...

INDEX

468